Exoplanetary Atmospheres

PRINCETON SERIES IN ASTROPHYSICS

Edited by David N. Spergel

Exoplanetary Atmospheres

Theoretical Concepts and Foundations

Kevin Heng

PRINCETON UNIVERSITY PRESS

PRINCETON AND OXFORD

Copyright © 2017 by Princeton University Press

Published by Princeton University Press
41 William Street, Princeton, New Jersey 08540

In the United Kingdom: Princeton University Press
6 Oxford Street, Woodstock, Oxfordshire, OX20 1TR

press.princeton.edu

ISBN 978-0-691-16697-1 (cloth)
ISBN 978-0-691-16698-8 (paperback)

Library of Congress Control Number: 2016950505

British Library Cataloging-in-Publication Data is available

This book has been composed in LaTeX

The publisher would like to acknowledge the author of this volume for providing the print-ready files from which this book was printed

Printed on acid-free paper. ∞

Printed in the United States of America

10 9 8 7 6 5 4 3 2 1

Dedications

Dick McCray, for showing me that science is to be enjoyed
Helmer Aslaksen, who believed in me when no one else did
Scott Tremaine, for teaching me how to read the literature
Rashid Sunyaev, for giving me a chance to learn from greatness
Sara Seager, for inspiring me to think big
Willy Benz, for teaching me the business of science
Ray Pierrehumbert, for being a fountain of energy and creativity
Claudia, my little black cat, the dearest creature I have ever known
Felix Hetzenecker-Heng, who forced the first draft 7 months before deadline
Stefanie Hetzenecker, my wife and best friend, who stood by me in all times

Contents

Foreword by Sara Seager

The field of research of exoplanet atmospheres is flourishing in observation and theory. Both the quality and quantity of observations are increasing rapidly—for a variety of planet types including transiting planets and directly imaged giant planets. With the James Webb Space Telescope on the horizon, the promise of large numbers of a huge variety of exoplanet atmospheres with observations at high-precision, high-spectral resolution will finally be fulfilled. As a result, more and more researchers are entering the field, especially students from many disciplines. Observations alone are not enough—theory to interpret the observations and to guide future observing strategy is key.

The theory of exoplanets atmospheres must draw on a vast body of physics, chemistry, and atmospheric science. Yet, exoplanet atmospheres is truly its own discipline, requiring the topics to be tied together in a way not normally taught in any standard physics or atmospheric science class. Further, remote sensing of distant exoplanets means we will always be limited in data extent and quality, as compared to Solar System planets which can be visited directly by orbiters and landers. So, different techniques and different applications of theory are needed over traditional planetary science. But more importantly, exoplanets have a huge diversity, appearing in nearly all masses, sizes, and orbits physically plausible. We anticipate this diversity will extend to the planet atmospheres— and indeed already have evidence in this regard for hot Jupiters observed with the Hubble Space Telescope. But just how does one tie all of the topics together?

Dr. Heng succeeds in providing an insightful and comprehensive treatise threading together key elements of exoplanet atmospheres, starting with radiative transfer and through opacities, chemistry, fluid dynamics, convection, and touching on atmospheric escape. Both deep and broad, Dr. Heng's book goes beyond a standard graduate-level textbook to provide a very thorough foundation for those wanting to perform research in the field. For example, early in the book Dr. Heng treats the two-stream approximation to the radiative transfer equation in meticulous detail and later, over three chapters, gives an impressive treatment of atmospheric fluid dynamics. The best way to learn and gain intuition is to apply—in this field by implementing equations into one's own computer code and then experimenting with it. To this end—and a fresh ingredient of many chapters—Dr. Heng outlines a recipe checklist for implementation, including pitfalls. One of my favorite features of the book are comments throughout on how to simplify a complicated equation, when simplification is appropriate, and what the dangerous caveats to the simplification are.

The book is the most thorough text on exoplanet atmospheres to date. It dives deeper and therefore builds upon my book "Exoplanet Atmospheres," the first on the subject and published in 2010 also by Princeton University Press. Indeed, the author intends his new book to be a logical continuation of the first. The book is complementary to the range of existing scholarly books on exoplanets in general.

Dr. Heng is one of the world's foremost experts on exoplanet atmospheres and has encapsulated his expertise into the book. His book will take you through the complexities and simplicities, the elegance and challenges of exoplanet atmospheres. Enjoy the journey!

Sara Seager
Professor of Planetary Science
Professor of Physics
Massachusetts Institute of Technology
Cambridge, MA, U.S.A.

Preface

Why did I feel the need to write this book in the first place? I started in astrophysics studying supernova remnants for my doctoral thesis. Almost immediately after obtaining my doctorate, I dropped the topic and embarked on a period of exploration, eventually chancing upon the subject of exoplanetary atmospheres in the same year that exoplanet pioneer Sara Seager's "Exoplanet Atmospheres" textbook was published. The topic intrigued me, because studying and understanding the atmospheres of exoplanets is an indispensable step on the path towards addressing one of the oldest questions posed by humanity: Are we alone? And how do we scan, from afar, this myriad of worlds to find out?

It became readily apparent that atmospheric science is an unavoidably interdisciplinary endeavor. Undeniably, it has its roots in the Earth sciences—and why would it not, since we actually *live* in this particular atmosphere? Yet it has long had an eye towards the stars, as the Solar System provides several examples of atmospheres that have been carefully scrutinized by planetary scientists. To understand the structure and appearance of an atmosphere requires a working understanding of fluid dynamics, radiation and its passage through matter, chemistry, the influence of aerosols and clouds, thermodynamics and phenomenology. To place the atmosphere of an exoplanet in a broader context, beyond the confines of our Solar System, requires that we understand the birth and death of stars other than our Sun, the properties of stellar populations and how substellar objects known as brown dwarfs bridge the continuum between exoplanets and stars [230]. To do my own work, I had to reconcile techniques, terminology and modeling philosophies from numerous monographs and papers, across the atmospheric, oceanic and climate sciences, astronomy and astrophysics, planetary science, geophysics and chemistry. Even the term "model" means different things to different communities. Thus, the idea was planted in my mind: the need for a single book that explained the concepts and principles needed to embark on theoretical research in the atmospheres of exoplanets.

The original intention was to write it in the style of "exoplanetary atmospheres for dummies," based on the notion that one would not have to wade through pages of equations or prose to get at the salient features of an idea. This has evolved into something slightly more sophisticated: to explain a concept using the shortest and/or most intuitive approach possible and relegating the standard or classical explanations to the problem sets. This approach has borne unexpected fruit, as some of the alternative derivations are either novel or cleaner, but without necessarily sacrificing on technique. It is my firm belief that one does not really understand physics until one has to apply the concepts acquired to examples; about a sixth of the textbook is devoted to problem sets.

In a departure from the traditional approach, some of the puzzles described in the problem sets are qualitative and quasi-open-ended, designed to provoke thought and encourage debate.

To tackle a vast and interdisciplinary field of inquiry like exoplanetary atmospheres, it is not enough to check that its theoretical foundations are not Earth-centric—and, if so, to generalize them. It also helps to understand the history behind some of these ideas and appreciate what the different communities of scientists accept as being standards of proof. Thus, I return to my earlier worry: what is a model? To purists, a model is an approximate description of Nature based on a governing equation that is itself derived from a conservation law—of mass, momentum, energy or some other, more generalized quantity such as the potential vorticity. I favor this approach. To others, it may be a mathematical function describing how some aspect of the atmosphere behaves as the different parameters describing it are varied—but derived entirely from fitting this ad hoc function to experimental data; the function itself is "pulled out of a hat," unconstrained by any law of Nature. To compound the issue, the use of simulations to address scientific questions is coming into its own as a third way of establishing scientific truth, bringing with it an entire set of epistemological and metaphysical concerns [95]. To engage constructively in conversations about exoplanetary atmospheres, across the spectrum of disciplines, requires that one is perpetually aware of what commonly-used terms mean to different scientists.

The way that data are measured, collected and interpreted in the Earth and planetary sciences differs somewhat from astronomy. Data of the Earth's atmosphere are abundant and rich in detail. Measurements of the atmospheric properties of the planets and moons of our Solar System are a little less rich in detail, but still of a quality that would make any exoplanet astronomer envious. It is important to recognize that these fields of inquiry are collecting a wealth of information on only a handful of objects. Astronomers studying exoplanetary atmospheres are in the other regime: the data are sparse, usually incomplete, and reveal fragmentary information about any given atmosphere, but the number of objects being studied is vastly larger. The challenge is to reconcile the possible benchmarking opportunities offered by the Solar System with the statistical trends revealed by exoplanets. In this debate, it is wise to not mistake precision for accuracy. We will always know more about the Solar System in detail, but it remains unclear if we are stuck in the situation of looking for our metaphorical keys under the nearest lamp—the astronomical data are already hinting at this possibility. Assigning importance to scientific questions based on the cosmic proximity of the object being studied distracts from the bigger, broader questions exoplanetary science is attempting to address. I do not see how one can conduct research in exoplanetary science without a working understanding of the data collected by astronomers (rather than by Earth or planetary scientists). It is no accident that the first chapter contains a broad review of observational techniques and the data they produce.

This brings us to a technique loved by astrophysicists and loathed by the other disciplines that are used to enjoying more precision: understanding a concept or principle using order-of-magnitude arguments. A more complex model is not necessarily more correct. Order-of-magnitude arguments often reveal salient behavior, which may be used to develop intuition before embarking on more sophisticated calculations. I have peppered the textbook throughout with such arguments, as I believe they are a useful way of attaining understanding without being bogged down in mathematical drudgery—and this is coming from someone who enjoys mathematics. Yet, I have insisted on maintaining a level of discipline about what the various mathematical symbols mean: "on the order of" ($\sim$), "approximately" ($\approx$), "proportional to" ($\propto$), "equal to" ($=$) and "defined as" ($\equiv$). At no point do I use some subset of these symbols interchangeably.

In this first edition, I have chosen to focus on the theoretical framework for understanding atmospheres—what I call the foundations of the subject, since I do not expect them to ever change. The Navier-Stokes and radiative transfer equations will probably continue to be studied long after I am gone. The basic principles behind hydrostatic equilibrium are expected to remain invariant. I do not expect the basic laws of thermodynamics or the concept of Gibbs free energy to be altered anytime soon. What will change are the measurement and interpretation techniques used by astronomers, which will continue to rush ahead at a dazzling pace. In the second edition, I expect to write about biosignature gases, geochemical cycles and habitability, and expand upon the material on chemistry, escape and inversion techniques. This first edition is about the "bread and butter" (or "nuts and bolts") of the theory of exoplanetary atmospheres and was written in about two years. I wish to thank Mark Marley, Sara Seager, Daniel Kitzmann, Ruth Murray-Clay, James Owen, Sébastien Fromang, Eric Agol and two anonymous referees for constructive comments that improved the clarity and quality of the manuscript. I am especially thankful to Jim Lyons for first teaching me the basics of atmospheric chemistry.

Throughout the textbook, I have interlaced the mathematical formalism with narrative prose meant to impart physical intuition and, whenever possible, convey different ways of thinking about familiar ideas. In this age of information overload, it is not enough to provide the student with facts—it is also the job of the teacher to convince the student why something is interesting and worth learning. To a large extent, the chapters may be read independently of one another. It has been a great joy to be able to combine my rigorous training as a scientist with my love of writing (and a brief few years of experience gleaned from being a professional journalist), which I have done for the *American Scientist* magazine, and assembling things in the same manner as a cook preparing a recipe (having gone to culinary school and trained as a chef). In the best-case scenario, theory aims to predict natural phenomena before they are measured by astronomers. A more modest goal—one which I aspire to and the central tenet of this textbook—is to provide a robust theoretical scaffolding upon which to place into context the interpretation of fragmented data. If this book inspires the next generation of exoplanet scientists to perform sounder interpretations of their data, then it would have served its intended purpose.

Chapter One

Observations of Exoplanetary Atmospheres: A Theorist's Review of Techniques in Astronomy

1.1 THE BIRTH OF EXOPLANETARY SCIENCE

It is no exaggeration that the close of the twentieth century was a time of great discovery. For centuries, if not millennia, humanity speculated upon the existence of other worlds—it is an ancient question that transcends cultures and nations. We debated the existence of eight or nine planets[1] (depending on if one counts Pluto) orbiting our Sun: four rocky planets (Mercury, Venus, Earth and Mars) and four gas/ice giants (Jupiter, Saturn, Neptune and Uranus), flanked by the asteriod and Kuiper belts and encompassed by the Oort cloud. In the mid-1990s, this Solar System-centric view was shattered with the first discoveries of *exoplanets*—planets orbiting stars other than our Sun, beyond the Solar System—first around a pulsar [253, 254] and a few years later around a Sun-like star [166]. In the intervening years, the confirmed discoveries of exoplanets have increased exponentially and—at the time of publication of this textbook—number in the thousands.

It appears as if Nature is infinitely more creative than us at forming these new worlds [92]. The reported discovery of 51 Peg b in 1995 was the first example of a *hot Jupiter*, a Jupiter-sized exoplanet orbiting its star at a small fraction of the distance (~ 0.01–0.1 AU) between Jupiter and our Sun (≈ 5 AU) and experiencing intense starlight that causes its temperatures to reach ~ 1000 K. It was also the first instance of many surprises, including the discovery that *super Earths*—exoplanets with radii between that of our Earth and Neptune, which do not exist in the Solar System—are common and that Nature has a talent for making closely-packed, multi-exoplanet systems [145] and *circumbinary exoplanets* [51]—those orbiting a pair of stars. These discoveries have forced us to rethink our hypotheses and views on how the Solar System and other exoplanetary systems formed. If one may make a statement that will stand the test of time, it is that Nature will undoubtedly continue to surprise and challenge us in the future.

What is the myriad of worlds Nature has formed in the broader Universe? What is the chemical inventory of these worlds? And how many of them are capable of, and are, harboring life? Planetary science, which is the study of

---

[1]Intriguingly, Batygin & Brown [15] have suggested the existence of a yet undiscovered planet in the outer Solar System.

our Solar System, will continue to advance, but it is clear that the answers to these questions will ultimately come from the data gathered by astronomers harnessing telescopes to scan the heavens for worlds orbiting other stars. As exoplanet scientists, we have to develop a working knowledge of the nature and types of data gathered by astronomers in order to construct models that may be confronted by the data—or, at least, have some relevance to them. It is the purpose of this chapter to summarize the various techniques used by astronomers to observe and study exoplanets and their atmospheres.

1.2   TRANSITS AND OCCULTATIONS

The transit method is the workhorse of detecting and characterizing exoplanets and has been described as the "royal road" of exoplanetary science [252]. It exploits the fact that for a population of exoplanet-hosting stars, some fraction of them have exoplanets that reside on nearly edge-on orbits, which causes a diminution of the total light from the system when the exoplanet passes in front of its star. This event may be recorded by a distant observer, i.e., an astronomer. If a sufficient number of stars are monitored, then a harvest of exoplanet transits may be reaped.

1.2.1   Basics: transit probability and duration

A first, natural question to ask is: how likely is it for an exoplanet to transit its star [22]? If we visualize *half* the angle subtended by the star (θ), then it is given by

$$\tan \theta = \frac{R_\star}{a}, \tag{1.1}$$

where $R_\star$ is the stellar radius and a is the exoplanet-star separation in distance. Since we expect this angle to be small, we have

$$\theta \approx \frac{R_\star}{a}. \tag{1.2}$$

If we integrate over the entire celestial sphere, then we obtain a "celestial band" within which a distant observer may record a transit,

$$\int \int d\theta \ d\phi \approx \frac{4\pi R_\star}{a}. \tag{1.3}$$

The transit probability obtains from dividing the coverage of the celestial band by the total solid angle subtended (4π steradians),

$$\mathcal{P}_{\text{transit}} = \frac{1}{4\pi} \int \int d\theta \ d\phi \approx \frac{R_\star}{a}. \tag{1.4}$$

For hot Jupiters, the transit probability is relatively high: about 5–50% (around Sun-like stars) [37]. This explains why these were among the first type of exoplanets found by the astronomers. By comparison, the transit probability of

Earth and Jupiter are 0.5% and 0.09%, respectively. Since the transit probability is linearly proportional to the stellar radius, it provides the motivation to hunt for smaller exoplanets around later-type[2] stars [40, 41]. There is an understandable tension between planetary scientists, who are interested in the science of habitability, wishing to exclusively study Earth-like exoplanets and astronomers who worry about detecting high-probability events.

The second, natural question to ask is: if an exoplanet transits its star, how *long* does the event last? We expect that short transits are infeasible to observe, where "short" may mean that it is less than the typical cadence associated with an observing strategy or instrument. We may obtain a rough estimate by considering exoplanets that transit exactly at the equator of the star and reside on circular orbits. The circumference of the orbit is $2\pi a$. The diameter of the star is $2R_\star$, which implies that the fraction of the orbit during which the exoplanet transits is $R_\star/\pi a$. If the period of the orbit is given by t_{period}, then the transit duration is [252]

$$t_{\text{transit}} = \frac{R_\star \, t_{\text{period}}}{\pi a}. \tag{1.5}$$

We may eliminate t_{period} for a via Kepler's third law (assuming the stellar mass far exceeds the mass of the exoplanet),

$$t_{\text{period}} = \frac{2\pi a^{3/2}}{(GM_\star)^{1/2}}, \tag{1.6}$$

and obtain

$$t_{\text{transit}} = 2R_\star \left(\frac{a}{GM_\star}\right)^{1/2}, \tag{1.7}$$

where G is Newton's gravitational constant and $M_\star$ is the mass of the star. It is worth noting that $\sqrt{GM_\star/a}$ is simply the circular speed. For a hot Jupiter, Earth and Jupiter, we have $t_{\text{transit}} \approx 1$–4, 13 and 30 hours, respectively. The weak scaling of t_{transit} with a means that even for a Jupiter-sized exoplanet located at $a = 100$ AU, one obtains $t_{\text{transit}} \approx 130$ hours, although this neglects the fact that $t_{\text{period}} \approx 10^3$ years and $\mathcal{P}_{\text{transit}} \approx 0.005\%$.

Several considerations relegate these algebraic formulae to being accurate only at the order-of-magnitude level. If the exoplanet resides on an eccentric and/or inclined orbit, then correction factors on the order of unity need to be added to $\mathcal{P}_{\text{transit}}$ and t_{transit} [234]. More interestingly, if multiple exoplanets exist within a single system, then they exert mutual gravitational forces on one another and each orbit deviates from having a Keplerian period [60]. Specifically, these *transit timing variations* (TTVs) may be used to infer the masses of these exoplanets [3, 103].

---

[2]Astronomers have developed a spectral classification for stars, labeling them by the alphabets O, B, A, F, G, K and M, which inspired the mnemonic, "Oh be a fine girl/guy, kiss me." O stars are early-type stars that live short lives ($\sim 10^6$ years), while M stars may last for the age of the Universe ($\sim 10^{10}$ years). Within each stellar type, there is a further subdivision by numbers. For example, an M0 star is an early-type red dwarf, while M9 is a late-type one.

1.2.2 Stellar density and limb darkening

The *ingress* and *egress* of the transit of an exoplanet are the moments when it first obscures the star and exits the transit, respectively. It turns out that the ingress and egress encode information about the density of the star. By assuming the star to be spherical,

$$M_\star = \frac{4\pi}{3}\rho_\star R_\star^3, \tag{1.8}$$

where $\rho_\star$ is the stellar mass density, we may rewrite Kepler's third law as [206]

$$\rho_\star = \frac{3\pi}{G}\left(\frac{a}{R_\star}\right)^3 t_{\text{period}}^{-2}. \tag{1.9}$$

By observing repeated transits, one may measure the orbital period of the exoplanet. The quantity $R_\star/a$ may be directly measured using transit photometry, as it is related to the duration of ingress/egress and the transit depth [206, 252].

As you will learn in Chapter 2, the *optical depth* (τ) is the correct measure of whether any medium is transparent ($\tau \ll 1$) or opaque ($\tau \gg 1$). We see an "edge" to the Sun because we are detecting its $\tau \sim 1$ boundary. Since the optical depth is a wavelength-dependent quantity, the radius of the Sun varies with wavelength. Similarly, we expect the radii of stars to be wavelength-dependent. Furthermore, the two-dimensional projection, on the sky, of the star, which we term the *stellar disk*, does not have a sharp edge. A combination of these effects causes the phenomenon of *limb darkening*, which smooths the ingress and egress of the transit light curves and causes them to have rounded bottoms [158]. To correctly extract the radius of the exoplanet from the light curve requires that one account for limb darkening correctly.

1.2.3 Transmission and emission spectra

To lowest order, the radius of an exoplanet does not depend on wavelength. To the next order, it does because its atmosphere (if it has one) will tend to absorb starlight or its own thermal emission differently across wavelength [23, 205]. Let the transit radius of the exoplanet at some reference wavelength be denoted by R_0; let the transit radius at a general wavelength be R, such that $R = R_0 + \delta R$ and δR is the (small) deviation from the reference radius across wavelength. If we denote the *change* in transit depth by δ_{transit}, then it is [23]

$$\delta_{\text{transit}} = \frac{\pi\left(R^2 - R_0^2\right)}{\pi R_\star^2} \approx \frac{2R_0\,\delta R}{R_\star^2}. \tag{1.10}$$

It is essentially the annulus of the atmosphere ($2\pi R_0 \delta R$), seen in transit, divided by the area of the stellar disk ($\pi R_\star^2$). The change in the transit radius is typically some multiple of the (isothermal) *pressure scale height* of the atmosphere, which is given by

$$H = \frac{k_{\text{B}}T}{mg}, \tag{1.11}$$

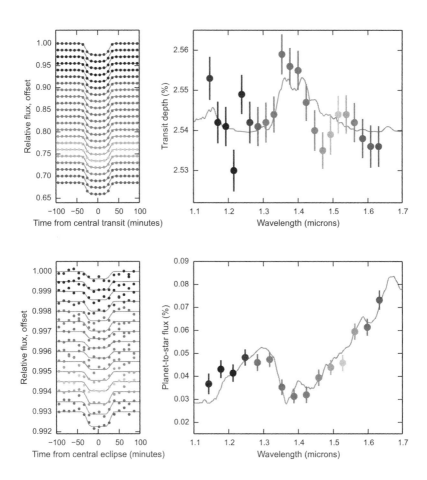

Figure 1.1: Transit light curves (top-left panel) of the hot Jupiter WASP-43b, measured at near-infrared wavelengths, and the corresponding transmission spectrum (top-right panel), taken using the Wide Field Camera 3 (WFC3) on the Hubble Space Telescope [132]. The spectral feature peaking at about 1.4 μm is attributed to the water molecule. Notice how the limb darkening of the star alters the *widths* of the transit light curves across wavelength, while the changing opaqueness of the atmosphere varies their *depths*. Also shown are the secondary-eclipse light curves (bottom-left panel) and the corresponding emission spectrum (bottom-right panel). Courtesy of Laura Kreidberg and Jacob Bean.

where k_B is Boltzmann's constant, T is the temperature, m is the mean molecular mass and g is the surface gravity of the exoplanet. If we plug in typical numbers ($T = 1000$ K, $g = 1000$ cm s^{-2}), then we obtain $H \approx 400$ km for a hydrogen-dominated atmosphere. For a hot Jupiter, we obtain $\delta_{transit} \sim 10^{-4}$; for an Earth-sized exoplanet with the same pressure scale height, we obtain $\delta_{transit} \sim 10^{-5}$. If we wish to characterize a twin of the Earth orbiting a Sun-like star, the change in transit depth drops to $\delta_{transit} \sim 10^{-7}$, making it a formidable technological feat. A plausible way forward is to scrutinize exoplanets around smaller stars. For example, if the star had a radius 10% that of the Sun, this would increase the transit depth by a factor of 100 for any atmosphere.

Generally, starlight filters through the atmosphere of a transiting exoplanet along a chord [62, 97]. The location of this chord, and thus the size of the transit radius, depends on wavelength. Specifically, δR and $\delta_{transit}$ are wavelength-dependent quantities. By measuring $\delta_{transit}$ across a range of wavelengths, one may constrain the composition of the atmosphere (Figure 1.4) [38]. A thought experiment illustrates this: imagine an exoplanetary atmosphere that is composed purely of water vapor. At some wavelengths, the water molecule absorbs radiation strongly and the transit radius of the exoplanet becomes larger. At other wavelengths, it is transparent to radiation and the transit radius becomes smaller. Scanning the transit radius across wavelength produces a *transmission spectrum*, which encodes the opacity function of the atmosphere and allows us to identify the constituent atoms and molecules. Figure 1.1 provides an example of such observations by astronomers. In the ultraviolet, transmission spectra probe the upper atmospheres of exoplanets and provide constraints on atmospheric escape [246].

A complementary and harder way of scrutinizing the atmosphere of an exoplanet is to measure its *occultation* or *secondary eclipse* (Figure 1.1). This occurs when the exoplanet is obscured by its star, which is usually a smaller effect than a transit. At this moment in time, the system shines only in starlight. Just before secondary eclipse, one may measure photons from the dayside of the atmosphere by substracting out the stellar contribution [39, 46]. The secondary eclipse depth is

$$D_{eclipse} = \frac{F}{F + F_\star} \approx \frac{F}{F_\star}, \qquad (1.12)$$

where F and $F_\star$ are the wavelength-dependent fluxes from the exoplanet and the star, respectively. To extract the *secondary-eclipse spectrum* (F as a function of wavelength) requires that we understand the stellar spectrum ($F_\star$ as a function of wavelength) and the peculiarities of the star (activity, flares, starspots, etc.).

1.2.4   Geometric albedos

If one records the secondary eclipse depth in the range of wavelengths dominated by reflected starlight (usually in the visible or optical), then the *geometric albedo*

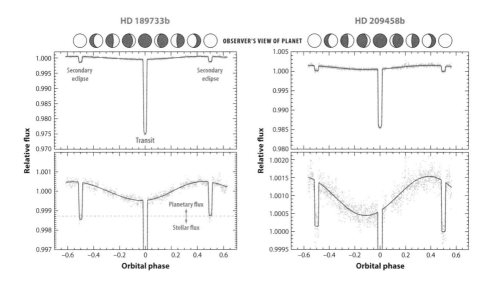

Figure 1.2: Infrared phase curves of the hot Jupiters HD 189733b [121, 122] and HD 209458b [262]. Courtesy of Heather Knutson and Robert Zellem.

may be measured [208],

$$A_g = D_{\mathrm{eclipse}} \left(\frac{a}{R} \right)^2. \tag{1.13}$$

It is the albedo of the exoplanet at zero phase angle. Chapter 2 describes its relationship to other types of albedo in more detail.

1.2.5 Phase curves

As the exoplanet orbits its star, it exposes different sides of itself to the astronomer, who may measure its light as a function of the orbital phase. Such a light curve is known as the *phase curve* and was first[3] measured for the hot Jupiter HD 189733b [121]. It is a one-dimensional "map," since the phase curve contains both longitudinal (east-west) and geometric information on the exoplanetary atmosphere (Figure 1.2). If phase curves are measured at different wavelengths, then the atmosphere is being probed at slightly different altitudes or pressures, thus providing limited two-dimensional information [236]. In rare cases, one may obtain both latitudinal and longitudinal information by sampling the transit light curve with high cadence at ingress and egress, a technique known as *eclipse mapping* [156]. The method requires high precision also in the

---

[3]To be fair, Harrington et al. [82] detected phase variations for the *non-transiting* hot Jupiter v Andromedae b in 2006.

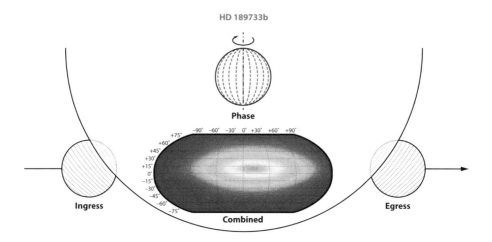

Figure 1.3: Eclipse mapping of the hot Jupiter HD 189733b [50]. The phase curve and ingress/egress of a transit provide complementary information in two dimensions. Courtesy of Julien de Wit and Sara Seager.

measured orbital parameters (e.g., eccentricity), since their uncertainties may easily mimic the timing offsets associated with atmospheric flux variations [50].

1.2.6 Putting it all together: A wealth of information from transits and occultations

Figure 1.4 unifies the information we have discussed regarding transits and secondary eclipses into a single schematic. It focuses on exoplanets that are tidally locked and have permanent daysides and nightsides, but the principles highlighted in the schematic apply generally. Specifically, it describes how the peak offset of the phase curve is expected to occur at secondary eclipse in the absence of atmospheric circulation. Measuring a peak offset in the phase curve thus implies the presence of atmospheric winds [216].

1.3 RADIAL VELOCITY MEASUREMENTS

One often visualizes the planets of our Solar System orbiting a static Sun. In reality, the Sun and the planets orbit a common center of gravity. As the star wobbles about this center of gravity, its light is blue- or redshifted relative to the astronomer, enabling its *radial velocity* signal to be measured. This was how the first exoplanet orbiting a Sun-like star was discovered [166]. Specifically, the

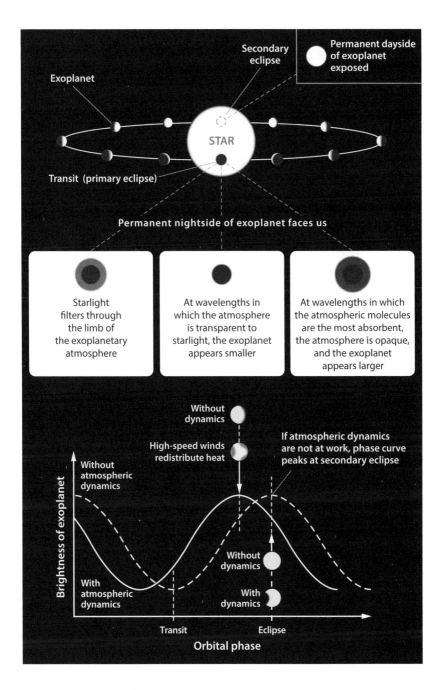

Figure 1.4: Schematic depicting the phases of a transiting exoplanet, how this maps onto a one-dimensional phase curve and the influence of atmospheric dynamics [96].

velocity semi-amplitude is measured [150, 252],

$$K_\star = \left(\frac{2\pi G}{t_{\text{period}}} \right)^{1/3} \frac{M \sin i}{(M_\star + M)^{2/3}} \left(1 - e^2 \right)^{-1/2}, \qquad (1.14)$$

where M is the mass of the exoplanet and i and e are the inclination and eccentricity of its orbit, respectively.

The preceding expression describes the well-known degeneracy associated with radial-velocity measurements, which is that it is $M \sin i$, and not M, that is being measured. The stellar mass $(M_\star)$ is usually inferred from models of stellar evolution. If the exoplanet transits its star, then $\sin i \approx 1$ and M may be measured. For a hot Jupiter, we have $K_\star \sim 0.1$ km s^{-1}, nearly an order of magnitude faster than a professional sprinter. For the 12-year orbit of Jupiter, this drops to $K_\star \sim 10$ m s^{-1}. For the Earth, we have $K_\star \sim 10$ cm s^{-1}. By comparison, humans tend to walk at speeds ~ 1 m s^{-1}. One may imagine that the technology needed to make these measurements is demanding as one moves towards lower masses.

Precise radial-velocity measurements may also be used to infer the tilt, between the rotational axis of the star and the orbital axis of the exoplanet, projected onto the plane of the sky. A star that rotates produces blueshifted light in the hemisphere rotating towards the line of sight of the astronomer; in the other hemisphere that is rotating away, the light is redshifted. If the axes of the star and the exoplanet are aligned, then as the exoplanet transits the stellar disk it first produces a deficit of blueshifted light, followed by an equal deficit of redshifted light (or vice versa). If the orbit of the exoplanet is misaligned with respect to the rotational axis of the star, then these deficits of blue- and redshifted light are unequal, which is known as the *Rossiter-McLaughlin effect* [167, 200]. It has been used to measure the spin-orbit misalignments of hot Jupiters [197, 241].

More recently, astronomers figured out that high-resolution spectrographs may also be used to study the atmospheres of exoplanets. Since these spectrographs are typically mounted on ground-based telescopes, astronomers have to apply clever techniques to subtract both the stellar and telluric[4] lines from the total spectrum. The resulting ultra-high-resolution[5] spectrum may be used to produce the transmission spectrum [257] or be cross-correlated with theoretical templates of various molecules to identify their presence or absence in the atmosphere of the exoplanet [227, 228].

---

[4]Meaning the spectral lines associated with the Earth's atmosphere.

[5]The resolution is defined as $\lambda/\Delta\lambda$, where λ is the wavelength and $\Delta\lambda$ is the increment in wavelength. In this instance, we are discussing resolutions $\sim 10^5$ compared to the $\sim 10^2$–10^3 usually encountered.

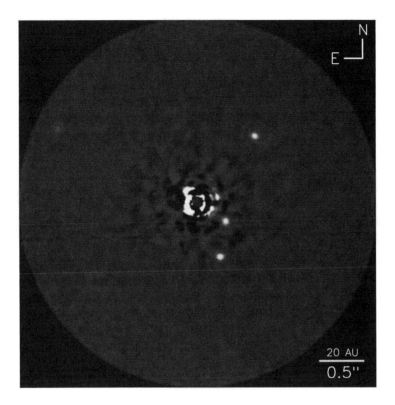

Figure 1.5: The iconic image of the four young ($\sim 10^7$–10^8 years), gas-giant exoplanets orbiting the HR 8799 star, whose light has been suppressed [163, 164]. Courtesy of the National Research Council of Canada, Christian Marois and Keck Observatory.

1.4 DIRECT IMAGING

As its name suggests, direct imaging aims to directly take a photograph (and spectrum) of an exoplanet and its atmosphere. This is a formidable challenge, since the light from the exoplanet versus the star is typically between one part in a million to a billion, depending on whether one observes in the visible or infrared range of wavelengths. It is easier to directly image an exoplanet when it is young and hot, flushed with its remnant heat of formation, in which case the contrast between it and its star may be as high as one part in ten thousand. A technique is needed to block out the light from the star such that the light of the exoplanet may be isolated and the emission spectrum (F) may be measured. Such a feat was achieved for the four gas giants orbiting the HR 8799 star (Figure 1.5) [163, 164], which allowed water and carbon monoxide to be identified in its atmosphere via spectroscopy [13, 128]. The disadvantage of direct imaging is that the radius and mass of the exoplanet cannot be directly measured, as they need to be inferred from spectral and evolutionary models and thus have model-dependent values [13, 154, 161].

1.5   GRAVITATIONAL MICROLENSING

The Polish and Princeton astronomer Bohdan Paczyński was the first to propose that unseen objects in the halo of our Galaxy may act as gravitational lenses, which cause the transient brightening of a background star [183]. He estimated the probability of such an event occurring to be one in a million, implying that a million stars need to be monitored in order for it to be detected. Such reasoning was later applied to the detection of exoplanets [159]. A foreground star lensing a background or source star would produce a smoothly rising and falling light curve, as the light from the latter becomes magnified when the stars move past each other.[6] If an exoplanet is orbiting the foreground star, then it acts to distort this smooth curve in a manner analogous to astigmatism. The relatively high probability ($\sim 10\%$) of detecting this effect arises from the numerical coincidence between the orbital distance of Jupiter-like exoplanets ($a \approx 5$ AU) and the *Einstein radius* of Sun-like stars [75]. The Einstein radius is the characteristic length scale at which gravitational lensing occurs,

$$l_{\text{Einstein}} \sim \frac{(GM_\star l_{\text{lens}})^{1/2}}{c}, \qquad (1.15)$$

and is the geometric mean of the Schwarzschild radius,

$$l_{\text{Schwarzschild}} \sim \frac{GM_\star}{c^2}, \qquad (1.16)$$

and the distance to the lens or foreground star (l_{lens}). Despite being an elegant technique, gravitational lensing is of utility only in detecting exoplanets and not in characterizing their atmospheres. Furthermore, the detection is unrepeatable and it is typically challenging to photometrically distinguish the lens star from the source star.

1.6   FUTURE MISSIONS AND TELESCOPES

Given the dynamism of the field of astronomy, it is somewhat pointless to try and exactly predict the types of space missions and telescopes that will be built and mobilized in the future, but one may perhaps describe the general investments that will be made by astronomers in their hunt for exoplanets. Regardless of its exact configuration, there is clearly a need for a space-based telescope that

---

[6]Yes, you read it correctly: stars in our Galaxy are not static and appear to move with speeds ~ 10–100 km s^{-1}.

will record spectra of exoplanetary atmospheres, from the ultraviolet to the infrared range of wavelengths, and perhaps build a statistical sample[7] of them in order to search for trends in the data. From the ground, telescopes will feature ever bigger mirrors (or arrays of mirrors) and more sophisticated spectrographs with the goal of measuring light from ever fainter sources. Eventually, when the technology is available and mature, humanity may mobilize a fleet of space telescopes, flying in formation, to image a twin of Earth, using a technique known as *interferometry*.

---

[7]At the time of writing, a trio of space-based telescopes dedicated to exoplanet detection (CHEOPS, TESS and PLATO) were being designed and built for the main purpose of greatly increasing the sample size of confirmed exoplanets orbiting nearby, bright stars [98].

Chapter Two

Introduction to Radiative Transfer

2.1 THE OPTICAL DEPTH: THE MOST FUNDAMENTAL QUANTITY IN RADIATIVE TRANSFER

It is possible to sit through an entire radiative transfer class and not develop an intuition for the subject at all, since it is a technical one that threatens to drown the student in formalism. Thus, my approach is not to derive the governing equation of radiative transfer first, but rather to introduce you to the most fundamental quantity one needs to appreciate in order to understand the subject: the optical depth, which we shall denote by τ. The optical depth quantifies if a medium is transparent or opaque, thereby informing us on the approximation that we may use to describe the passage of radiation. It offers us a vocabulary to describe why and how the size of a star or exoplanet varies across wavelength. It identifies the surface of an object from which photons emanate and travel to our telescopes. It allows us to define the transit radius. To develop any intuition for radiative transfer, one needs to thoroughly understand what the optical depth is.

If you stare at clouds in the sky, you will realize that there are large, transparent clouds; there are also small, opaque clouds. Such an observation already informs us that physical size alone is a poor indicator of whether an object is transparent or opaque to radiation, because we are missing two other key ingredients: how loosely or tightly the material is being packed (i.e., the number or mass density) and how absorbent the constituent atoms or molecules are (i.e., the cross section or opacity). Formally, the optical depth is defined as

$$\tau \equiv \int n\sigma \, dx. \tag{2.1}$$

It is the product of the number density (n), a macroscopic quantity, multiplied by the cross section (σ), a microscopic quantity, and integrated across the spatial extent of the object. Other equivalent definitions of the optical depth exist, as we shall see shortly.

When a photon travels through a medium, we may ask how far it will propagate before it is absorbed by an atom or molecule. The answer is that it travels a distance corresponding to an optical depth on the order of unity,

$$\tau \sim 1. \tag{2.2}$$

In Problem 2.8.2, you will learn that τ is roughly the number of interactions the photon has with the medium. The precise value of the optical depth per interaction depends on the detailed properties of the atmosphere, but it currently suffices to understand that it is $\tau \sim 1$. The photon would travel a long distance if the medium was tenuous or poorly absorbent—the formulation of the optical depth automatically takes these two possibilities into consideration. If we return to the thought experiment of clouds in the sky, we now have a vocabulary for describing them: transparent clouds have $\tau \ll 1$, while opaque ones have $\tau \gg 1$. In other words, photons traveling through transparent clouds do not suffer a single absorption[1] event on average, while those penetrating opaque clouds incur multiple absorptions and re-emissions. Another familiar example is when one walks through fog or a snowstorm: the visibility only extends out to a distance corresponding to $\tau \sim 1$.

Nuclear astrophysics teaches us that the Sun has a nuclear reactor residing at its core with temperatures reaching $\sim 10^6$–10^7 K. Applying Wien's law informs us that this corresponds to radiation in the X-ray range of wavelengths, invisible to the naked eye. Yet, the rays of the Sun reach our eyes as visible light. How do we resolve this paradox? It turns out that the core of the Sun resides at $\tau \gg 1$. Instead, it is the surface corresponding to $\tau \sim 1$, known as the *photosphere*, that our eyes see, at least in visible light. It is the surface of last absorption or scattering before the photons stream across space into our telescopes.

Another observed and interesting fact about the Sun is that it has different sizes, depending on whether one observes it at X-ray, ultraviolet, visible or radio wavelengths. This illustrates another important concept regarding the optical depth—it is wavelength dependent.[2] What we call the "radius" of the Sun is its surface corresponding to $\tau \sim 1$. Since the location of this surface depends on wavelength, $\tau \sim 1$ at different wavelengths corresponds to different radii. Thus, the optical depth provides a useful vocabulary for describing why and how the size of an object changes across wavelength.

Now, consider the intense light of a star impinging upon a close-in exoplanet (e.g., a hot Jupiter). How far does the starlight penetrate? Again, it corresponds to the surface where the optical depth is on the order of unity (at a specific wavelength), because such a concept applies symmetrically to both emission from an object and absorption by it. When an exoplanet transits its host star, the measured *transit radius* always picks out the chord, cutting across the limb of its atmosphere, corresponding to a chord optical depth on the order of unity (Figure 2.1).

Finally, we note that the *magnitude* of the optical depth determines how the radiation behaves as it passes through the medium. When $\tau \ll 1$, radiation is in the *freely streaming limit* and propagates at the speed of light. When $\tau \gg 1$, it is absorbed and re-emitted multiple times within a short distance, causing its

---

[1] The same arguments also apply to scattering events.

[2] As another example, we note that styrofoam is opaque to visible light, but it is transparent in the microwave.

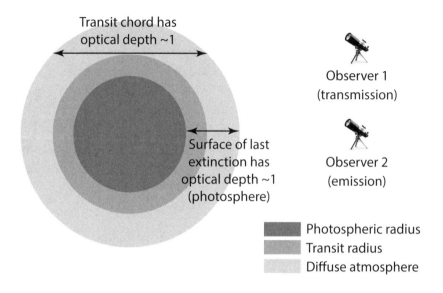

Figure 2.1: Schematic depicting the photospheric and transit radii. Observer 1 measures the transit radius of the exoplanet, which is the radius corresponding to a transit chord of optical depth ~ 1. Observer 2 measures photons emanating from the photosphere, which is the surface of last absorption or scattering. The optical depth between this surface and Observer 2 is ~ 1. The transit radius is generally larger than the photospheric radius [25], although not to the extent it has been exaggerated in this schematic.

overall passage through the medium to resemble diffusion. In Chapter 3, we will discuss one of the simplest implementations of radiative transfer known as the *two-stream approximation*, which is mostly designed to work when $\tau \lesssim 1$.

2.2   BASIC QUANTITIES IN RADIATIVE TRANSFER

2.2.1   Opacity, cross section and extinction coefficient

There are three ways of expressing the optical depth, because there is an equal number of ways to quantify how absorbent a medium is. We will use the term *extinction* to collectively describe both absorption and scattering.

The first measure of extinction is the cross section (σ), which we have already described. It may be visualized as the area of the "target" provided by an atom, molecule or particle—if a collidant passes within this target area, a collision will occur. The cross section per unit mass is the *opacity*, which we will generally denote by κ. A column of atmosphere has $\tilde{m}$ of mass per unit area—it is aptly termed the *column mass*. Thus, another definition of the optical depth, which is more appropriate for atmospheric studies, is

$$\tau \equiv \int \kappa \, d\tilde{m}. \tag{2.3}$$

The relationship between the pressure and the column mass is essentially New-ton's second law (cast in a per-unit-area form): $P = \tilde{m}g$, where g is the surface gravity of the exoplanet.

Yet another way of defining the optical depth comes from using the *extinction coefficient* (α_e).

$$\tau \equiv \int \alpha_e \, dx. \tag{2.4}$$

Its reciprocal is the mean free path traveled by the photon ($l_{\mathrm{mfp}} = 1/\alpha_e = 1/n\sigma$).

A source of confusion originates from the fact that some references refer to the cross section, opacity and extinction coefficient collectively as the "extinction coefficient,"[3] despite the fact that σ, κ and α_e have physical units of cm^2, cm^2 g^{-1} and cm^{-1}, respectively. It is my hope that you will use these terms more judiciously.

2.2.2 The extinction efficiency

Consider a spherical particle with a radius of r. Its geometric cross section is πr^2. However, its actual cross section for extinction is generally larger than πr^2,

$$\sigma = Q_e \pi r^2. \tag{2.5}$$

The factor Q_e is known as the *extinction efficiency* [191]. Its functional form and value depend on whether a particle is small or large. It may be separated into its absorption (Q_a) and scattering (Q_s) components: $Q_e = Q_a + Q_s$.

The beauty about physics is that whether something is "small" or "large" is not a subjective judgment based on human emotion. For the interaction of radiation with matter, it is in comparison to the wavelength (λ). Specifically, a particle is small if $2\pi r/\lambda \ll 1$ and large if $2\pi r/\lambda \gg 1$. For example, a micron-sized particle is (radiatively) large when it interacts with radiation in the visible range of wavelengths, but it is small when the interaction is with photons in the far infrared.

To derive the exact functional form of Q_e requires a full-blown calculation of absorption and scattering by a given particle [214]. In practice, we, as atmospheric scientists, look up tables of $Q_e(\lambda, r)$ and apply them to our models [52, 53, 136]. Roughly speaking, the extinction efficiency has a somewhat universal form,

$$Q_e \sim \min\left\{\left(\frac{2\pi r}{\lambda}\right)^4, 1\right\}, \tag{2.6}$$

---

[3]For example, see Appendix 2 of the textbook by Goody & Yung [74].

with a transition from *Rayleigh scattering* ($2\pi r/\lambda \ll 1$) to scattering by large particles ($2\pi r/\lambda \gg 1$) occurring at a value of the normalized particle radius that depends on its composition. With the exception of resonant transitions that are specific to each material, the overall shape of the extinction efficiency is similar for different members of, for example, the same silicate family (e.g., the olivine group).

2.2.3   The single-scattering, geometric, spherical and Bond albedos

If we now distinguish between the absorption (σ_a) and scattering (σ_s) cross sections, then the *single-scattering albedo* is defined as

$$\omega_0 \equiv \frac{\sigma_s}{\sigma_s + \sigma_a}. \tag{2.7}$$

It quantifies the fraction of light reflected or scattered by a particle during a single scattering event. Another quantity needed to fully quantify scattering by a single particle is the *scattering asymmetry factor* (g_0), which describes how isotropic or anisotropic the scattering is; we will discuss g_0 in greater detail in Chapter 3. In practice, both $\omega_0(\lambda, r)$ and $g_0(\lambda, r)$ are tabulated quantities one obtains from detailed calculations [136]. If we have $g_0 = 0$, then the incident light is scattered equally in all directions (i.e., isotropically).[4] If we have $g_0 > 0$, then it is peaked in the forward direction; if it is $g_0 < 0$, then backscattering prevails. Even without engaging in first-principles calculations of g_0, it is interesting to note its asymptotic behavior [191],

$$g_0 \rightarrow \begin{cases} 0, & \frac{2\pi r}{\lambda} \ll 1, \\ 1, & \frac{2\pi r}{\lambda} \gg 1. \end{cases} \tag{2.8}$$

Evidently, small particles scatter isotropically, while large particles preferentially deflect radiation into the same direction from which it came. For example, a sub-micron-sized particle will scatter infrared radiation isotropically, but engage in the forward scattering of X-rays.

   An atmosphere is composed of an enormous number of particles of different species, each with its own absorption and scattering properties. Collectively, they determine the fraction of incident light that is reflected out of the atmosphere at a given wavelength. As has been discussed by Seager [208], a historical and commonly used form of the albedo is the *geometric albedo* (A_g), which is the brightness of a planet or moon compared to an illuminated Lambertian disk[5] at a specific wavelength and full phase. This somewhat awkward definition means that it is possible to have $A_g > 1$. The geometric albedo is *not* the fraction of incident starlight scattered by an atmosphere.

---

   [4]Strictly speaking, $g_0 = 0$ corresponds to a symmetric phase function. For example, Rayleigh scattering has $g_0 = 0$ but also exhibits a slight departure from isotropy. Conversely, the Henyey-Greenstein scattering phase function displays perfect isotropy when $g_0 = 0$.
   [5]Isotropically scattering disk.

As the exoplanet orbits around its star, it shows different faces of its scattering surface to the astronomer, corresponding to different orbital *phases*. Integrating over all solid angles yields the *spherical albedo* [160, 204, 208, 237],

$$A_s = A_g \mathcal{P}_g, \tag{2.9}$$

with $\mathcal{P}_g$ being the *phase integral*. In other words, the geometric albedo is the spherical albedo at zero phase angle, which is also the moment of secondary eclipse or superior conjunction for an astronomer monitoring a transiting exoplanet. The relationship between the geometric and spherical albedos depends on the details of the scattering surface. For example, a Lambertian sphere (i.e., isotropic scattering) has $\mathcal{P}_g = 3/2$ [204], while Rayleigh scattering yields $\mathcal{P}_g = 4/3$ [208].

If the spherical albedo is in turn integrated over all wavelengths, we obtain the *Bond albedo* [160, 237],

$$A_{\mathrm{B}} \equiv \frac{\int I_\star A_s \, d\lambda}{\int I_\star \, d\lambda}, \tag{2.10}$$

where $I_\star$ is the wavelength-dependent intensity of the star. Generally, $I_\star$ is not a Planck function. Essentially, the spherical albedo is integrated over all wavelengths, weighted by the stellar spectrum. The Bond albedo is a true measure of the incident energy that is scattered, and not absorbed, by the atmosphere.

In an astronomical context, the geometric albedo for a transiting exoplanet is directly obtained from measuring its secondary eclipse [28, 208],

$$A_g = \frac{F}{F_\star} \left(\frac{a}{R} \right)^2, \tag{2.11}$$

where $F/F_\star$ is the ratio of the flux of the exoplanet to its star *as observed at Earth*, a is the exoplanet-star separation and R is the radius of the exoplanet. In practice, this is done via photometry for a range of visible wavelengths, rather than at a single wavelength, which yields the wavelength-integrated geometric albedo. By assuming[6] the form of $\mathcal{P}_g$, one obtains the wavelength-integrated spherical albedo. If the range of wavelengths over which the secondary eclipse is measured encompasses the blackbody function of the star, then the wavelength-integrated spherical albedo is the Bond albedo. An important caveat is that the exoplanetary atmosphere is cool enough that its thermal emission is not contributing significantly in the same range of visible wavelengths [91]. This caveat may be violated for very hot exoplanets orbiting Sun-like stars or temperate exoplanets around cool red dwarfs.

In Chapter 4, we will see that there is a relationship between A_{B}, ω_0 and g_0 within the two-stream approximation of radiative transfer. Generally, this relationship is non-trivial to elucidate.

---

[6]At least, until astronomical techniques are advanced enough to measure the phase integral.

2.3   THE RADIATIVE TRANSFER EQUATION

We now derive the governing equation of radiation traveling through matter. Sophisticated derivations of the radiative transfer equation exist [74, 172], but I will opt for the shortest possible derivation. Namely, instead of working with length or spatial distance, we will directly work with the optical depth, since we are now convinced that it is the correct measure of transparency or opaqueness. Let the *intensity*[7] (energy per unit area, time, wavelength and solid angle subtended) of radiation be given by I. Its physical units are erg cm^{-3} s^{-1} sr^{-1}.

In general, the intensity impinges upon the surface of an object at an angle θ; we shall characterize this angle by its cosine, $\mu \equiv \cos\theta$. We choose the sign convention for μ to be such that reflection from this surface corresponds to $\mu > 0$, while penetration past the surface and into the object has $\mu < 0$. With this convection, the loss in intensity due to extinction is given by I/μ. This loss may be compensated by the fact that the object generally has a finite temperature and thus emits its own thermal emission $(-S/\mu)$. The quantity S is aptly termed the *source function*. Thus, the change in intensity (ΔI), across the optical depth corresponding to the extent of the object $(\Delta\tau)$, is

$$\Delta I = \frac{\Delta\tau}{\mu}\left(I - S\right). \tag{2.12}$$

If we take $\Delta\tau$ to be infinitesimally small, then we obtain the *radiative transfer equation*,

$$\mu\frac{\partial I}{\partial\tau} = I - S. \tag{2.13}$$

It looks deceivingly simple, because all of the complexity has been hidden within the source function. Besides thermal emission, S may include flux scattered from other directions into the intensity beam under consideration. It is often forgotten that the radiative transfer equation was first derived for stars by astrophysicists and later made its way into the atmospheric and climate sciences [191].

2.4   SIMPLE SOLUTIONS OF THE RADIATIVE TRANSFER EQUATION

2.4.1   Beer's law

The simplest solution of the radiative transfer equation comes from discarding the source function $(S = 0)$, which yields

$$\int_{I_0}^{I}\frac{dI}{I} = \int_{0}^{\tau}\frac{d\tau}{\mu}. \tag{2.14}$$

---

[7] A potential source of confusion comes from the fact that atmospheric scientists sometimes prefer the terms *radiance* and *irradiance* when referring to the intensity and flux, respectively. I categorically avoid the use of these terms.

The integration is performed from the surface of the object, which we marked by $\tau = 0$, to some arbitrary depth or distance within it. The incident or initial intensity impinging upon the surface is given by I_0.

If we integrate the preceding expression, we obtain

$$I = I_0 e^{\tau/\mu}. \tag{2.15}$$

This is *Beer's law*, which tells us that the intensity passing through the object is being diminished exponentially. In this limit, $\tau = 1$ corresponds to one e-folding of the intensity, i.e., about 37% of the intensity beam is removed when an optical depth of unity is traversed. Note again that $\mu < 0$ in our convention.

In Chapter 4, we will derive a generalization of Beer's law that includes non-isotropic scattering. To a very good approximation, Beer's law describes how starlight is absorbed when it is incident upon an exoplanetary atmosphere. It is less appropriate for describing how the thermal emission of an exoplanet behaves, because in this case radiation has both incoming and outgoing components and the source function cannot be neglected.

2.4.2   Direct solution (pure absorption only)

Beyond Beer's law, the radiative transfer equation may be solved directly and analytically only in the limit of pure absorption. By "direct," we mean that one is solving for the intensity, rather than integrating the equation first over μ and solving for the angle-integrated quantities. To prove this point, we include coherent scattering in the radiative transfer equation and demonstrate that it is soluble only when scattering is absent. By "coherent," we mean that scattering does not alter the wavelength of a photon. The opposite limit is that of complete redistribution (or complete non-coherence), where the wavelength information is randomly distributed over some range [172]. Strictly speaking, spectral lines are poorly described by coherent scattering and better approximated by complete redistribution over the line profiles. However, if one is examining a wavelength interval or bin that is wide compared to the widths of individual lines, then coherent scattering is a decent approximation.

In the limit of coherent, isotropic scattering, the source function becomes [172],

$$S = \frac{\omega_0 J}{4\pi} + (1 - \omega_0) B, \tag{2.16}$$

where B is the blackbody or Planck function. The *total intensity* is

$$J \equiv \int_0^{4\pi} I \, d\Omega, \tag{2.17}$$

where $d\Omega = d\mu \, d\phi$ is the solid angle subtended and ϕ is the azimuthal angle. The limiting values of the source function appear to make sense: $S = B$ for pure absorption ($\omega_0 = 0$) and $S = J/4\pi$ (the mean intensity) for pure scattering ($\omega_0 = 1$).

With this choice of S, the radiative transfer equation becomes

$$\mu\frac{\partial I}{\partial \tau} = I - \frac{\omega_0 J}{4\pi} - (1 - \omega_0) B. \tag{2.18}$$

Note that we have intentionally not labeled the intensity, source function and Planck function with a subscript, because our formulation applies regardless of whether these quantities are cast in per wavelength, per frequency or per wavenumber units. In practice, they may also apply to intervals or bins in wavelength, frequency or wavenumber. An argument for using frequency or wavenumber units is that spectral lines are more evenly distributed across them, rather than across wavelength.

Consider the transfer of radiation between two points labeled by 1 and 2. By integrating the equation between these points, we obtain

$$I_2 e^{-\tau_2/\mu} - I_1 e^{-\tau_1/\mu} = -\frac{1}{\mu}\int_{\tau_1}^{\tau_2}\left[\frac{\omega_0 J}{4\pi} + (1 - \omega_0) B\right] e^{-\tau/\mu}\, d\tau. \tag{2.19}$$

Written in this form, it becomes apparent why one cannot proceed unless $\omega_0 = 0$, because one does not know the functional form of $J(\tau)$ a priori.

We now specialize to an atmosphere—we account for radiation traveling either upwards or downwards and subscript the quantities by $\uparrow$ and $\downarrow$, respectively.[8] We integrate equation (2.19) over all angles, while assuming that I_1, I_2 and B are constant. By writing the *fluxes* (energy per unit area, time and wavelength) as $F = \pi I$ (with the appropriate subscripts), we obtain

$$\begin{aligned}F_{\uparrow_1} &= F_{\uparrow_2}\mathcal{T} + \pi B\left(1 - \mathcal{T}\right),\\ F_{\downarrow_2} &= F_{\downarrow_1}\mathcal{T} + \pi B\left(1 - \mathcal{T}\right).\end{aligned} \tag{2.20}$$

The *transmission function* takes the form,

$$\mathcal{T} \equiv 2\int_0^1 \mu e^{-\Delta\tau/\mu}\, d\mu, \tag{2.21}$$

where $\Delta\tau \equiv \tau_2 - \tau_1$ is the difference in optical depth between the two points and we demand $\tau_2 > \tau_1$ as a property of the coordinate system. The transmission function has a convenient range of values between 0 and 1. When $\mathcal{T} = 1$, no radiation is absorbed and blackbody emission is not produced. The incident flux simply passes right through the pair of layers—we have a completely transparent medium. When $\mathcal{T} = 0$, nothing passes through and the pair of layers radiates purely as a blackbody.

Writing the solutions in this form allows us to describe the transfer of radiation from Points 1 to 2 and vice versa, using either location as a starting point or boundary condition. In Chapter 3, we will see that formulating the problem as being radiative transfer between pairs of layers is the basis of the two-stream treatment.

---

[8]The directions of up and down correspond to $0 \le \theta \le \pi/2$ and $\pi/2 \le \theta \le \pi$, respectively.

2.5   A PRACTICAL CHECKLIST FOR RADIATIVE TRANSFER CALCULATIONS

Now that we have developed a basic intuition for radiative transfer and the various quantities associated with it, it is time to list all of the necessary ingredients required to perform a numerical calculation. They include:

- Specifying the appropriate form of the radiative transfer equation and any approximations taken, checking that the assumptions made are plausible and selecting the numerical method. In Chapter 3, we examine the *two-stream solutions* of the radiative transfer equation. Solutions of the radiative transfer equation are sometimes known as *forward models*, as they take a set of chemical abundances and opacities, compute forward, and translate them into a synthetic spectrum.

- Setting up a model atmosphere with a finite number of discrete layers, each with its own temperature, pressure and column mass, and calculating the fluxes passing through each layer. For example, equation (2.20) is a forward model in the limit of pure absorption and isothermal layers. To use it, we need the opacities as inputs, which allows us to compute the optical depths, transmission functions and fluxes associated with each layer. Knowledge of the fluxes allows the temperature of each layer to be calculated self-consistently.

- Deciding upon the chemical composition of the atmosphere being investigated (Chapter 7) and obtaining the opacities of the atoms, molecules and aerosols or condensates involved, typically from pre-computed tables (Chapter 5). Combining all of the opacities, weighted by the relative abundance of each species, across temperature, pressure and wavelength, yields the *opacity function* of the atmosphere.

- Numerically iterating to obtain a converged solution, since the opacities, optical depths and temperature are all interdependent quantities. The solution is judged to have converged once the temperature in each layer ceases to evolve, which in practice means that the change in temperature, between time steps, is less than a stated numerical threshold.

The converged solution yields the flux emerging from the top of the atmosphere, which is its synthetic spectrum. The temperatures in all of the layers collectively yield the temperature-pressure profile of the model atmosphere.

How do we relate the temperature and fluxes associated with each layer of the atmosphere? In the absence of atmospheric dynamics, one hopes to solve for *radiative equilibrium*, which occurs when the heat entering and exiting each layer vanishes. It is a statement of *local* energy conservation and originates from the first law of thermodynamics (Chapter 9),

$$\rho c_P \frac{DT}{Dt} = \frac{DP}{Dt} + \rho Q, \qquad (2.22)$$

where ρ is the mass density, c_P is the specific heat at constant pressure and Q is the term associated with heating. If we focus just on radiative heating (and ignore convection and conduction), we have

$$\rho Q = -\nabla.\mathcal{F}_-, \tag{2.23}$$

where $\mathcal{F}_-$ is the wavelength-integrated net flux. The net flux of a layer is the *difference* between the flux entering and exiting it, which explains why I have labeled it with a minus sign. Geometrically, we can see that the preceding expression makes sense: when radiation emanates from the layer, the divergence of the net flux is positive, which leads to cooling ($Q < 0$). The opposite happens when radiation enters the layer. The minus sign is crucial in this respect—if it is missing, one gets the unphysical situation of runaway cooling or heating occurring.

If we ignore work done on the system ($\frac{DP}{Dt} = 0$) and atmospheric dynamics ($\frac{D}{Dt} = \frac{\partial}{\partial t}$), then we obtain the equation that is typically used to solve for radiative equilibrium in a one-dimensional model of an atmosphere,

$$\frac{\partial T}{\partial t} = -\frac{1}{\rho c_P}\frac{\partial \mathcal{F}_-}{\partial z}. \tag{2.24}$$

The computational iteration between the fluxes and temperatures, which depend on each other, is performed until the temperature stops changing in time,[9] at which point radiative equilibrium is formally obtained. Note that the spatial coordinate z increases with altitude.

Strictly speaking, if a calculation does not enforce radiative equilibrium, then we should not consider it to be a bona fide radiative transfer calculation. Radiative equilibrium is often confused with *global energy conservation*, which is the statement that the energy entering and exiting the atmosphere, as a whole, must equate. The latter is a necessary but insufficient condition for the former, a fact that is often unappreciated. We will cast this statement in precise mathematical terms in Chapter 4.

2.6   CLOUDS

The formation of aerosols or condensates in a gaseous environment is one of the outstanding puzzles of modern astrophysics and affects several of its branches, including that of exoplanetary atmospheres [162]. It is also a puzzle for climate scientists on Earth. In this textbook, I will use the terms *cloud, haze, aerosol* and *condensate* interchangeably, while being aware that there are subtle differences in their meaning related to their formation mechanisms. To compound matters, the usage of terminology differs between the various disciplines, so care must be

---

[9]In practice, one simply needs to get $\frac{\partial T}{\partial t}$ to reach a value that is below some numerical threshold (e.g., 10^{-6}). Obtaining a precise zero is impossible in such an iterative numerical calculation.

taken when reading the published literature. For example, hazes are typically associated with photochemical products,[10] while clouds follow the condensation curves set by thermodynamics. Forming clouds from first principles, out of the atmospheric gas, is a daunting theoretical challenge as it requires that we understand the following steps:

- Elucidating the details of *nucleation*, whereby the precursor seed particles form and onto which the further growth of the cloud particles may proceed [84]. A true theory should predict the shape, composition and size distribution of cloud particles formed.

- Calculating the gaseous and solid phases of the atmospheric chemistry self-consistently, meaning that as the gas gets depleted to form solids, its atomic and molecular abundances are adjusted accordingly.

- Treating the radiative transfer and opacities of the atmospheric gas and solid cloud particles self-consistently, including getting the details of absorption and scattering, associated with each component, correct. In other words, the opacities for both the gas and cloud are calculated from first principles (using quantum mechanics), across wavelength, temperature and pressure, rather than being described by ad hoc parameters.

- Performing this set of calculations self-consistently[11] against the dynamical background of the atmosphere. To be kept aloft, the aerosol needs to be dynamically supported, meaning that its terminal velocity, due to the action of gravity, is cancelled out by upwelling atmospheric flow. *Sedimentation* occurs when the condensed solids sink deeper into the atmosphere, out of reach of the photosphere.

It is difficult to *form* the cloud particles from first principles, but once they do it is hardly controversial to model their effects on the computed spectrum. For our purposes, the term *aerosol* is the most neutral one—it refers to any particle that may be described by its size, composition and index of refraction (or extinction efficiency).

To lowest order, the main effect of the aerosol is to introduce an extinction efficiency into the radiative transfer calculation. Crudely speaking, it may be approximated by this fitting formula [139],

$$Q_e = \frac{5}{Q_0 x^{-4} + x^{0.2}}, \qquad (2.25)$$

where $x \equiv 2\pi r / \lambda$, r is the radius of the aerosol (which is assumed to be spherical) and λ is the wavelength. When $x \gg 1$, the extinction efficiency asymptotes to

---

[10]This is the accepted view in the planetary sciences, but the term "haze" in the Earth sciences is used as a measure of particle size.

[11]Rather than being parametrized by an *eddy mixing coefficient* (usually denoted by K_{zz}), which is essentially the use of a diffusion coefficient to crudely describe large-scale atmospheric circulation.

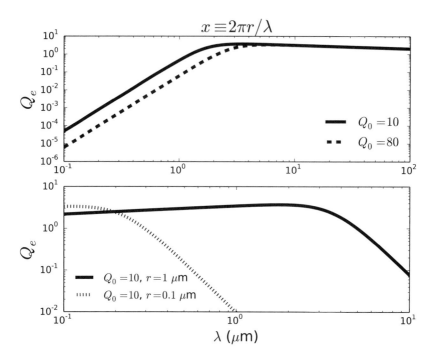

Figure 2.2: The extinction efficiency of spherical cloud particles versus the normalized radius (top panel) and wavelength (bottom panel). It is important to note that the curves in the top panel represent *all* particle radii. In other words, the extinction efficiency as a function of the normalized radius (x) is size-independent to lowest order. The different values of $Q_0 = 10$ and 80 mimic the effects of cloud composition, with the lower and higher values representing refractory and volatile species, respectively. The bottom panel shows the extinction efficiency associated with refractory cloud particles of two different radii: $r = 0.1$ and 1 μm.

a roughly constant value. When $x \ll 1$, it describes Rayleigh scattering by small particles. The dimensionless parameter Q_0 is a proxy for composition: refractory species (e.g., silicates) tend to have $Q_0 \approx 10$, while volatile ones have $Q_0 \approx 40$–80 (e.g., water).

Figure 2.2 shows the extinction efficiency associated with cloud particles of different sizes and compositions. The particle radius exerts a somewhat larger influence on the extinction efficiency than the composition. Generally, the influence of a cloud particle on the absorption and scattering of radiation is only felt at $\lambda \lesssim 2\pi r$. This insight generalizes straightforwardly to a cloud with a size distribution of particles.

To lowest order, these simple estimates allow us to glean some intuition for what clouds would do to the spectrum of an exoplanetary atmosphere. When the optical depth approaches unity, spectral features and lines become muted and "less sharp" for $\lambda \lesssim 2\pi r$ [23]. However, this effect may be negated by increasing the abundances of the atomic and molecular species responsible for these features and lines. In this way, one may understand how clouds introduce a degeneracy to the interpretation of chemical abundances in an atmosphere. To the next order, cloud particles are described by a single-scattering albedo and scattering asymmetry factor. A full-fledged radiative transfer calculation would take the wavelength- and layer-dependent nature of these quantities into account when implementing, for example, the two-stream solutions. More accurate calculations would employ *multi-stream* treatments [35].

2.7 ATMOSPHERIC RETRIEVAL

Traditionally, when one builds a model of an atmosphere, one starts with a governing equation—either of fluid dynamics, radiation, chemistry or some combination of them—and a set of assumptions and computes. Ideally, the calculation predicts some measurable quantity that may be confronted by the astronomical observations. A judgment may then be made on whether the model has successfully represented Nature in some way, but only against the preconceived notions we have about an atmosphere.

Atmospheric retrieval tries to shed itself of these preconceived notions by providing a framework in which one may invert a set of observables to obtain the properties of an atmosphere, such as its thermal structure or chemical abundances [18, 138, 139, 143, 144, 151, 152]. It has its origin in the atmospheric sciences, where inversion techniques were used to analyze remote sensing observations of Earth [199]. It has also been used in the planetary sciences to analyze the atmospheres of the Solar System bodies [109]. Typically, there is a combination of satellite and in-situ data to provide some ground truth to anchor the retrieval. As my colleagues in these fields of inquiry like to say, "We can simply let the data do the talking."

This faith in the data alone becomes less viable in the study of exoplanetary atmospheres, because we are essentially dealing with the remote sensing of unresolved point sources and in-situ measurements are out of the question. Furthermore, the data are often sparse and fragmented in the sense that one only sees incomplete facets of the exoplanetary atmosphere (e.g., its dayside but not its nightside, insufficient wavelength coverage) that do not allow for unique interpretations. Letting our interpretations be guided entirely by the data—for example, without enforcing radiative or chemical equilibrium—becomes foolhardy when the number of model parameters exceeds the number of data points. Our interpretation needs to be anchored by the laws of physics and chemistry to some degree. This is an ongoing debate.

In this section, we will discuss the basic ingredients and computational procedure that go into constructing and implementing an atmospheric retrieval model. We will then discuss the pitfalls involved in using these inversion techniques.

2.7.1   Basic ingredients for constructing a retrieval model

Earlier in the chapter, we already summarized the basic ingredients needed for radiative transfer. Here, we augment that list for atmospheric retrieval.

- The first ingredient one needs is a way of describing the temperature-pressure profile of the atmosphere using a set of parameters. One may either use an ad hoc fitting function [109, 151] or a simplified solution of the radiative transfer equation [18, 143, 144]. Chapter 4 discusses the latter.

- Next, one needs to specify the *identities* of the atoms and molecules being included in the retrieval (e.g., water, methane, carbon monoxide). The relative abundances of these atoms and molecules may be treated as being free parameters of the retrieval model. Alternatively, one may wish to enforce chemical equilibrium, where the molecular abundances are determined by the elemental abundances. When chemical equilibrium is enforced, the only free parameters present are the elemental abundances. Chapter 7 discusses simple models for computing the relative abundances of molecules in chemical equilibrium.

- For a fixed combination of parameter values for the temperature-pressure profile and chemical abundances, one may use a forward model to convert a set of opacities into a synthetic spectrum. In retrieval models, this conversion is performed once for each set of parameter values. There is typically no attempt to solve for radiative equilibrium, as the numerical iteration previously described is not performed.

Millions, if not billions, of combinations of these parameters are considered. Each synthetic spectrum is then compared against the observed one and a goodness of fit is computed for each model. In this manner, a range of best-fit models are found given the uncertainties on the data and the posterior distributions of the molecular abundances may be computed. In practice, a minimization algorithm is required to efficiently and thoroughly scan this large, multi-dimensional parameter space.

It is beyond the scope of the current edition of this textbook to cover the minimization algorithms used in atmospheric retrieval, but we will briefly mention them. The *optimal estimation method* exploits the property that if the prior distribution of an input quantity takes the form of a Gaussian function, then the expressions used for minimization are simplified [199]. If one desires a uniform prior, then the width of the Gaussian is taken to be very large to mimic this property [138]. *Monte Carlo methods* allow one to consider more general functional forms for the priors and have been implemented in various flavors [18, 143, 144, 151, 152].

2.7.2   Pitfalls

A retrieval model is only as good as the assumptions and techniques used. Here, we list some of the possible pitfalls.

2.7.2.1   Degeneracies

The spectral lines associated with different molecules inevitably overlap across wavelength, which naturally leads to degeneracies between their relative abundances as multiple combinations of relative abundances may lead to the same net absorption at a given wavelength. Aerosols or clouds introduce another form of degeneracy, as they act to diminish the strength of spectral lines, an effect that may be compensated by increasing the abundances of the molecules associated with these lines. Any good retrieval model will quantify these degeneracies in a formal and systematic way.

2.7.2.2   Being misled by your priors

The posterior distributions of the properties of the atmosphere inferred from the retrieval calculation are dependent on the *priors* assumed. As an example, the carbon-to-oxygen ratio (C/O) may be estimated using,

$$C/O \approx \frac{\tilde{n}_{CO} + \tilde{n}_{CH_4} + \tilde{n}_{CO_2}}{\tilde{n}_{CO} + \tilde{n}_{H_2O} + 2\tilde{n}_{CO_2}}, \tag{2.26}$$

where $\tilde{n}$ generally denotes the *mixing ratio* of a molecule, which is its abundance normalized by that of molecular hydrogen (for a hydrogen-dominated atmosphere); self-explanatory subscripts indicate the specific molecule being referred to. If carbon monoxide (CO) is the dominant molecule, then we have $C/O \approx 1$. If carbon dioxide (CO_2) is dominant, then we have $C/O \approx 0.5$. If methane (CH_4) or water (H_2O) are dominant, then we have $C/O \gg 1$ and $C/O \ll 1$, respectively. In other words, even if we assume the prior distributions of these molecules to be uniform, we produce two artificial peaks in the prior distribution of C/O [143].

2.7.2.3   Local versus global energy conservation

In the ensemble of models being computed by a retrieval calculation, it is natural to discard those where the emergent thermal emission from the exoplanetary atmosphere exceeds the energy input from the star (assuming that the exoplanet produces negligible internal heat) [151]. This is simply a statement of global energy conservation—you cannot produce more than what you put in. What is less obvious is *local* energy conservation, which is radiative equilibrium—the net heating or cooling of each layer of the model atmosphere must be the same across layers (see Chapter 4) in the absence of atmospheric circulation.

2.7.2.4   Not everything is chemically possible

When an atmosphere is in chemical equilibrium, not every combination of the relative abundances is possible. As each layer of the model atmosphere has its own values of temperature and pressure, the mixing ratios of molecules take on specific values given the elemental abundances. One is not free to specify the mixing ratios as free parameters. When atmospheric dynamics is present and the dynamical mixing time scale is less than the chemical time scale, it is possible for the mixing ratios to depart from chemical equilibrium given the local values of temperature and pressure—but again, not every combination of mixing ratios is possible. If we invoke the quenching approximation (Chapter 6), then the mixing ratios originate from a deeper part of the atmosphere that is in chemical equilibrium given its local values of temperature and pressure. As another example, consider carbon dioxide in hot ($\gtrsim$ 1000 K), hydrogen-dominated atmospheres, which is never more abundant than carbon monoxide or water unless the metallicity is implausibly high [100].

2.7.2.5   Opacities: The devil is in the details

The task of calculating atmospheric opacities is a subtle and detail-oriented business. Across the ultraviolet to infrared range of wavelengths, one has to compute the shapes of millions to billions of lines—and this is for one molecule at a specific pairing of temperature and pressure. *Line-by-line calculations* require that one sample *all* of these lines with no approximation taken, which is a formidable task when the atmospheric temperatures attain $\sim$ 1000 K. Binning techniques such as the *k-distribution method* may be employed to speed up the calculation, but this comes at the cost of invoking a set of approximations (see Chapter 5). Hidden in all of these details is the issue of the pressure broadening of the far wings of spectral lines, which remains an unsolved physics problem. Furthermore, photometric observations of exoplanetary atmospheres record radiation integrated over a broad bandpass, implying that many combinations of molecules could produce the same opacity across this wavelength range. When you read published papers that omit these details, you should exercise one of the most precious instruments in your toolkit as a scientist: skepticism.

2.7.2.6   Numerical convergence

Running a retrieval model is a technically demanding task, as one needs to discretize space (via the temperature-pressure profile) and wavelength (the opacity function) and also consider a set of atoms and molecules. To complete the calculation within a reasonable amount of time, one often has to use a coarse numerical resolution across one of these axes. To determine if the resolution is sufficient requires a set of resolution tests to be performed—it is a thankless but necessary step. And as a rule of thumb, one should always adopt the *second-highest* resolution if the highest one attains convergence.[12]

2.8   PROBLEM SETS

2.8.1   The Sun

(a) The photosphere of the Sun has a temperature of about 5800 K. Using Wien's law, what is the peak wavelength of its blackbody function? What color of light does this peak correspond to?

(b) The *chromosphere* resides *above* the photosphere and has temperatures $\sim 10^4$ K. Why do we generally not see the chromosphere in visible light? At what wavelengths do we detect the chromosphere?

2.8.2   The random walk of a photon

Photons created at the center of the Sun must travel a distance of $R_\odot$ to escape from it, where $R_\odot \approx 700,000$ km is the solar radius. Each photon travels a distance of about the mean free path (l_{mfp}) before being absorbed and re-emitted (or scattered).

(a) If the Sun were completely transparent, estimate the time it would take a photon to escape.

(b) Imagine if the Sun were not too opaque such that the center-to-surface optical depth (τ) were between 1 and 10. Convince yourself that the number of absorption/scattering events would be about $R_\odot/l_{mfp}$. Show that the number of events would be $N \sim \tau$.

(c) Imagine if the Sun were very opaque such that the center-to-surface optical depth greatly exceeded unity. On average, the displacement of the photon would be zero. But the mean of the square of the displacement would be $N l_{mfp}^2$. Thus, show that $N \sim \tau^2$.

(d) Consider the case of the opaque Sun. Taking the number density to be $n \sim 10^{24}$ cm^{-3} (pure hydrogen) and the cross section to be $\sigma \sim 10^{-24}$ cm^2 (Thomson scattering), estimate the value of l_{mfp}. Estimate the time taken for a photon to leak out of the Sun from its center.

(e) What is a key weakness of this analysis? (Hint: it is *not* that the numbers are imprecise.)

2.8.3   The greenhouse effect on Earth

A naive estimate of the temperature on Earth would lead us to the conclusion that our atmosphere is warmer than expected. This greenhouse warming is caused by the variation in optical depth, across wavelength, of the Earth's atmosphere.

---

[12]A piece of advice I attribute to Scott Tremaine.

(a) On average, the Earth orbits the Sun at a distance of a. Let the effective temperature of the Sun be $T_\star$ and its radius be $R_\star$. Let the Bond albedo of the Earth's atmosphere be A_B. If we take $A_B = 0.3$, estimate the equilibrium temperature,

$$T_{eq} = T_\star \left(\frac{R_\star}{2a}\right)^{1/2} (1 - A_B)^{1/4}, \tag{2.27}$$

at Earth. Convince yourself that the equilibrium temperature corresponds to the incident solar flux averaged over the entire surface of the Earth. Is T_{eq} less or greater than the freezing point of water?

(b) What is the optical depth of the Earth's atmosphere to sunlight?

(c) Sunlight absorbed by the Earth is reprocessed into infrared emission (such that the second law of thermodynamics is obeyed). What is the optical depth of Earth's atmosphere to the infrared emission?

2.8.4 The various forms of the Planck function

A common pitfall for novices of radiative transfer is the failure to properly elucidate the various forms of the Planck function. When written in per wavelength units, it is

$$B_\lambda = \frac{2hc^2}{\lambda^5} \left(e^{hc/\lambda k_B T} - 1\right)^{-1}. \tag{2.28}$$

where h is the Planck constant, c is the speed of light, k_B is the Boltzmann constant and T is the temperature. (Note that we usually reserve ν for the kinematic viscosity, but in this problem we will use it to denote the frequency.)

(a) Let the Planck function in per frequency units be denoted by B_ν. By enforcing energy conservation,

$$\int B_\nu \, d\nu = \int B_\lambda \, d\lambda, \tag{2.29}$$

derive the functional form of B_ν.

(b) Next, let the Planck function in per wavenumber ($\tilde{\nu} \equiv 1/\lambda$) units be denoted by $B_{\tilde{\nu}}$ and derive its expression as well.

(c) Show that

$$\pi \int_0^\infty B_\lambda \, d\lambda = \pi \int_0^\infty B_\nu \, d\nu = \pi \int_0^\infty B_{\tilde{\nu}} \, d\tilde{\nu} = \sigma_{SB} T^4. \tag{2.30}$$

What is the expression for σ_{SB}? (Hint: you may require the use of Riemann zeta functions.) Hence, the Stefan-Boltzmann constant is not a fundamental constant, but is composed of other physical constants.

2.8.5 Direct solutions of the radiative transfer equation

(a) For the treatment of isothermal atmospheric layers discussed in Section 2.4.2, show that

$$\mathcal{T} = (1 - \Delta\tau) e^{-\Delta\tau} + (\Delta\tau)^2 \mathcal{E}_1, \tag{2.31}$$

where $\mathcal{E}_1(\Delta\tau)$ is the exponential integral of the first order, defined as [2, 7]

$$\mathcal{E}_i(\Delta\tau) \equiv \int_1^\infty x^{-i} e^{-x\Delta\tau} \, dx. \tag{2.32}$$

(b) Generally, the Planck function varies across optical depth. Even in this non-isothermal situation, it is possible to obtain a direct solution of the radiative transfer equation. Consider the following functional form for B,

$$B = \begin{cases} B_1 + B'\,(\tau - \tau_1), & \downarrow \text{ direction,} \\ B_2 + B'\,(\tau - \tau_2), & \uparrow \text{ direction,} \end{cases} \tag{2.33}$$

where

$$B' \equiv \frac{\partial B}{\partial \tau} = \frac{B_2 - B_1}{(\tau_2 - \tau_1)} \tag{2.34}$$

is the gradient of the Planck function across the medium bounded by Points 1 and 2. Quantities subscripted by 1 and 2 are evaluated at their respective locations. Mathematically, this functional form is the Taylor series expansion of B, about the point τ_1 or τ_2, truncated at the linear term. Show that this choice of B ensures that $B = B_1$ and $B = B_2$ when $\tau = \tau_1$ and $\tau = \tau_2$, respectively. Furthermore, show that the solutions of the radiative transfer are

$$F_{\uparrow_1} = F_{\uparrow_2}\mathcal{T} + \pi B_2\,(1 - \mathcal{T}) + \pi B'\left[\frac{2}{3}\left(1 - e^{-\Delta\tau}\right) - \Delta\tau\left(1 - \frac{\mathcal{T}}{3}\right)\right],$$

$$F_{\downarrow_2} = F_{\downarrow_1}\mathcal{T} + \pi B_1\,(1 - \mathcal{T}) - \pi B'\left[\frac{2}{3}\left(1 - e^{-\Delta\tau}\right) - \Delta\tau\left(1 - \frac{\mathcal{T}}{3}\right)\right], \tag{2.35}$$

where we again have $\Delta\tau \equiv \tau_2 - \tau_1$.

2.8.6 The transit radius and the optical depth

Consider an exoplanet, with an atmosphere, transiting its star. Starlight cuts across the limb of the atmosphere along a chord. Let the spatial coordinate along this chord be x. Let the transit radius be R.

(a) Let x and R form the sides of a right-angled triangle with a hypotenuse given by $R + z$, where z is the vertical coordinate. By asserting that $z \ll R$, show that

$$z \approx \frac{x^2}{2R}. \tag{2.36}$$

(b) Assume an isothermal and hydrostatic atmosphere. Derive the number density (n) as a function of z and a reference value of the number density (n_{ref}) corresponding to $z = 0$. By integrating along the transit chord, show that the chord optical depth is [62]

$$\tau = n_{\text{ref}} \sigma \sqrt{2\pi H R}, \tag{2.37}$$

where H is the (isothermal) pressure scale height.

(c) Obtain an alternative and approximate derivation of the optical depth by asserting that $\tau = n_{\mathrm{ref}}\sigma l$ and the length scale l must be the mean of H and R. (Hint: think about the *type* of averaging involved.)

(d) Now, consider a non-isothermal atmosphere with the temperature being expressed as a Taylor series expansion truncated at the linear term [97],

$$T = T_{\mathrm{ref}} + \frac{\partial T}{\partial z} z, \tag{2.38}$$

where T_{ref} is a reference value of the temperature. Derive the number density profile, $n(z)$. What is the expression for the non-isothermal scale height? Finally, derive the expression for the chord optical depth.

Chapter Three

The Two-Stream Approximation of Radiative Transfer

3.1   WHAT IS THE TWO-STREAM APPROXIMATION?

The passage of radiation through an atmosphere generally occurs in many directions—photons are absorbed (and re-emitted) or scattered multiple times before they escape it. These interactions with matter may also alter their wavelengths. Thus, we are faced with a formidable problem spanning three dimensions and across a broad range of wavelengths or frequencies. To make any progress in understanding, we need to reduce the complexity of the problem down to a level suitable for pedagogy.

In Chapter 10, we will discuss the utility of the shallow water models for understanding exoplanetary atmospheres and point out that "shallow" has a subtle meaning in the context of fluid dynamics. Within the context of radiative transfer, it is more straightforward—atmospheres with extents much less than the radius of the exoplanet are considered shallow. In this case, the horizontal propagation of radiation may be neglected and it appears as if radiative transfer occurs only in two directions: upwards and downwards. This is the physical basis of the *two-stream approximation*. It is an acceptable approximation for Earth, Mars, Titan and Venus. It is decent even for calculations of hot Jupiters with model atmospheres that are still considerably thinner than their radii, because the extent of the model domain is set by the depth of penetration of stellar heating.

The mathematical basis of the two-stream approximation is a little subtler. The radiative transfer equation has two independent variables: optical depth and propagation angle. Rather than solve this partial differential equation, it is easier to first integrate over the outgoing (upwards) and incoming (downwards) hemispheres separately, before solving the resulting pair of ordinary differential equations that depend only on the optical depth. The problem with this approach is that one always ends up with one more dependent variable than the number of equations available, which renders the system under-determined. To close the system of equations requires that we make assumptions about the ratios of these dependent variables, which is the origin of the *Eddington coefficients*. The freedom to choose different values for a set of Eddington coefficients leads to an ambiguity in the types of *closures* used. The choice of closure is in turn related to fundamental issues of energy conservation.

It is worth mentioning that it is possible to perform separate sets of two-stream calculations for incident starlight and thermal emission from the exoplanetary atmosphere, which implies that separate sets of Eddington coefficients are used in each waveband [33, 240]. This is precisely what is done in Chapter 4 when we derive the temperature-pressure profiles. One may also elect to use *multi-stream* methods [35], especially when the details of scattering are important [117].

This is the upshot of the two-stream approximation: there is a price to pay for simplicity and, if one is not careful, this might be the self-consistency of physics in one's two-stream calculations. And we will also see that, despite its apparent simplicity, the formalism behind setting up and solving for the two-stream solutions is formidable.

3.2   THE RADIATIVE TRANSFER EQUATION AND ITS MOMENTS

To begin, we state without proof the radiative transfer equation with non-isotropic but coherent scattering [35, 74, 172],

$$\mu \frac{\partial I}{\partial \tau} = I - \omega_0 \int_0^{4\pi} \mathcal{P} I \, d\Omega' - (1 - \omega_0) B, \qquad (3.1)$$

where θ is the polar angle in spherical coordinates (or co-latitude), $\mu \equiv \cos \theta$, I is the intensity, τ is the optical depth, ω_0 is the single-scattering albedo and B is the Planck function. As mentioned in Chapter 2, "coherent" refers to the assumption that scattering does not alter the wavelength of a photon. The quantity $\mathcal{P}$ is the *scattering phase function*, which we will discuss later in the chapter. It is integrated over all incident polar (θ') and azimuthal (ϕ') angles; we have defined $\mu' \equiv \cos \theta'$. The solid angle subtended by the incident beam is $d\Omega' \equiv d\mu' d\phi'$.

We wish to integrate equation (3.1) over all angles to get rid of one of the independent variables (μ). To do so requires that we first define what the moments of the intensity are. We recognize the total intensity as also being the zeroth moment of the intensity,

$$J \equiv \int_0^{2\pi} \int_{-1}^1 I \, d\mu \, d\phi. \qquad (3.2)$$

It may be separated into its outgoing ($\uparrow$) and incoming ($\downarrow$) components,

$$J_\uparrow \equiv \int_0^{2\pi} \int_0^1 I \, d\mu \, d\phi, \; J_\downarrow \equiv \int_0^{2\pi} \int_{-1}^0 I \, d\mu \, d\phi, \qquad (3.3)$$

such that $J \equiv J_\uparrow + J_\downarrow$. We may do the same for the first moments of the intensity,

$$F_\uparrow \equiv \int_0^{2\pi} \int_0^1 \mu I \, d\mu \, d\phi, \; F_\downarrow \equiv \int_0^{2\pi} \int_{-1}^0 \mu I \, d\mu \, d\phi, \qquad (3.4)$$

which are the *outgoing* and *incoming fluxes*, respectively. We have adopted terminology that is specific to an atmosphere. The *total flux* is given by $F_+ \equiv F_\uparrow + F_\downarrow$, while the *net flux* is $F_- \equiv F_\uparrow - F_\downarrow$. We will see later that distinguishing between the total and net fluxes has profound physical implications.

It is easier and cleaner to demonstrate the method of moments[1] in the isotropic scattering limit ($\mathcal{P} = 1/4\pi$), such that the integral in equation (3.1) simply becomes $J/4\pi$. We will start with this limit and deal with the complications associated with non-isotropic scattering later in the chapter. The radiative transfer equation reduces to

$$\mu \frac{\partial I}{\partial \tau} = I - \frac{\omega_0 J}{4\pi} - (1 - \omega_0) B. \tag{3.5}$$

We begin with the outgoing hemisphere. Instead of the optical depth, we use the *slant optical depth*,

$$\tau' \equiv \frac{\tau}{\bar{\mu}_\uparrow}, \tag{3.6}$$

where $\bar{\mu}_\uparrow$ is a characteristic value of μ in the outgoing hemisphere ($0 \leq \mu \leq 1$). As we will see shortly, this step is a mathematical book-keeping trick to keep track of minus signs associated with $\mu < 0$, which is really only relevant for the incoming hemisphere. Integrating over $d\Omega \equiv d\mu d\phi$, we obtain

$$\frac{1}{\bar{\mu}_\uparrow} \frac{\partial F_\uparrow}{\partial \tau'} = J_\uparrow - \frac{\omega_0 J}{2} - 2\pi (1 - \omega_0) B. \tag{3.7}$$

At this point, you will notice that we have one equation and three unknowns ($F_\uparrow$, $J_\uparrow$ and $J_\downarrow$). If we execute the same procedure for the other hemisphere, we will obtain another equation, but we will still have three unknowns. Our system of equations is mathematically under-determined.

To proceed, we have to somehow relate the total intensities and fluxes. The simplest approach is to assume that their ratios are constant,

$$\epsilon_\uparrow \equiv \frac{F_\uparrow}{J_\uparrow}, \quad \epsilon \equiv \frac{F_+}{J}, \tag{3.8}$$

which are quantities known as the *Eddington coefficients*. Physically, there is no reason why the Eddington coefficients should be constant in general, but it does provide a clear mathematical path forward, because we may now reduce the number of unknowns by one,

$$\frac{\partial F_\uparrow}{\partial \tau'} = \bar{\mu}_\uparrow F_\uparrow \left(\frac{1}{\epsilon_\uparrow} - \frac{\omega_0}{2\epsilon} \right) - \frac{\bar{\mu}_\uparrow \omega_0}{2\epsilon} F_\downarrow - 2\pi \bar{\mu}_\uparrow (1 - \omega_0) B. \tag{3.9}$$

Instead of a governing equation for the intensity, we now have one for the outgoing flux. Note that all of the quantities in the preceding equation are monochromatic (i.e., they are defined at a specific wavelength, frequency or wavenumber).

---

[1] This approach is not so mysterious when one realizes that fluid quantities are the moments of the Boltzmann equation.

Turning to the incoming hemisphere $(-1 \leq \mu \leq 0)$, we similarly define the *slant optical depth*,

$$\tau' \equiv \frac{\tau}{\bar{\mu}_\downarrow}. \tag{3.10}$$

Here, the purpose of this step becomes clear, because we have $\bar{\mu}_\downarrow < 0$ and explicitly stating $\bar{\mu}_\downarrow$ helps us to keep track of the minus signs. If we define an additional Eddington coefficient,

$$\epsilon_\downarrow \equiv \frac{F_\downarrow}{J_\downarrow}, \tag{3.11}$$

and integrate the radiative transfer equation over all incoming angles, then we obtain

$$\frac{\partial F_\downarrow}{\partial \tau'} = \bar{\mu}_\downarrow F_\downarrow \left(\frac{1}{\epsilon_\downarrow} - \frac{\omega_0}{2\epsilon} \right) - \frac{\bar{\mu}_\downarrow \omega_0}{2\epsilon} F_\uparrow - 2\pi \bar{\mu}_\downarrow \left(1 - \omega_0 \right) B. \tag{3.12}$$

This equation has a form that is identical to its outgoing counterpart, but with the difference that $\bar{\mu}_\downarrow < 0$. This difference will introduce an asymmetry into the pair of governing equations for the fluxes.

We now appeal to physics—we assume that no asymmetry exists between the hemispheres for the characteristic cosines of the co-latitude and the Eddington coefficients,

$$\bar{\mu} = \bar{\mu}_\uparrow = -\bar{\mu}_\downarrow, \quad \epsilon' = \epsilon_\uparrow = \epsilon_\downarrow. \tag{3.13}$$

With these assumptions, the pair of governing equations may be cast into a more illuminating form,

$$\begin{aligned} \frac{\partial F_\uparrow}{\partial \tau'} &= \gamma_a F_\uparrow - \gamma_s F_\downarrow - \gamma_B B, \\ \frac{\partial F_\downarrow}{\partial \tau'} &= -\gamma_a F_\downarrow + \gamma_s F_\uparrow + \gamma_B B, \end{aligned} \tag{3.14}$$

which has the coefficients,

$$\begin{aligned} \gamma_a &\equiv \bar{\mu} \left(\frac{1}{\epsilon'} - \frac{\omega_0}{2\epsilon} \right), \\ \gamma_s &\equiv \frac{\bar{\mu}\omega_0}{2\epsilon}, \\ \gamma_B &\equiv 2\pi \bar{\mu} \left(1 - \omega_0 \right). \end{aligned} \tag{3.15}$$

When scattering is absent $(\gamma_s = 0)$, the equations in (3.14) become decoupled from each other—outgoing fluxes remain outgoing and incoming ones remain incoming. Scattering has the effect of transforming outgoing rays into incoming ones and vice versa.

Written in this form, it is easy to see that we now have a well-determined problem: two equations and two unknowns. Furthermore, these equations behave like a pair of coupled, first-order, ordinary differential equations with constant coefficients and with the independent variable being the slant optical depth. Before obtaining the general solutions for the two-stream equations, it is instructive to examine them in the purely scattering and absorbing limits.

3.3   TWO-STREAM SOLUTIONS WITH ISOTROPIC SCATTERING

3.3.1   Enforcing energy conservation in purely scattering limit

Towards the end of Chapter 2, we showed that radiative equilibrium—which is a statement of local energy conservation between atmospheric layers—is attained when the temperature, described by the following equation,

$$\frac{\partial T}{\partial t} = -\frac{1}{\rho c_P}\frac{\partial \mathcal{F}_-}{\partial z},\tag{3.16}$$

ceases to evolve, where ρ is the mass density and c_P is the specific heat capacity at constant pressure. In other words, heating or cooling is a response to the spatial variation of the net flux across altitude. We may now cast the wavelength-integrated net and total fluxes in terms of our two-stream formalism,

$$\mathcal{F}_{\pm} \equiv \int F_{\uparrow} \pm F_{\downarrow}\, d\lambda.\tag{3.17}$$

One may also choose to perform the integration across frequency or wavenumber. It is the wavelength-integrated net flux that is involved, and not the monochromatic net flux, because temperature is a wavelength-independent quantity.[2]

Consider the limit of pure scattering, where the single-scattering albedo becomes unity ($\omega_0 = 1$). In this limit, we have $\gamma_B = 0$ and the blackbody emission vanishes. If we take the difference of the pair of equations in (3.14), we obtain

$$\frac{\partial F_-}{\partial \tau'} = (\gamma_a - \gamma_s)\, F_+.\tag{3.18}$$

Since we have assumed that the passage of radiation through the atmosphere is coherent in wavelength, we demand a stricter criterion: that the vertical gradient (which is proportional to the gradient with respect to the slant optical depth) vanishes at each wavelength. Generally, the total flux must be non-zero, which means

$$\frac{\partial F_-}{\partial \tau'} = 0 \implies \gamma_a = \gamma_s \implies \epsilon = \epsilon'.\tag{3.19}$$

Physically, a purely scattering atmosphere is unable to absorb radiation and thus cannot be heated.

We end up with a single Eddington coefficient. Since there appears to be no consensus on how to order them, we will number the Eddington coefficients in the order in which they appear in this chapter. In this spirit, we will name ϵ the *first Eddington coefficient.*

---

[2]This is a common source of confusion, because an often-used term is the *brightness temperature*, which is the effective temperature corresponding to the blackbody flux measured at a specific wavelength. This is not really a temperature. Rather, it is a convenient way to visualize the measured flux. Mathematically, one inverts the Planck function to obtain brightness temperature.

3.3.2   Enforcing correct blackbody flux in purely absorbing limit

To determine the value of the first Eddington coefficient, we need to examine the opposite limit: that of pure absorption. For an isothermal atmosphere, it is possible to show that (Problem 3.8.1),

$$
\begin{aligned}
F_{\uparrow_1} &= F_{\uparrow_2} e^{\gamma_a(\tau_1' - \tau_2')} + \frac{\gamma_B B}{\gamma_a}\left[1 - e^{\gamma_a(\tau_1' - \tau_2')}\right], \\
F_{\downarrow_2} &= F_{\downarrow_1} e^{\gamma_a(\tau_1' - \tau_2')} + \frac{\gamma_B B}{\gamma_a}\left[1 - e^{\gamma_a(\tau_1' - \tau_2')}\right],
\end{aligned}
\tag{3.20}
$$

where we have considered the transfer of radiation between two points labeled by 1 and 2. Our convention is always to let Point 1 sit above Point 2 in altitude. It suffices to let these two points span the entire atmosphere. At this juncture, we will remark that the optical depth is a coordinate; the *difference* in optical depths between points is the actual quantity of interest. Let the difference in slant optical depth spanning the atmosphere be $\Delta\tau' \equiv \tau_2' - \tau_1'$.

You will notice that these solutions look a lot like the ones we derived in equation (2.20) of Chapter 2, except that the transmission function takes on a simpler form,

$$
\mathcal{T} = e^{-\gamma_a \Delta\tau'} = e^{-\Delta\tau/\epsilon}.
\tag{3.21}
$$

Note that $\Delta\tau$ is the non-slanted or "normal" optical depth of the atmosphere.

Now, consider the physical situation of a purely-absorbing, isothermal and opaque atmosphere. When the atmosphere is opaque, we have $\Delta\tau \to \infty$, $\mathcal{T} = 0$ and

$$
F_{\uparrow_1} = \frac{\gamma_B B}{\gamma_a}, \quad F_{\downarrow_2} = \frac{\gamma_B B}{\gamma_a}.
\tag{3.22}
$$

In an isothermal atmosphere, we expect each hemisphere to contribute πB of flux.[3] This implies that $\gamma_B/\gamma_a = \pi$ and thus

$$
\epsilon = \frac{1}{2}.
\tag{3.23}
$$

Therefore, if we wish to enforce the correct blackbody flux in the limit of a purely-absorbing, isothermal and opaque atmosphere, there is no freedom in choosing the value of the first Eddington coefficient. If we update the coefficients in the governing equations for the outgoing and incoming fluxes, we get

$$
\begin{aligned}
\gamma_a &\equiv \bar{\mu}\left(2 - \omega_0\right), \\
\gamma_s &\equiv \bar{\mu}\omega_0, \\
\gamma_B &\equiv 2\pi\bar{\mu}\left(1 - \omega_0\right).
\end{aligned}
\tag{3.24}
$$

Another misconception is the belief that one may independently vary the preceding coefficients and the first Eddington coefficient, when in fact they depend on one another.

---

[3]Evaluate $\int \mu B \, d\Omega$ over each hemisphere and this statement will become clear.

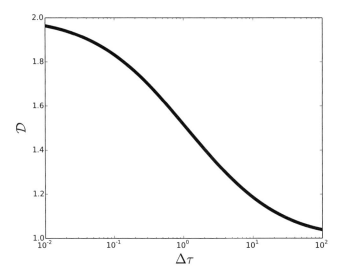

Figure 3.1: The diffusivity factor as a function of the difference in optical depth across a pair of layers. Note that this calculation can only be done for purely-absorbing, isothermal atmospheres. When scattering is present, all bets are off.

The first Eddington coefficient appears as a coefficient of the optical depths in the transmission function, and its reciprocal is often termed the *diffusivity factor*,

$$\mathcal{D} = \frac{1}{\epsilon}. \tag{3.25}$$

But since we may derive the direct solution of the radiative transfer equation, in the purely-absorbing limit, and the transmission function associated with it, we may do a comparison with the two-stream solution and derive the actual functional form of $\mathcal{D}$. Problem 2.8.5 teaches you how to derive the transmission function of the direct solution. By equating it to the two-stream transmission function, we obtain

$$\mathcal{D} = -\frac{1}{\Delta\tau} \ln\left[(1 - \Delta\tau)\, e^{-\Delta\tau} + (\Delta\tau)^2\, \mathcal{E}_1\right], \tag{3.26}$$

where $\mathcal{E}_1$ is the exponential integral of the first order [2, 7]. Note that such a comparison and derivation of $\mathcal{D}$ may only be performed in the limit of pure absorption.

Instead of an entire atmosphere, let Points 1 and 2 now represent two layers of an atmosphere. The value of the diffusivity factor, and thus of the first Eddington coefficient, generally varies for different pairs of atmospheric layers, depending on the optical depth of the path between each pair. Figure 3.1 shows calculations of the diffusivity factor as a function of $\Delta\tau$. Notice that we have $\mathcal{D} \approx 2$ when $\Delta\tau \ll 1$. Thus, the two-stream approximation is in good agreement with the direct solution of the radiative transfer equation—at least in the purely-absorbing, isothermal limit—if a sufficient number of layers are used to describe an atmosphere such that the difference in optical depth between each pair of layers is small.

Generally, as we will see later, the diffusivity factor, in the two-stream approximation, takes the form,

$$\mathcal{D} = \frac{[(\gamma_a + \gamma_s)(\gamma_a - \gamma_s)]^{1/2}}{\bar{\mu}}. \tag{3.27}$$

As we have already remarked in Chapter 2, we may not obtain a direct solution of the radiative transfer equation when isotropic scattering is present. Non-isotropic scattering exacerbates the situation, as we now have to deal with a scattering phase function of arbitrary form. Therefore, it is impossible to *generally* derive how the diffusivity factor depends on $\Delta\tau$. Curiously, some researchers use $\mathcal{D}$ as a free parameter and tune its value in two-stream calculations to match more sophisticated calculations. A commonly-used value is $\mathcal{D} = 1.66$, despite the fact that it was estimated via comparing calculations for the atmosphere of Earth [8]. I would urge you to proceed with caution—to be self-consistent within the two-stream framework, one should just use $\mathcal{D}$ without treating it like a free parameter.

3.3.3 General solutions (isotropic scattering)

If we add and substract the pair of equations in (3.14) in turn, we end up with governing equations for the net and total fluxes,

$$\begin{aligned} \frac{\partial F_+}{\partial \tau'} &= (\gamma_a + \gamma_s) F_-, \\ \frac{\partial F_-}{\partial \tau'} &= (\gamma_a - \gamma_s) F_+ - 2\gamma_B B. \end{aligned} \tag{3.28}$$

This pair of equations can be combined into a single, second-order differential equation for the net flux,

$$\frac{\partial^2 F_+}{\partial \tau'^2} - (\bar{\mu}\mathcal{D})^2 F_+ = -2\gamma_B (\gamma_a + \gamma_s) B, \tag{3.29}$$

where the diffusivity factor was previously stated in equation (3.27). It is important to note that no assumptions have been made on the functional form of B at this point. The preceding equation behaves like an *ordinary* (rather than partial) and inhomogeneous differential equation.

Generally, we would like to apply the two-stream solutions to pairs of atmospheric layers. We expect that each layer would have its own temperature and pressure and thus its own value of the opacity at a given wavelength. Any path traversing a pair of layers would necessarily experience a change in temperature, pressure and opacity. Since the Planck function depends on temperature, it cannot be constant along this path. Therefore, we need a treatment that allows for B to depend on τ'. Mathematically, we may use either Point 1 or 2 as a reference and Taylor-expand the Planck function about it,

$$B = B_j + \frac{\partial B}{\partial \tau} (\tau - \tau_j) + \frac{\partial^2 B}{\partial \tau^2} \frac{(\tau - \tau_j)^2}{2} + \text{higher-order terms}, \qquad (3.30)$$

where $j = 1$ or 2. Note that the factors of $\bar{\mu}$ appear both in the numerator and denominator of the terms involving derivatives and thus cancel out, allowing us to use the non-slanted optical depth in the preceding expression.

For simplicity, we retain only the first two terms—it is always possible to retain higher-order terms if one desires better accuracy. We write the first derivative of the Planck function as

$$B' \equiv \frac{\partial B}{\partial \tau} \approx \frac{B_2 - B_1}{\tau_2 - \tau_1}, \qquad (3.31)$$

regardless of whether one uses Point 1 or 2 as the reference. I will let you convince yourself that when $\tau = \tau_1$, we obtain $B \approx B_1$, the value of the Planck function evaluated at Point 1, in both series expansions for B. Likewise, when $\tau = \tau_2$, we obtain $B \approx B_2$, the value of the Planck function evaluated at Point 2.

In what follows, I will highlight the most crucial mathematical steps en route to obtaining the two-stream solutions and relegate the tedious parts of the algebra to Problem 3.8.2, which I encourage you to work through. We first solve the second-order ordinary differential equation for the total flux using the method of undetermined coefficients. From the solution for the total flux, we may use equation (3.28) to obtain the solution for the net flux. With the net and total fluxes in hand, we may construct the outgoing and incoming fluxes,

$$
\begin{aligned}
F_\uparrow &= \mathcal{A}_1 \zeta_+ e^{\alpha \tau'} + \mathcal{A}_2 \zeta_- e^{-\alpha \tau'} + \pi B + \frac{\pi B'}{2}, \\
F_\downarrow &= \mathcal{A}_1 \zeta_- e^{\alpha \tau'} + \mathcal{A}_2 \zeta_+ e^{-\alpha \tau'} + \pi B - \frac{\pi B'}{2},
\end{aligned}
\qquad (3.32)
$$

where the coefficients $\mathcal{A}_1$ and $\mathcal{A}_2$ are determined by imposing the boundary conditions $F_{\uparrow_2}$ and $F_{\downarrow_1}$. The *coupling coefficients* are defined as

$$\zeta_\pm \equiv \frac{1}{2} \left[1 \pm (1 - \omega_0)^{1/2} \right], \qquad (3.33)$$

so named because when scattering is absent, we have $\zeta_+ = 1$ and $\zeta_- = 0$ and the boundary conditions do not simultaneously appear in each of the two-stream solutions.

By imposing the boundary conditions, we obtain

$$
\begin{aligned}
F_{\uparrow_2} &= \mathcal{A}_1 \zeta_+ e^{\alpha \tau'_2} + \mathcal{A}_2 \zeta_- e^{-\alpha \tau'_2} + \pi \mathcal{B}_{2+}, \\
F_{\downarrow_1} &= \mathcal{A}_1 \zeta_- e^{\alpha \tau'_1} + \mathcal{A}_2 \zeta_+ e^{-\alpha \tau'_1} + \pi \mathcal{B}_{1-},
\end{aligned}
\qquad (3.34)
$$

where we have found it convenient to define the quantities,

$$\mathcal{B}_{j-} \equiv B_1 + B'\left(\tau_j - \tau_1\right) - \frac{B'}{2},$$

$$\mathcal{B}_{j+} \equiv B_2 + B'\left(\tau_j - \tau_2\right) + \frac{B'}{2}. \tag{3.35}$$

By writing the transmission function as

$$\mathcal{T} \equiv e^{-\mathcal{D}\Delta\tau}, \tag{3.36}$$

with $\Delta\tau \equiv \tau_2 - \tau_1$, we may express the non-isothermal two-stream solutions as

$$
\begin{aligned}
F_{\uparrow_1} =&\frac{1}{\left(\zeta_-\mathcal{T}\right)^2 - \zeta_+^2}\left\{\left(\zeta_-^2 - \zeta_+^2\right)\mathcal{T}F_{\uparrow_2} - \zeta_-\zeta_+\left(1 - \mathcal{T}^2\right)F_{\downarrow_1}\right.\\
&\left. + \pi\left[\mathcal{B}_{1+}\left(\zeta_-^2\mathcal{T}^2 - \zeta_+^2\right) + \mathcal{B}_{2+}\mathcal{T}\left(\zeta_+^2 - \zeta_-^2\right) + \mathcal{B}_{1-}\zeta_-\zeta_+\left(1 - \mathcal{T}^2\right)\right]\right\},\\
F_{\downarrow_2} =&\frac{1}{\left(\zeta_-\mathcal{T}\right)^2 - \zeta_+^2}\left\{\left(\zeta_-^2 - \zeta_+^2\right)\mathcal{T}F_{\downarrow_1} - \zeta_-\zeta_+\left(1 - \mathcal{T}^2\right)F_{\uparrow_2}\right.\\
&\left. + \pi\left[\mathcal{B}_{2-}\left(\zeta_-^2\mathcal{T}^2 - \zeta_+^2\right) + \mathcal{B}_{1-}\mathcal{T}\left(\zeta_+^2 - \zeta_-^2\right) + \mathcal{B}_{2+}\zeta_-\zeta_+\left(1 - \mathcal{T}^2\right)\right]\right\}.
\end{aligned}
\tag{3.37}
$$

The two-stream solutions in equation (3.37) allow us to transfer radiation from the first (subscripted by 1) to the second (subscripted by 2) layer, and vice versa, given the boundary conditions. (As before, our convention is to sit Layer 1 above Layer 2 in altitude, such that $\tau_2 > \tau_1$.) They have been cast in a convenient form for computation, because all of the non-flux quantities (transmission function, coupling coefficients) are dimensionless and have values that are bounded between 0 and 1. The coupling coefficients determine the interplay between the boundary conditions on the flux and the blackbody emission— they are essentially weights for determining what fraction of each component an observer will measure. Generally, a model atmosphere is divided into a finite number of layers and subjected to a pair of boundary conditions at the top (stellar heating) and bottom (interior heating) of the computational domain. The two-stream solutions are applied to each pair of layers in turn, and the boundary conditions are propagated throughout the atmosphere until radiative equilibrium is attained.

One may further gain intuition by going to the purely absorbing, albeit non-isothermal, limit ($\zeta_+ = 1$, $\zeta_- = 0$),

$$
\begin{aligned}
F_{\uparrow_1} =&\mathcal{T}F_{\uparrow_2} + \pi\left(\mathcal{B}_{1+} - \mathcal{B}_{2+}\mathcal{T}\right),\\
F_{\downarrow_2} =&\mathcal{T}F_{\downarrow_1} + \pi\left(\mathcal{B}_{2-} - \mathcal{B}_{1-}\mathcal{T}\right).
\end{aligned}
\tag{3.38}
$$

When a path is opaque ($\mathcal{T} = 0$), the boundary conditions do not contribute to the emergent flux and one sees only the blackbody emission from either Point 1 (for the outgoing flux) or 2 (for the incoming flux). When the path is transparent ($\mathcal{T} = 1$), one fully sees the boundary condition and an extra contribution due to non-isothermal behavior. If the slab is isothermal, then this extra contribution vanishes.

3.4   THE SCATTERING PHASE FUNCTION

A treatment of non-isotropic scattering requires us to deal with the scattering phase function in the radiative transfer equation. Fortunately, we may employ a trick: by assuming two general properties of the scattering phase function, we may manipulate the radiative transfer equation into its two-stream form even without explicitly knowing the functional form of $\mathcal{P}$.

The first assumption we make is that $\mathcal{P}$ only depends on θ'', which is the *change* in the polar angle or co-latitude of a photon due to scattering,

$$\theta'' \equiv \theta' - \theta. \tag{3.39}$$

Let $\mu'' \equiv \cos\theta''$. For two arbitrary points in spherical geometry, the dot product of their position vectors yields

$$\mu'' = \mu'\mu + \left(1 - \mu'^2\right)^{1/2}\left(1 - \mu^2\right)^{1/2}\cos\left(\phi' - \phi\right). \tag{3.40}$$

The preceding equation informs us that μ'' possesses a symmetry property,

$$\mu'' \to \mu'' \text{ if } \mu' \to -\mu' \text{ and } \mu \to -\mu. \tag{3.41}$$

Since μ'' is single-valued for $0 \leq \theta'' \leq \pi$, we may conclude that θ'' is invariant under double sign-flips of μ and μ'. If $\mathcal{P} = \mathcal{P}(\theta'')$ only, then the scattering phase function is also invariant under this transformation.

The second assumption is that integrating over $\mathcal{P}$ yields the same answer regardless of whether we are in the (θ, ϕ), (θ', ϕ') or (θ'', ϕ'') coordinate systems,

$$\int_0^{4\pi} \mathcal{P}\,d\Omega = \int_0^{4\pi} \mathcal{P}\,d\Omega' = \int_0^{4\pi} \mathcal{P}\,d\Omega'' = 1, \tag{3.42}$$

where we have $d\Omega'' \equiv d\mu''d\phi''$ and $\phi'' \equiv \phi' - \phi$. You should convince yourself that this is trivially true when $\mathcal{P} = 1/4\pi$. An implication of this assumption is that we can always replace $\mathcal{P}$ by some other function, involving itself and an additional function that only depends on θ'', and the same normalization symmetry[4] applies. We will put this property to use later when deriving the governing equations involving non-isotropic scattering.

Typically, one uses a fitting function known as the *Henyey-Greenstein* scattering phase function [101] in one's model,

$$\mathcal{P} = \frac{1 - g_0^2}{4\pi\left(1 + g_0^2 - 2g_0\cos\theta''\right)^{3/2}}, \tag{3.43}$$

where g_0 is the *scattering asymmetry factor*. Notice how we recover $\mathcal{P} = 1/4\pi$ when $g_0 = 0$. In such a treatment, g_0 is the input parameter to $\mathcal{P}$. By contrast, we will see shortly that, in our derivation, g_0 emerges as an integral involving $\mathcal{P}$.

---

[4]In Heng, Mendonça & Lee [94], we stated that integrating $\mathcal{P}$ over all angles yields 4π. A more commonly-used normalization constant is unity, as we have stated here. The choice of normalization constant does not affect our derivation of either the two-stream governing equations or its solutions.

3.5   TWO-STREAM SOLUTIONS WITH NON-ISOTROPIC SCATTERING

3.5.1   Derivation of governing equations

We need to set ourselves up to take moments of the radiative transfer equation with non-isotropic scattering [191]. Consider an arbitrary function $\mathcal{H} = \mathcal{H}(\theta)$; we multiply equation (3.1) by it and integrate over all angles,

$$
\frac{1}{\bar{\mu}} \frac{\partial}{\partial \tau'} \int_0^{2\pi} \left(\int_0^1 \mu \mathcal{H} I \, d\mu - \int_{-1}^0 \mu \mathcal{H} I \, d\mu \right) d\phi
$$
$$
= \int_0^{4\pi} \mathcal{H} I \, d\Omega - \mathcal{I} - (1 - \omega_0) \int_0^{4\pi} \mathcal{H} B \, d\Omega,
\tag{3.44}
$$

where we have defined the integrals,

$$
\mathcal{I} \equiv \omega_0 \int_0^{4\pi} \mathcal{G} I \, d\Omega',
$$
$$
\mathcal{G} \equiv \int_0^{4\pi} \mathcal{H} \mathcal{P} \, d\Omega.
\tag{3.45}
$$

The minus sign in the integral involving $\bar{\mu}$ in equation (3.44) comes from the characteristic value of μ being positive and negative in the outgoing and incoming hemispheres, respectively.

The problem is now reduced to evaluating the integrals $\mathcal{I}$ and $\mathcal{G}$ for different functional forms of $\mathcal{H}$. In the simplest case of $\mathcal{H} = 1$, we obtain $\mathcal{G} = 1$, $\mathcal{I} = \omega_0 J$ and

$$
\frac{\partial F_-}{\partial \tau'} = \bar{\mu} \left(1 - \omega_0 \right) \left(J - 4\pi B \right).
\tag{3.46}
$$

This yields the first of the governing equations for non-isotropic scattering.

The next natural guess is $\mathcal{H} = \mu$, because it allows us to introduce another Eddington coefficient into the problem [191]. By writing $\mu = \cos(\theta' - \theta'')$ and using the trigonometric identity,

$$
\cos \left(\theta' - \theta'' \right) = \cos \theta' \cos \theta'' + \sin \theta' \sin \theta'',
\tag{3.47}
$$

we may show that

$$
\mathcal{G} = \int_0^{2\pi} \int_{-1}^1 \left[\mu' \mu'' + \left(1 - \mu'^2 \right)^{1/2} \left(1 - \mu''^2 \right)^{1/2} \right] \mathcal{P} \, d\mu \, d\phi.
\tag{3.48}
$$

Using the symmetry property described in equation (3.41), we see that the second term in the preceding equation is an odd integral, which vanishes.

Since $\mathcal{G}$ is the integrand of $\mathcal{I}$ and we are evaluating it at fixed values of μ', the normalization symmetry previously discussed applies to the integrand $\mu' \mu'' \mathcal{P}$, which means we can integrate $\mathcal{G}$ over $d\Omega''$ rather than $d\Omega$,

$$
\mathcal{G} = \mu' \int_0^{4\pi} \mu'' \mathcal{P} \, d\Omega''.
\tag{3.49}
$$

We now invoke the definition of the scattering asymmetry factor [191],

$$g_0 \equiv \int_0^{4\pi} \mu'' \mathcal{P} \, d\Omega''. \tag{3.50}$$

When $\mathcal{P} = 1/4\pi$, we may easily verify that $g_0 = 0$. There is an analogy between the flux and the scattering asymmetry factor, since they are the first moments of the intensity and scattering phase function, respectively. Instead of describing scattering by a function $\mathcal{P}$, we may now quantify it using a number g_0, which generally depends on wavelength. It follows that $\mathcal{G} = g_0 \mu'$ and $\mathcal{I} = \omega_0 g_0 F_+$.

If we define the second moments of the intensity as

$$K_\uparrow \equiv \int_0^{2\pi} \int_0^1 \mu^2 I \, d\mu \, d\phi, \;\; K_\downarrow \equiv \int_0^{2\pi} \int_{-1}^0 \mu^2 I \, d\mu \, d\phi, \tag{3.51}$$

and let $K_- \equiv K_\uparrow - K_\downarrow$, then we are now in a position to evaluate equation (3.44) with $\mathcal{H} = \mu$,

$$\frac{\partial K_-}{\partial \tau'} = \bar{\mu} F_+ \left(1 - \omega_0 g_0 \right). \tag{3.52}$$

The preceding expression is the second governing equation. The problem is that we now have two equations and four unknowns (J, F_-, F_+ and K_-). Two of them may be related by the first Eddington coefficient ($\epsilon = F_+/J = 1/2$), which means we need one more Eddington coefficient. Physically, $K_\uparrow/c$ (and its counterparts) is the *radiation pressure*.

A necessary constraint is that equations (3.46) and (3.52) must reduce to the expressions we previously derived for isotropic scattering, which is given by equation (3.28). A simple inspection will reveal that equation (3.46), which governs the net flux, already does—what appears to be missing is a governing equation for the total flux, which equation (3.52) must somehow provide. Thus, the starting point for the second Eddington coefficient must be

$$\epsilon_2 \equiv \frac{K_-}{F_+}, \tag{3.53}$$

which transforms equation (3.52) into

$$\epsilon_2 \frac{\partial F_+}{\partial \tau'} = \bar{\mu} F_+ \left(1 - \omega_0 g_0 \right). \tag{3.54}$$

But for it to reduce to its isotropic counterpart, we must have

$$\epsilon_2 = \frac{F_+}{2F_-}, \tag{3.55}$$

which finally yields

$$\frac{\partial F_+}{\partial \tau'} = 2\bar{\mu} F_- \left(1 - \omega_0 g_0 \right). \tag{3.56}$$

By comparing equations (3.46) and (3.56) with the expressions in (3.28), we may obtain the generalized coefficients,

$$
\begin{aligned}
\gamma_{\mathrm{a}} &= \bar{\mu}\left[2 - \omega_0\left(1 + g_0\right)\right], \\
\gamma_{\mathrm{s}} &= \bar{\mu}\omega_0\left(1 - g_0\right), \\
\gamma_{\mathrm{B}} &= 2\pi\bar{\mu}\left(1 - \omega_0\right).
\end{aligned}
\tag{3.57}
$$

In other words, the governing equations for the net and total fluxes, for non-isotropic scattering, are identical to those stated in (3.28), but with more general expressions for their coefficients as given above. By selecting specific values and expressions for the Eddington coefficients, we have chosen a *closure*. We will see later that we have essentially rederived an existing and commonly-used closure known as the *hemispheric closure*. Ironically, the closure named after the British astronomer Arthur Eddington produces unphysical results, as we will see.

3.5.2 General solutions (non-isotropic scattering)

If you have worked through Problem 3.8.2, you have already done the hard work of learning the technique for deriving the two-stream solutions with isotropic scattering. Fortunately, these techniques may be applied, without modification, to the two-stream solutions with non-isotropic scattering. Specifically, equations (3.28) and (3.29) may be solved to obtain

$$
\begin{aligned}
F_{\uparrow 1} &= \frac{1}{\left(\zeta_- \mathcal{T}\right)^2 - \zeta_+^2}\Bigg\{\left(\zeta_-^2 - \zeta_+^2\right)\mathcal{T}F_{\uparrow 2} - \zeta_-\zeta_+\left(1 - \mathcal{T}^2\right)F_{\downarrow 1} \\
&\quad + \frac{\gamma_{\mathrm{B}}}{\gamma_{\mathrm{a}} - \gamma_{\mathrm{s}}}\left[\mathcal{B}_{1+}\left(\zeta_-^2\mathcal{T}^2 - \zeta_+^2\right) + \mathcal{B}_{2+}\mathcal{T}\left(\zeta_+^2 - \zeta_-^2\right) + \mathcal{B}_{1-}\zeta_-\zeta_+\left(1 - \mathcal{T}^2\right)\right]\Bigg\}, \\
F_{\downarrow 2} &= \frac{1}{\left(\zeta_- \mathcal{T}\right)^2 - \zeta_+^2}\Bigg\{\left(\zeta_-^2 - \zeta_+^2\right)\mathcal{T}F_{\downarrow 1} - \zeta_-\zeta_+\left(1 - \mathcal{T}^2\right)F_{\uparrow 2} \\
&\quad + \frac{\gamma_{\mathrm{B}}}{\gamma_{\mathrm{a}} - \gamma_{\mathrm{s}}}\left[\mathcal{B}_{2-}\left(\zeta_-^2\mathcal{T}^2 - \zeta_+^2\right) + \mathcal{B}_{1-}\mathcal{T}\left(\zeta_+^2 - \zeta_-^2\right) + \mathcal{B}_{2+}\zeta_-\zeta_+\left(1 - \mathcal{T}^2\right)\right]\Bigg\},
\end{aligned}
\tag{3.58}
$$

where the transmission function is again $\mathcal{T} \equiv e^{-\mathcal{D}\Delta\tau}$, the diffusivity factor is again described by equation (3.27) and the expression for $\mathcal{B}_{j\pm}$ is now generalized to

$$
\begin{aligned}
\mathcal{B}_{j-} &\equiv B_1 + B'\left(\tau_j - \tau_1\right) - \frac{\bar{\mu}B'}{\gamma_{\mathrm{a}} + \gamma_{\mathrm{s}}}, \\
\mathcal{B}_{j+} &\equiv B_2 + B'\left(\tau_j - \tau_2\right) + \frac{\bar{\mu}B'}{\gamma_{\mathrm{a}} + \gamma_{\mathrm{s}}}.
\end{aligned}
\tag{3.59}
$$

Furthermore, the coupling coefficients take on more general forms,

$$
\zeta_\pm \equiv \frac{1}{2}\left[1 \pm \left(\frac{\gamma_{\mathrm{a}} - \gamma_{\mathrm{s}}}{\gamma_{\mathrm{a}} + \gamma_{\mathrm{s}}}\right)^{1/2}\right] = \frac{1}{2}\left[1 \pm \left(\frac{1 - \omega_0}{1 - \omega_0 g_0}\right)^{1/2}\right].
\tag{3.60}
$$

Although $\bar{\mu}$ appears in the expression for the diffusivity factor, γ_a and γ_s contain factors of $\bar{\mu}$ as well. In the end, the two-stream solutions do not depend on specifying a value of $\bar{\mu}$ if one is dealing with the non-slanted optical depth. As I have already mentioned, $\bar{\mu}$ is used as a mathematical book-keeping trick for keeping track of errant minus signs. Physically, it plays no role in the end.

There is a final and important caveat to the solutions in equation (3.58) that needs to be mentioned and discussed. They are invalid in the limit of pure scattering ($\omega_0 = 1$), because they only produce pure reflection. Pure reflection and pure scattering are mutually exclusive behaviors—an atmosphere may only scatter (and not absorb), but this is not the same as saying that it always needs to be opaque. It is also possible to be purely scattering and transparent, in which case the incident radiation simply passes through. This is an important distinction that cannot be over-emphasized. Problem 3.8.4 works through the purely-scattering two-stream solutions in detail.

3.6 DIFFERENT CLOSURES OF THE TWO-STREAM SOLUTIONS

Table 3.1: Comparison of closures for the two-stream approximation

Name	$\gamma_a/\bar{\mu}$	$\gamma_s/\bar{\mu}$	$\gamma_B/\bar{\mu}$
Hemispheric	$2 - \omega_0(1+g_0)$	$\omega_0(1-g_0)$	$2\pi(1-\omega_0)$
Eddington	$\frac{1}{4}[7 - \omega_0(4+3g_0)]$	$-\frac{1}{4}[1 - \omega_0(4-3g_0)]$	$2\pi(1-\omega_0)$
Quadrature	$\frac{\sqrt{3}}{2}[2 - \omega_0(1+g_0)]$	$\frac{\sqrt{3}\omega_0}{2}(1-g_0)$	$\sqrt{3}\pi(1-\omega_0)$

We now understand the price to pay for the simplicity of the two-stream approximation: the Eddington coefficients. Our first-principles derivation allows us to define ϵ and ϵ_2 and to demonstrate that $\epsilon = 1/2$ is a plausible choice, which ultimately leads to the set of coefficients listed in equation (3.57). It turns out that this set of expressions for γ_a, γ_s and γ_B has a name—it is termed the *hemispheric* or *hemi-isotropic closure* [168, 191, 240]. We have essentially proved that the hemispheric closure has a sound physical and mathematical basis. Curiously, there are several closures used in the published literature [35, 74, 168, 172, 173, 191, 240]. Besides the hemispheric closure, the most common of these are the *Eddington* and *quadrature closures*. We will now explore the implications of adopting different types of closures.

The first test that we subject the Eddington and quadrature closures to is the conservation of energy in the limit of pure scattering. One may verify that

$$\frac{\partial F_-}{\partial \tau} \propto (\gamma_a - \gamma_s) \propto (1 - \omega_0), \qquad (3.61)$$

which implies that radiative equilibrium is attained for all three closures in the purely-scattering limit.

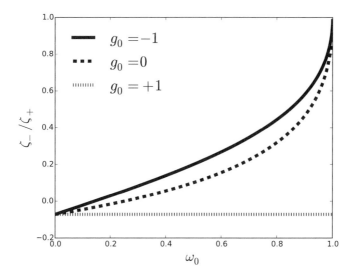

Figure 3.2: Spurious enhancement or diminution factor associated with using the Eddington closure. Both reflection and blackbody emission are artificially enhanced or diminished by a factor of ζ_-/ζ_+ when $\mathcal{T} = 0$.

The second test is to check that the correct amount of blackbody flux is produced in the limit of a purely-absorbing, isothermal and opaque atmosphere. For the hemispheric and quadrature closures, we have $\gamma_s = 0$, $\zeta_+ = 1$ and $\zeta_- = 0$ when $\omega_0 = 0$. We may define a dimensionless figure of merit,

$$f_\infty \equiv \frac{\gamma_B}{\pi \left(\gamma_a - \gamma_s\right)}, \qquad (3.62)$$

and check that it is unity for these closures. In other words, $f_\infty = 1$ corresponds to each hemisphere producing πB of flux. It originates from the coefficient of the terms associated with the Planck function in the two-stream solutions.

Subjecting the Eddington closure to the second test is a little trickier. Strangely enough, we have

$$\gamma_s = -\frac{\bar{\mu}}{4} \neq 0 \qquad (3.63)$$

even when $\omega_0 = 0$ (pure absorption). This is unphysical behavior within the context of the two-stream approximation, because it means that outgoing rays may transform into incoming ones, and vice versa, even when scattering is absent. Essentially, reflection is produced where there should be none. It is already the first hint that the Eddington closure should not be used.[5]

---

[5]We note that the Eddington closure is used by the textbook of Goody & Yung [74].

Another consequence of having $\gamma_s \neq 0$ even when scattering is absent is that we have

$$f_\infty \equiv \frac{\gamma_B}{\pi \left(\gamma_a - \gamma_s\right)} \left(1 - \frac{\zeta_-}{\zeta_+}\right). \qquad (3.64)$$

If we plug in $\omega_0 = 0$, then we end up with $f_\infty \approx 1.1$. It implies that the Eddington closure is causing a spurious enhancement of the blackbody flux, by about 10%, in the limit of pure absorption.

Generally, the Eddington closure introduces spurious enhancements or diminutions of both the reflected and blackbody fluxes, depending on the exact values of ω_0 and g_0 (Figure 3.2). To compound the problem, the errors are heterogeneous unless one is in the limit of pure forward scattering. The size of the errors is non-negligible: $\sim 10\%$. As the errors propagate into calculations involving different pairs of layers, the final error incurred is difficult to quantify. If the errors ultimately cancel, it can only be a lucky accident—one that no sober researcher should blindly hope for.

The problems associated with the Eddington closure highlight an important difference in scientific culture between astrophysics and the Earth and planetary sciences. We are primarily interested in building robust models to scan broad regimes of parameter space, and it is of fundamental importance that they are general and physically consistent—it is insufficient that they work just in a narrow parameter regime or for specific case studies.

3.7 THE DIFFUSION APPROXIMATION FOR RADIATIVE TRANSFER

Deep within an atmosphere, the mean free path of a photon becomes very small. If the mean free path is much smaller than the pressure scale height of the atmosphere, then the photon typically suffers an enormous number of absorption and scattering events as it propagates across the atmospheric layers. (If you have worked through Problem 2.8.2, you will convince yourself that the number of such events is $\sim \tau^2$.) In such a limit, the transfer of radiation resembles diffusion [172]. We expect that the radiative transfer equation should reduce to the governing equation for diffusion.

In the diffusion limit, the two-stream approximation breaks down, and it makes little sense to distinguish between the net and total fluxes. Hence, we will simply remove the subscripts associated with the various moments of the intensity. Furthermore, in such an optically thick situation, the radiation field becomes Planckian (i.e., thermalized) and scattering is isotropic,

$$J = 4\pi B, \quad g_0 = 0. \qquad (3.65)$$

We invoke the third Eddington coefficient,

$$\epsilon_3 \equiv \frac{K}{J} = \frac{1}{3}. \qquad (3.66)$$

Its value comes from setting $I = B$ and evaluating the zeroth and second moments of the intensity.

We use equations (3.46) and (3.52) as our starting points, which now take the form,

$$\frac{\partial F}{\partial \tau} = 0, \ F = 4\pi\epsilon_3 \frac{\partial B}{\partial \tau}. \tag{3.67}$$

It is interesting to note that the deep parts of an atmosphere are in radiative equilibrium—by definition. The flux itself is constant across optical depth and is driven by the gradient of the Planck function—a distinctly non-isothermal effect.

Let the wavelength-integrated flux be $\mathcal{F}$. By writing $d\tau = -\rho\kappa \, dz$, we obtain

$$\mathcal{F} = -\frac{4\pi\epsilon_3}{\rho} \int_0^\infty \frac{1}{\kappa} \frac{\partial B}{\partial z} \, d\lambda. \tag{3.68}$$

We imagine that there is a wavelength-integrated version of this equation mediated by some wavelength-integrated opacity (κ_R), which looks like [172]

$$\mathcal{F} = -\frac{4\pi\epsilon_3}{\rho\kappa_R} \frac{\partial \mathcal{B}}{\partial z}, \tag{3.69}$$

where we have defined

$$\mathcal{B} \equiv \int_0^\infty B \, d\lambda = \frac{\sigma_{SB} T^4}{\pi} \tag{3.70}$$

and σ_{SB} is the Stefan-Boltzmann constant. By equating the two expressions for $\mathcal{F}$, we obtain the *Rosseland mean opacity*,

$$\kappa_R = \frac{4\sigma_{SB} T^3}{\pi} \left(\int_0^\infty \frac{1}{\kappa} \frac{\partial B}{\partial T} \, d\lambda \right)^{-1}. \tag{3.71}$$

As always, the integration can also be performed over frequency or wavenumber.

The Rosseland mean opacity is an example of a *harmonic mean*, rather than the arithmetic mean that we attribute to "common sense." To illustrate this concept, we consider the three different ways[6] of averaging the numbers 1 and 100,

$$\frac{1 + 100}{2} = 50.5, \ (1 \times 100)^{1/2} = 10, \ \frac{2}{1 + 1/100} \approx 2, \tag{3.72}$$

which are the arithmetic, geometric and harmonic means, respectively. The arithmetic mean attributes equal weight to each number, whereas the geometric mean tries to find the average between them in a logarithmic sense. The latter is the correct quantity to use when one is taking the mean between a set of quantities with values spanning many orders of magnitude. For example, the Einstein radius in gravitational lensing is, to within a factor of unity, the

---

[6]Sometimes known as the Pythagorean means.

geometric mean of the Schwarzschild radius—a typically tiny quantity on astronomical scales—and the distance to the lens (which typically spans galactic or cosmic scales). By contrast, the harmonic mean attributes almost all of the weight to the smallest quantity.

Physically, we can see why the harmonic mean is the correct mean to use in an optically thick situation, because the transfer of radiation is controlled predominantly by the wavelengths at which the opacity is small and the atmosphere is transparent. This is the reason why the Rosseland mean is designed to give disproportionate weight to the lowest opacities.

If we return to the expression for $\mathcal{F}$ that involves κ_{R} and invoke hydrostatic equilibrium, we end up with

$$\mathcal{F} = \frac{16\sigma_{\mathrm{SB}}gT^3}{3\kappa_{\mathrm{R}}} \frac{\partial T}{\partial P}, \tag{3.73}$$

where g is the surface gravity of the exoplanet. The preceding equation takes the same form as *Fick's law of diffusion*, where the flux is proportional to the gradient of an internal quantity and the coefficient is interpreted as the *diffusion coefficient*.

By equating our version of Fick's law to an internal heat flux ($\sigma_{\mathrm{SB}}T_{\mathrm{int}}^4$) and performing the integration over pressure, we may obtain the temperature-pressure profile of the deep atmosphere,

$$T = \left[\frac{3}{4}\left(\tau_{\mathrm{R}} + \mathcal{C}\right)\right]^{1/4} T_{\mathrm{int}}, \tag{3.74}$$

where $\mathcal{C} \sim 1$ is a constant of integration and τ_{R} is the optical depth associated with the Rosseland mean opacity. This is known as *Milne's solution*, after the British astrophysicist Edward Milne [173]. Originally, it was used to describe self-luminous objects such as stars. When $\tau_{\mathrm{R}} \sim 1$, the observer measures $T \approx T_{\mathrm{int}}$. In the case of the Sun, $T_{\mathrm{int}} \approx 5800$ K is the photospheric temperature.

3.8 PROBLEM SETS

3.8.1 Alternative method for obtaining two-stream solutions

(a) Derive the two-stream solutions, in the limit of pure absorption and an isothermal atmosphere, by directly and separately solving each first-order differential equation. This involves realizing that

$$\begin{aligned}
\frac{\partial}{\partial \tau'}\left(F_\uparrow e^{-\gamma_{\mathrm{a}}\tau'}\right) &= e^{-\gamma_{\mathrm{a}}\tau'}\frac{\partial F_\uparrow}{\partial \tau'} - \gamma_{\mathrm{a}}F_\uparrow e^{-\gamma_{\mathrm{a}}\tau'}, \\
\frac{\partial}{\partial \tau'}\left(F_\downarrow e^{\gamma_{\mathrm{a}}\tau'}\right) &= e^{\gamma_{\mathrm{a}}\tau'}\frac{\partial F_\downarrow}{\partial \tau'} + \gamma_{\mathrm{a}}F_\downarrow e^{\gamma_{\mathrm{a}}\tau'}.
\end{aligned} \tag{3.75}$$

(b) Does this approach work when scattering is present? Why or why not?

3.8.2 Two-stream solutions with isotropic scattering

In deriving the non-isothermal two-stream solutions, we have to find expressions for $\mathcal{A}_1 \zeta_+ e^{\alpha \tau_1'}$ and $\mathcal{A}_2 \zeta_- e^{-\alpha \tau_1'}$ when constructing $F_{\uparrow_1}$. For $F_{\downarrow_2}$, we need expressions for $\mathcal{A}_1 \zeta_- e^{\alpha \tau_2'}$ and $\mathcal{A}_2 \zeta_+ e^{-\alpha \tau_2'}$.

(a) By manipulating the pair of equations in (3.34), which incorporate the boundary conditions, show that

$$\mathcal{A}_2 e^{-\alpha \tau_1'} = \frac{1}{\zeta_-^2 \mathcal{T} - \zeta_+^2 \mathcal{T}^{-1}} \left[\zeta_- F_{\uparrow_2} - \zeta_+ \mathcal{T}^{-1} F_{\downarrow_1} - \pi \left(\zeta_- \mathcal{B}_{2+} - \zeta_+ \mathcal{T}^{-1} \mathcal{B}_{1-} \right) \right] .$$

(3.76)

(b) Similarly, show that

$$\mathcal{A}_2 e^{-\alpha \tau_2'} = \frac{1}{\zeta_-^2 \mathcal{T} - \zeta_+^2 \mathcal{T}^{-1}} \left[\zeta_- \mathcal{T} F_{\uparrow_2} - \zeta_+ F_{\downarrow_1} - \pi \left(\zeta_- \mathcal{T} \mathcal{B}_{2+} - \zeta_+ \mathcal{B}_{1-} \right) \right] . \quad (3.77)$$

(c) By again using equation (3.34), show that

$$\begin{aligned}
\mathcal{A}_1 \zeta_+ e^{\alpha \tau_1'} &= \mathcal{T} F_{\uparrow_2} - \mathcal{A}_2 \zeta_- \mathcal{T}^2 e^{-\alpha \tau_1'} - \pi \mathcal{B}_{2+} \mathcal{T}, \\
\mathcal{A}_1 \zeta_- e^{\alpha \tau_2'} &= \mathcal{T}^{-1} F_{\downarrow_1} - \mathcal{A}_2 \zeta_+ \mathcal{T}^{-2} e^{-\alpha \tau_2'} - \pi \mathcal{B}_{1-} \mathcal{T}^{-1}.
\end{aligned}$$

(3.78)

(d) Thus, derive the solutions in equation (3.37).

3.8.3 Henyey-Greenstein scattering phase function

Does the Henyey-Greenstein scattering phase function obey the normalization symmetry stated in equation (3.41)?

3.8.4 Two-stream solutions in the limit of pure scattering

(a) Return to the pair of equations in (3.28) and consider them in the limit of $\omega_0 = 1$. Apply the boundary conditions and show that

$$\begin{aligned}
F_{\uparrow_1} &= F_{\uparrow_2} - \frac{(F_{\uparrow_2} - F_{\downarrow_1})(\gamma_a + \gamma_s)(\tau_2' - \tau_1')}{2 + (\gamma_a + \gamma_s)(\tau_2' - \tau_1')}, \\
F_{\downarrow_2} &= F_{\downarrow_1} + \frac{(F_{\uparrow_2} - F_{\downarrow_1})(\gamma_a + \gamma_s)(\tau_2' - \tau_1')}{2 + (\gamma_a + \gamma_s)(\tau_2' - \tau_1')}.
\end{aligned}$$

(3.79)

(b) If the preceding solutions describe a slab, what do they reduce to when the slab is transparent or opaque?

(c) What do the two-stream solutions in equation (3.58) asymptote to when $\omega_0 \rightarrow 1$?

3.8.5 The diffusion approximation

(a) Recall our form of Fick's law and rewrite it in the form,

$$\mathcal{F} = -l_{\mathrm{diff}} \frac{\partial}{\partial z} \left(\sigma_{\mathrm{SB}} T^4 \right) ,$$

(3.80)

where l_{diff} is the diffusion length scale. Derive the expression for l_{diff}. Consider the situation deep within a hot exoplanetary atmosphere. Using $\kappa_{\text{R}} = 0.01$ cm^2 g^{-1}, $P = 100$ bar, $T = 1000$ K and $m = 2m_{\text{H}}$, where m_{H} is the mass of a hydrogen atom, compute l_{diff}.

(b) Using the same values of T and m, as well as $g = 1000$ cm s^{-2}, compute the pressure scale height $H = k_{\text{B}}T/mg$, which is the characteristic height in an atmosphere. What is the ratio H/l_{diff}? Is the diffusion approximation a good one in this situation?

(c) Repeat the calculation for the atmosphere of Earth. If $P = 1$ bar, what is the column mass? If the typical infrared optical depth is 2, what is the corresponding value of the opacity? Use this value as an estimate for the Rosseland mean opacity. Is the diffusion approximation a good one for Earth?

Chapter Four

Temperature-Pressure Profiles

4.1 A MYRIAD OF ATMOSPHERIC EFFECTS: GREENHOUSE WARMING AND ANTI-GREENHOUSE COOLING

Fundamentally, we would like to have full knowledge of the thermal structure of an atmosphere, as it has a major influence on computing its observed properties. Specifically, we would like to measure or infer how the temperature varies with altitude or pressure and how this *temperature-pressure profile* is altered by effects associated with the atmospheric opacities, aerosols or clouds, etc. We would like to know how a low or high albedo influences it. For Earth and the Solar System bodies, the temperature-pressure profile may actually be measured via satellite or in-situ probes. For exoplanets, this task is a lot harder and we need to resort to indirect methods.

In the absence of data on an exoplanetary atmosphere, astronomers typically label an exoplanet using its *equilibrium temperature*

$$T_{\rm eq} \equiv T_\star \left(\frac{R_\star}{2a} \right)^{1/2} (1 - A_{\rm B})^{1/4}, \tag{4.1}$$

which is the temperature equivalent to the incident starlight received by it and spread out over all 4π steradians. It is a quantity that may be computed *independent* of the properties of the atmosphere. Here, $T_\star$ is the stellar effective temperature, $R_\star$ is the stellar radius, a is the exoplanet-star separation and $A_{\rm B}$ is the Bond albedo. Our Solar System provides stunning counter-examples to this simplistic approach: Earth's equilibrium temperature is below the freezing point of water and Venus's is not much higher. In reality, the Earth hosts liquid water and Venus is an inferno with temperatures reaching about 700 K.

The reason why these equilibrium temperature estimates are so far off from reality is that atmospheres generally have a myriad of opacity sources that introduce warming and cooling effects. Absorption of radiation by atmospheric gases such as water, carbon dioxide and methane tend to produce *greenhouse warming*. Enhanced absorption or the scattering of starlight tends to produce *anti-greenhouse cooling*. Scattering of thermal emission of the exoplanet introduces the *scattering greenhouse effect*, where infrared radiation is redirected, via scattering, back towards the exoplanet [61, 117]. Aerosols or clouds complicate this picture as they possess some combination of these attributes and produce both cooling and heating, depending on their intrinsic properties and geometric configuration [191]. We would like to understand this richness of behavior without resorting to—or in advance of—a full-blown computer simulation. Like Ray Pierrehumbert [191], my belief is that true understanding starts from simple, clear models that isolate salient behavior.

A body of work exists in the astrophysical literature on temperature-pressure profiles of irradiated exoplanetary atmospheres [79, 81, 107, 208], including my own [88, 94]. In this chapter, we will use these analytical models to reproduce the effects previously mentioned and develop an intuition for them.

4.2 THE DUAL-BAND OR DOUBLE-GRAY APPROXIMATION

To render the radiative transfer equation amenable to analytical solution, we make the approximation that the starlight incident upon the exoplanet and its thermal emission occur at distinctly different wavelengths, known as the *dual-band* or *double-gray* approximation. This is not an unreasonable approximation, since Wien's law[1] tells us that the peak wavelength of blackbody emission depends on the temperature,

$$\lambda_{\text{peak}} \approx 0.5 \ \mu\text{m} \ \left(\frac{T}{6000 \ \text{K}} \right)^{-1}. \tag{4.2}$$

The preceding estimate applies to a Sun-like star. For Earth ($T \approx 300$ K), we have $\lambda_{\text{peak}} \approx 10 \ \mu$m, while a hot exoplanet ($T \approx 1500$ K) has $\lambda_{\text{peak}} \approx 2 \ \mu$m. The dual-band approximation appears to hold quite well.

We shall term the range of wavelengths corresponding to stellar heating the *shortwave* and subscript all associated quantities with an "S." The range of wavelengths associated with the thermal emission of the exoplanetary atmosphere is termed the *longwave*; the corresponding subscript is "L." Typically, the shortwave and longwave occur in the visible and infrared, respectively. However, it is possible for both of them to be in the infrared. For example, M stars or red dwarfs have effective temperatures of about 3000 K or lower, which means their blackbody emission peaks in the near-infrared.

The zeroth, first and second moments of the intensity are

$$J_{\text{S}} \equiv \int_{\text{S}} J \ d\lambda, \ F_{\text{S}} \equiv \int_{\text{S}} F_- \ d\lambda, \ K_{\text{S}} \equiv \int_{\text{S}} K_- \ d\lambda,$$
$$J_{\text{L}} \equiv \int_{\text{L}} J \ d\lambda, \ F_{\text{L}} \equiv \int_{\text{L}} F_- \ d\lambda, \ K_{\text{L}} \equiv \int_{\text{L}} K_- \ d\lambda. \tag{4.3}$$

Note the use of the total intensity versus the net flux and radiation pressure.

---

[1] Wien's law states that the peak wavelength of a blackbody curve and its temperature are related via a constant number: $\lambda_{\text{peak}} T = 2.8977729 \times 10^3 \ \mu$m K.

4.3   THE RADIATIVE TRANSFER EQUATION AND THE SCATTERING PARAMETER

We wish to set up the governing equations for the temperature-pressure profiles, but we do not need to start again from the radiative transfer equation. Instead, we will invoke the two-stream treatment, which takes the moments of the radiative transfer equation (Chapter 3). Specifically, we recall equations (3.46) and (3.52),

$$\frac{\partial F_-}{\partial \tau} = (1 - \omega_0)(J - 4\pi B),$$
$$\frac{\partial K_-}{\partial \tau} = F_+ (1 - \omega_0 g_0), \tag{4.4}$$

but written in terms of the non-slanted optical depth (τ). In other words, τ is the optical depth measured from the top of the atmosphere to some point within it along a straight path.

To study the effects of varying the atmospheric opacities, we need to define a new coordinate other than τ. If we express the single-scattering albedo in terms of the absorption (κ_a) and scattering (κ_s) opacities, we obtain

$$\omega_0 = \frac{\kappa_s}{\kappa_a + \kappa_s}. \tag{4.5}$$

We may show that the extinction opacity ($\kappa_e \equiv \kappa_a + \kappa_s$) takes the form,

$$\kappa_e = \frac{\kappa_a}{1 - \omega_0}. \tag{4.6}$$

Thus, the optical depth may be expressed in terms of

$$d\tau = \kappa_e \, d\tilde{m} = \frac{\kappa_a}{1 - \omega_0} \, d\tilde{m}, \tag{4.7}$$

where $\tilde{m}$ is the column mass of the atmosphere. We write it in this manner, because it allows for the absorption and scattering to be described by κ_a and ω_0, respectively. The absorption opacity and single-scattering albedo may now be used as input parameters, while the column mass serves as a coordinate. We have essentially separated out the opacities, which are microscopic quantities, from the pressure coordinate of the atmosphere.

It follows that the equations in (4.4) become

$$\frac{\partial F_-}{\partial \tilde{m}} = \kappa_a (J - 4\pi B),$$
$$\frac{\partial K_-}{\partial \tilde{m}} = \frac{\kappa_a F_+}{\beta_0^2}. \tag{4.8}$$

Written in this form, it becomes easy to see why we define the *scattering parameter* as

$$\beta_0 \equiv \left(\frac{1 - \omega_0}{1 - \omega_0 g_0} \right)^{1/2}. \tag{4.9}$$

This is an important point—the single-scattering albedo and scattering asymmetry factor do not appear as independent quantities in equation (4.8). Rather, they are always combined in the form of β_0. For this reason, we will always describe the temperature-pressure profiles by the scattering parameter. In the shortwave, we will relate the scattering parameter to the Bond albedo, a quantity that may be inferred from astronomical observations.

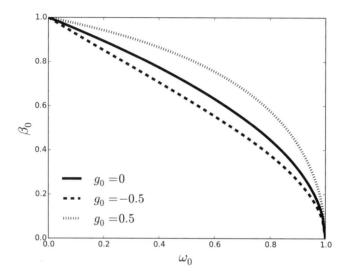

Figure 4.1: Scattering parameter as a function of the single-scattering albedo and scattering asymmetry factor. Pure absorption and scattering correspond to $\beta_0 = 1$ and 0, respectively. Isotropic, forward and backward scattering correspond to $g_0 = 0$, $g_0 > 0$ and $g_0 < 0$, respectively.

An atmosphere may be visualized as being a heat engine. The total amount of energy entering and leaving the atmosphere must be conserved. Integrated over wavelength, the energy content, per unit mass and time, of the atmosphere is

$$Q = \int_0^\infty \frac{\partial F_-}{\partial \tilde{m}} \, d\lambda = \frac{\partial \mathcal{F}_-}{\partial \tilde{m}} = \kappa_S J_S + \kappa_L \left(J_L - 4\sigma_{SB} T^4 \right), \qquad (4.10)$$

with $\mathcal{F}_-$ being the bolometric[2] net flux. The energy content, per unit area and time, of the atmosphere is derived from further integrating over the column mass,

$$\int_{\tilde{m}}^\infty \frac{\partial \mathcal{F}_-}{\partial \tilde{m}} \, d\tilde{m} = \tilde{Q} \left(\tilde{m}, \infty \right), \qquad (4.11)$$

---

[2]Meaning integrated over all wavelengths, frequencies or wavenumbers.

where we have defined, for any arbitrary pair of column masses $\tilde{m}_1$ and $\tilde{m}_2$,

$$\tilde{Q}\left(\tilde{m}_1, \tilde{m}_2\right) \equiv \int_{\tilde{m}_1}^{\tilde{m}_2} Q \, d\tilde{m}. \tag{4.12}$$

Equations (4.10) and (4.11) allow for an alternative expression of radiative equilibrium, which occurs when $Q = 0$. This in turn implies that $\tilde{Q}(0, \infty) = 0$. Thus, the energy per unit area and time (i.e., the flux), integrated over all wavelengths and pressures, entering and exiting the atmosphere is conserved. Radiative equilibrium is a necessary and sufficient condition for global energy conservation. The converse is not true—global energy conservation does not guarantee radiative equilibrium, a point of contention among even professional researchers. Mathematically, the gradient of the bolometric net flux may conspire to be non-zero at multiple locations, only to cancel out when summed over the entire atmosphere, such that $\tilde{Q}(0, \infty) = 0$ but $Q \neq 0$.

By following through on equation (4.11), energy conservation may be expressed as

$$\begin{aligned}
F_{\mathrm{L}} &= \mathcal{F}_\infty - F_{\mathrm{S}} - \tilde{Q}\left(\tilde{m}, \infty\right), \\
\epsilon_{\mathrm{L}} J_{\mathrm{L}_0} &= \mathcal{F}_\infty - F_{\mathrm{S}_0} - \tilde{Q}\left(0, \infty\right).
\end{aligned} \tag{4.13}$$

For convenience, we have defined $F_{\mathrm{S}_0} \equiv F_{\mathrm{S}}(\tilde{m} = 0)$, $F_{\mathrm{L}_0} \equiv F_{\mathrm{L}}(\tilde{m} = 0)$ and $J_{\mathrm{L}_0} \equiv J_{\mathrm{L}}(\tilde{m} = 0)$. The bolometric net flux from the deep interior (as $\tilde{m} \to \infty$) is

$$\mathcal{F}_\infty = \sigma_{\mathrm{SB}} T_{\mathrm{int}}^4, \tag{4.14}$$

where σ_{SB} is the Stefan-Boltzmann constant. The quantity T_{int} is the *interior* or *internal temperature* and $\mathcal{F}_\infty$ should be regarded as a boundary condition of our model atmosphere. It describes the remnant heat of formation of an exoplanet.

4.4 TREATMENT OF SHORTWAVE RADIATION

4.4.1 Relationship between Bond albedo and scattering parameter

In the shortwave, stellar heating dominates and thermal emission from the exoplanet is negligible. We recall our two-stream solutions with non-isotropic scattering in equation (3.58). By considering incident starlight ($F_{\downarrow_1} \neq 0$, $F_{\uparrow_2} = 0$) upon an opaque atmosphere ($\mathcal{T} = 0$) and ignoring thermal emission in the shortwave ($B = 0$), we may use the two-stream solutions to derive the spherical albedo[3] [88, 94],

$$A_s \equiv \frac{F_{\uparrow_1}}{F_{\downarrow_1}} = \frac{1 - \beta_0}{1 + \beta_0}. \tag{4.15}$$

---

[3] Recall that opaqueness and reflectivity are independent attributes.

By integrating the spherical albedo over all wavelengths, we obtain the Bond albedo,

$$A_B = \frac{1 - \beta_{S_0}}{1 + \beta_{S_0}}, \qquad (4.16)$$

where β_{S_0} is the mean or characteristic value of β_0 in the shortwave. Since our two-stream solutions and temperature-pressure profiles have a common mathematical origin, this derivation is self-consistent.

4.4.2 Generalization of Beer's law

We now seek to obtain solutions for the total intensity and net flux in the shortwave and both recover and generalize Beer's law. When the governing equations in (4.4) are integrated over the shortwave, they become

$$\frac{\partial F_S}{\partial \tilde{m}} = \kappa_S J_S,$$
$$\frac{\partial K_S}{\partial \tilde{m}} = \frac{\kappa'_S F_S}{\beta_{S_0}^2}, \qquad (4.17)$$

where the *absorption mean opacity* is [172]

$$\kappa_S \equiv \frac{\int_S \kappa_a J \, d\lambda}{\int_S J \, d\lambda}. \qquad (4.18)$$

The *flux mean opacity* is

$$\kappa'_S \equiv \frac{\int_S \kappa_a F_+ \, d\lambda}{\int_S F_- \, d\lambda}. \qquad (4.19)$$

We again have two equations and three unknowns (J_S, F_S and K_S). To proceed, we need a shortwave closure [79],

$$\epsilon_S \equiv \frac{K_S}{J_S} = \mu^2, \qquad (4.20)$$

where μ is the cosine of the polar angle or co-latitude. This form of the shortwave closure is chosen to correctly reproduce Beer's law in the limit of pure absorption and also generalize it to situations with scattering, as we will see shortly. In essence, we have guessed the form of the shortwave closure and validated our guess by reproducing Beer's law after the fact.

Drawing from the lessons of Chapter 3, we would like to combine the first-order differential equations in (4.17) into a single second-order differential equation, which behaves like an ordinary differential equation with $\tilde{m}$ being the independent variable. You will find that we cannot proceed unless we assume

$$\kappa_S = \kappa'_S. \qquad (4.21)$$

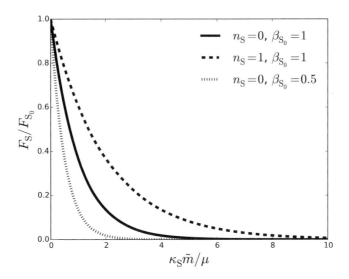

Figure 4.2: Generalization of Beer's law with non-isotropic scattering. The limit of constant shortwave opacity and pure absorption is given by $n_S = 0$ and $\beta_{S_0} = 1$. A steeper fall of the shortwave opacity with height allows for a deeper penetration of starlight ($n_S = 1$ and $\beta_{S_0} = 1$). Even when the shortwave opacity is constant, the presence of scattering causes a more rapid diminution of incident starlight ($n_S = 0$ and $\beta_{S_0} = 0.5$).

From this point onwards, we shall simply term κ_S the *shortwave opacity*. With this assumption, we obtain a pair of second-order ordinary differential equations,

$$\frac{\partial^2 J_S}{\partial \tilde{m}^2} - \frac{1}{\kappa_S}\frac{\partial \kappa_S}{\partial \tilde{m}}\frac{\partial J_S}{\partial \tilde{m}} - \left(\frac{\kappa_S}{\mu \beta_{S_0}}\right)^2 J_S = 0,$$
$$\frac{\partial^2 F_S}{\partial \tilde{m}^2} - \frac{1}{\kappa_S}\frac{\partial \kappa_S}{\partial \tilde{m}}\frac{\partial F_S}{\partial \tilde{m}} - \left(\frac{\kappa_S}{\mu \beta_{S_0}}\right)^2 F_S = 0. \tag{4.22}$$

When κ_S is constant, the pesky first derivatives in equation (4.22) vanish and the solutions are straightforward to obtain. What is surprising and less obvious is that a power-law form of the shortwave opacity,

$$\kappa_S = \kappa_{S_0}\left(\frac{\tilde{m}}{\tilde{m}_0}\right)^{n_S}, \tag{4.23}$$

where $\tilde{m}_0$ is a reference value of the column mass usually set to correspond to the bottom of atmosphere, admits the following analytical solutions [94],

$$J_S = J_{S_0} e^{\beta_S/\mu}, \quad F_S = F_{S_0} e^{\beta_S/\mu}, \tag{4.24}$$

with $J_{S_0} \equiv J_S(\tilde{m} = 0)$, $F_{S_0} \equiv F_S(\tilde{m} = 0)$ and

$$\beta_S \equiv \frac{\kappa_S \tilde{m}}{(n_S + 1)\,\beta_{S_0}}. \tag{4.25}$$

Since the column mass increases with pressure, a larger value of the index n_S means that starlight is able to penetrate deeper into the atmosphere. Mathematically, there are two solution branches, but we have picked the one that allows J_S and F_S to vanish as $\tilde{m} \to \infty$, because we expect starlight to be completely attenuated if it penetrates deeply enough into the atmosphere.

When the shortwave opacity is constant ($n_S = 0$) and scattering is absent ($\beta_{S_0} = 1$), we see that equation (4.24) correctly reproduces the classical version of Beer's law. Otherwise, equation (4.24) provides a generalization of Beer's law that includes the effects of non-isotropic scattering [94]. Figure 4.2 shows examples of the diminution of shortwave flux with different combinations of values for n_S and β_{S_0}. Generally, the presence of scattering ($\beta_{S_0} < 1$) causes a more rapid diminution of incident starlight within an atmosphere.

4.4.3 The photon deposition depth

With the generalization of Beer's law in hand, we are now in a position to locate the atmospheric layer at which most of the starlight is being absorbed. We shall term this layer the *photon deposition depth* [88, 94]. We seek a relationship between the pressure (P_D) corresponding to the photon deposition depth/layer and the Bond albedo.

The shortwave flux at $\tilde{m} = 0$ is interpreted as the incident stellar flux,

$$F_{S_0} = \mu F_\star, \tag{4.26}$$

where the *stellar constant*[4] is

$$F_\star \equiv \begin{cases} \sigma_{SB} T_{irr}^4, & 0 \le \phi \le \pi, \\ 0, & \pi \le \phi \le 2\pi, \end{cases} \tag{4.27}$$

and the *irradiation temperature* is

$$T_{irr} = T_\star \left(\frac{R_\star}{a}\right)^{1/2} (1 - A_B)^{1/4} = \sqrt{2}\,T_{eq}. \tag{4.28}$$

It is the temperature at the substellar point of an exoplanet, which is the point where the distance between itself and the star is the shortest.

It is important to note that $F_{S_0} < 0$ arises naturally from the fact that it is a *net flux with a vanishing outgoing component*. By using the expression for F_S from equation (4.24), we find that the mean value of the shortwave net flux is

$$\bar{F}_S \equiv \frac{1}{2\pi} \int_0^{2\pi} \int_{-1}^0 F_S \, d\mu \, d\phi = -\frac{\sigma_{SB} T_{irr}^4 \mathcal{E}_3}{2}, \tag{4.29}$$

---

[4]Generalized from the *solar constant*.

where $\mathcal{E}_3 = \mathcal{E}_3(\beta_S)$ and the exponential integral of the jth order is defined as [2, 7]

$$\mathcal{E}_j(\beta_S) \equiv \int_1^\infty x^{-j} e^{-x\beta_S}\, dx. \tag{4.30}$$

By denoting $\bar{F}_{S_0} \equiv \bar{F}_S(\tilde{m} = 0)$, it follows that

$$\frac{\bar{F}_S}{\bar{F}_{S_0}} = 2\mathcal{E}_3. \tag{4.31}$$

We define the photon deposition depth as the pressure level where $\bar{F}_S/\bar{F}_{S_0}$ suffers one e-folding, i.e., is equal to about 0.368, which occurs when $\beta_S \approx 0.63$. It follows that

$$
\begin{aligned}
P_D &= \left[\frac{0.63\,(n_S + 1)\,g P_0^{n_S}}{\kappa_{S_0}} \right]^{1/(n_S+1)} \left(\frac{1 - \omega_{S_0}}{1 - \omega_{S_0} g_{S_0}} \right)^{1/2(n_S+1)} \\
&= \left[\frac{0.63\,(n_S + 1)\,g P_0^{n_S}}{\kappa_{S_0}} \right]^{1/(n_S+1)} \left(\frac{1 - A_B}{1 + A_B} \right)^{1/(n_S+1)},
\end{aligned}
\tag{4.32}
$$

where ω_{S_0} and g_{S_0} are the mean or characteristic values of the single-scattering albedo and scattering asymmetry factor in the shortwave, respectively. The pressure corresponding to the bottom[5] of the atmosphere is $P_0 = \tilde{m}_0 g$ and g is the surface gravity of the exoplanet.

The photon deposition depth has the expected physical property that, as the scattering becomes more backward-peaked ($g_{S_0} < 0$), it resides higher up in the atmosphere. As $n_S \to \infty$, $P_D \to P_0$. When $n_S = 0$, the expression for P_D is particularly useful because it is independent of P_0. For pure forward scattering ($g_{S_0} = 1$), photon deposition behaves as if one is in the purely absorbing limit. Backward scattering ($g_{S_0} = -1$) tends to raise the photon deposition depth to higher altitudes (lower pressures).

4.5   TREATMENT OF LONGWAVE RADIATION

The governing equations in the longwave are

$$
\begin{aligned}
\frac{\partial F_L}{\partial \tilde{m}} &= \kappa_L J_L - 4\kappa_L'' \sigma_{SB} T^4, \\
\frac{\partial K_L}{\partial \tilde{m}} &= \frac{\kappa_L' F_L}{\beta_{L_0}^2},
\end{aligned}
\tag{4.33}
$$

where T is the temperature. Unlike in the shortwave, the longwave equations include blackbody emission from the exoplanetary atmosphere. The absorption, flux and Planck mean opacities are, respectively,

$$\kappa_L \equiv \frac{\int_L \kappa_a J\, d\lambda}{\int_L J\, d\lambda}, \quad \kappa_L' \equiv \frac{\int_L \kappa_a F_+\, d\lambda}{\int_L F_-\, d\lambda}, \quad \kappa_L'' \equiv \frac{\pi \int_L \kappa_a B\, d\lambda}{\sigma_{SB} T^4}. \tag{4.34}$$

---

[5] This really means the bottom of the model domain, since it is unclear if gas-giant exoplanets, for example, have a "bottom."

As before, we need to combine the pair of governing equations. In order to proceed, we have to assume that $\kappa_L = \kappa'_L = \kappa''_L$. We shall term κ_L the *longwave opacity*.

Now, we have two equations and *four* unknowns (J_L, F_L, K_L and T). Since we are ultimately interested in the temperature, we need two more Eddington coefficients,

$$\epsilon_L \equiv \frac{F_L}{J_L} = \frac{3}{8}, \ \epsilon_{L_3} \equiv \frac{K_L}{J_L} = \frac{1}{3}. \tag{4.35}$$

Problem 4.9.1 will teach you how to derive the values of these *longwave Eddington coefficients*, based on what we learned in Chapter 3. You will often encounter studies that assume the first longwave Eddington coefficient to be $\epsilon_L = 1/2$, which is inconsistent with the values we assume for the other Eddington coefficients.

4.6 ASSEMBLING THE PIECES: DERIVING THE GENERAL SOLUTION

We are finally ready to assemble the various pieces we painstakingly constructed and derive the temperature-pressure profile of an irradiated atmosphere with a finite remnant heat of formation. By combining the second equation in (4.33), the first equation in (4.13) and using the third longwave Eddington coefficient, we obtain

$$J_L = J_{L_0} + \frac{1}{\epsilon_{L_3}\beta_{L_0}^2} \int_0^{\tilde{m}} \kappa_L \left[\mathcal{F}_\infty - F_S - \tilde{Q}(\tilde{m}, \infty) \right] d\tilde{m}. \tag{4.36}$$

Eliminating the quantities J_L and J_{L_0} using equation (4.10) and the second equation in (4.13), respectively, yields

$$
\begin{aligned}
\sigma_{SB} T^4 = & \frac{\mathcal{F}_\infty}{4} \left(\frac{1}{\epsilon_L} + \frac{1}{\epsilon_{L_3}\beta_{L_0}^2} \int_0^{\tilde{m}} \kappa_L \, d\tilde{m} \right) + Q \\
& + \frac{1}{4} \left(-\frac{F_{S_0}}{\epsilon_L} + \frac{\kappa_S J_S}{\kappa_L} - \frac{1}{\epsilon_{L_3}\beta_{L_0}^2} \int_0^{\tilde{m}} \kappa_L F_S \, d\tilde{m} \right).
\end{aligned}
\tag{4.37}
$$

At this point, it is worth pausing and understanding the physical significance of each term in equation (4.37). The first term, involving $\mathcal{F}_\infty$, is the generalization of Milne's solution to a situation with non-isotropic scattering [94]. It is equivalent to the diffusion solution we derived in Chapter 3. The second term collects all of the terms associated with Q,

$$Q \equiv -\frac{1}{4} \left[\frac{Q}{\kappa_L} + \frac{\tilde{Q}(0, \infty)}{\epsilon_L} + \frac{1}{\epsilon_{L_3}\beta_{L_0}^2} \int_0^{\tilde{m}} \kappa_L \tilde{Q}(\tilde{m}, \infty) \, d\tilde{m} \right]. \tag{4.38}$$

The rest of the terms are associated with the stellar heating of the atmosphere.

Towards the end of Chapter 3, we saw that Milne's solution, derived in the diffusion limit, uses the Rosseland mean opacity. In the current derivation, we have assumed that the absorption, flux and Planck mean opacities are equal. That we end up with Milne's solution via both approaches can only imply that these three arithmetic means are equal to the harmonic mean, which is a strong assumption because they lend very different weights to the smallest and largest opacities across a wavelength range. In our bid to derive an analytical model, we expectedly paid the price of forgoing physical realism.

We obtain the global-mean temperature-pressure profile by integrating over all possible incident angles of starlight ($0 \leq \phi \leq 2\pi$ and $-1 \leq \mu \leq 0$) and dividing by 2π,

$$
\begin{aligned}
\bar{T}^4 = &\frac{T_{\text{int}}^4}{4} \left(\frac{1}{\epsilon_{\text{L}}} + \frac{1}{\epsilon_{\text{L}3}\beta_{\text{L}0}^2} \int_0^{\tilde{m}} \kappa_{\text{L}} \, d\tilde{m} \right) \\
&+ \frac{T_{\text{irr}}^4}{8} \left(\frac{1}{2\epsilon_{\text{L}}} + \frac{\kappa_{\text{S}}\mathcal{E}_2}{\kappa_{\text{L}}\beta_{\text{S}0}} + \frac{1}{\epsilon_{\text{L}3}\beta_{\text{L}0}^2} \int_0^{\tilde{m}} \kappa_{\text{L}}\mathcal{E}_3 \, d\tilde{m} \right) \\
&+ \frac{1}{2\pi} \int_0^{2\pi} \int_{-1}^0 \mathcal{Q} \, d\mu \, d\phi,
\end{aligned}
\tag{4.39}
$$

where we again have $\mathcal{E}_j(\beta_{\text{S}})$ being the exponential integral of the j-th order. The last term in equation (4.39) vanishes when radiative equilibrium *and* global energy conservation are attained.

To evaluate the integral involving the longwave opacity, we assume its functional form to be

$$
\kappa_{\text{L}} = \kappa_0 + \kappa_{\text{CIA}} \left(\frac{\tilde{m}}{\tilde{m}_0} \right),
\tag{4.40}
$$

where κ_0 and κ_{CIA} are constants. The physical significance of this functional form will be explained shortly. If we also assume a constant shortwave opacity ($n_{\text{S}} = 0$), then we end up with a fully explicit expression for the temperature-pressure profile,

$$
\begin{aligned}
\bar{T}^4 = &\frac{T_{\text{int}}^4}{4} \left[\frac{1}{\epsilon_{\text{L}}} + \frac{\tilde{m}}{\epsilon_{\text{L}3}\beta_{\text{L}0}^2} \left(\kappa_0 + \frac{\kappa_{\text{CIA}}\tilde{m}}{2\tilde{m}_0} \right) \right] \\
&+ \frac{T_{\text{irr}}^4}{8} \left[\frac{1}{2\epsilon_{\text{L}}} + \mathcal{E}_2 \left(\frac{\kappa_{\text{S}}}{\kappa_{\text{L}}\beta_{\text{S}0}} - \frac{\kappa_{\text{CIA}}\tilde{m}\beta_{\text{S}0}}{\epsilon_{\text{L}3}\kappa_{\text{S}}\tilde{m}_0\beta_{\text{L}0}^2} \right) \right. \\
&\left. + \frac{\kappa_0\beta_{\text{S}0}}{\epsilon_{\text{L}3}\kappa_{\text{S}}\beta_{\text{L}0}^2} \left(\frac{1}{3} - \mathcal{E}_4 \right) + \frac{\kappa_{\text{CIA}}\beta_{\text{S}0}^2}{\epsilon_{\text{L}3}\kappa_{\text{S}}^2\tilde{m}_0\beta_{\text{L}0}^2} \left(\frac{1}{2} - \mathcal{E}_3 \right) \right].
\end{aligned}
\tag{4.41}
$$

With this formula, we are now ready to explore the basic trends associated with temperature-pressure profiles.

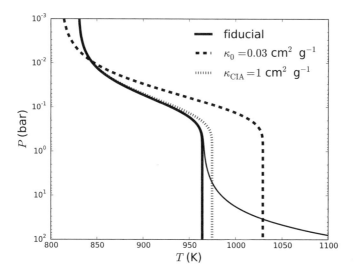

Figure 4.3: Elucidating the effects of greenhouse warming using temperature-pressure profiles in the purely absorbing limit. The fiducial model (thick, solid curve) has the following parameter values: $T_{\mathrm{int}} = 0$, $T_{\mathrm{irr}} = 1200$ K, $g = 10^3$ cm s^{-2}, $n_{\mathrm{S}} = 0$, $\kappa_{\mathrm{S}_0} = 0.01$ cm^2 g^{-1}, $\kappa_0 = 0.02$ cm^2 g^{-1}, $\kappa_{\mathrm{CIA}} = 0$, $\beta_{\mathrm{S}_0} = 1$ and $\beta_{\mathrm{L}_0} = 1$. The other models modify one of the parameter values as stated in the legend. The thin, solid curve is the fiducial model recomputed with $T_{\mathrm{int}} = 150$ K and shows Milne's solution.

4.7   EXPLORATION OF DIFFERENT ATMOSPHERIC EFFECTS

Our analytical model of the temperature-pressure profile of an atmosphere allows us to perform controlled experiments to understand the effects of varying the various parameters, much like how one does an experiment in the laboratory. To this end, we create a fiducial model with the following properties: constant opacities, no scattering in either the shortwave or longwave and the absence of collision-induced absorption. Its parameter values are listed in the caption of Figure 4.3. Based on this fiducial model, we selectively modify one parameter at a time in order to isolate its effect. Such an approach is the basis of good theory, at least in astrophysics.

4.7.1   Greenhouse warming

The gaseous component of a real atmosphere is composed of a soup of atoms and molecules, each with an ability to absorb radiation via the laws of quantum mechanics. Typically, molecules absorb photons via ro-vibrational transitions in the visible and infrared range of wavelengths, across an enormous array of spectral lines. We have swept this complexity under the carpet by parametrizing it in the form of the shortwave and longwave opacities. Nevertheless, we may learn valuable lessons from our analytical model.

The thick, solid curve in Figure 4.3 shows our "plain vanilla" fiducial model. Its shortwave opacity is less than its longwave counterpart, meaning that starlight penetrates beyond the infrared photosphere. At some point, it peters out and is fully absorbed by the atmosphere. The shape of the temperature-pressure profile, where temperatures generally increase with pressure until they become isothermal, is generic for irradiated atmospheres. If internal heat is present in the exoplanet, then a "tail" (Milne's solution) appears at great depths, as shown by the thin, solid curve. Later in the chapter, we will see that Milne's solution has an intimate connection to the absence or presence of convection, depending on how the infrared opacity varies with pressure.

If there was no atmosphere present, the temperature of the exoplanet would be about 850 K. That the temperature asymptotes to a value that is higher than 850 K is a manifestation of the *greenhouse effect*—starlight enters an atmosphere with relative ease, but thermal emission has greater difficulty escaping due to an enhanced infrared opacity, exactly like how a greenhouse functions.

The thick, dashed curve in Figure 4.3 shows that increasing the longwave or infrared opacity has the effect of strengthening the greenhouse warming. It is apparent that this warms the lower atmosphere and cools the upper atmosphere.

4.7.2   Collision-induced absorption

Symmetrically diatomic molecules such as hydrogen or nitrogen do not possess permanent dipole moments and thus are unable, to lowest order, to absorb radiation. However, at high pressures ($\gtrsim$ 1 bar) two molecules of hydrogen or nitrogen may come together and form a transient "super-molecule" with a weak quadrupole moment, which then absorbs radiation. This effect is known as *collision-induced absorption* (CIA). It is typically weak, but if it is the only game in town, then it becomes noticeable. In fact, CIA was first observed in the hydrogen-dominated atmospheres of Uranus and Neptune in the 1950s [102].

The functional form of the longwave opacity, as stated in equation (4.40), becomes clear. It consists of a constant component for the greenhouse gases (e.g., water, methane, carbon monoxide) and a linear component to mimic CIA associated with the inert buffer gas. Since CIA is a two-body process, we expect the optical depth to be proportional to the square of the number density of the atmospheric gas. This is reflected in our prescription that the optical depth associated with CIA is $\tau_{\mathrm{CIA}} \propto P^2$.

The thick, dotted curve in Figure 4.3 examines the effects of "switching on" CIA, which warms the atmosphere everywhere compared to the fiducial model and especially at high pressures. Unlike the pressure broadening of spectral lines, CIA is a continuum effect that occurs across a broad range of wavelengths. It

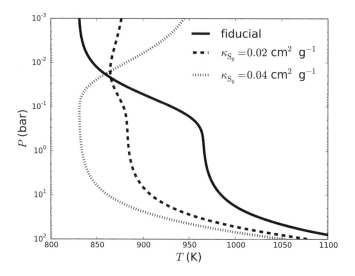

Figure 4.4: Elucidating the effects of anti-greenhouse cooling using temperature-pressure profiles in the purely absorbing limit. In this set of models, we have set $T_{\text{int}} = 150$ K. The thick, dashed curve has the shortwave and longwave opacities being equal. The thick, dotted curve has $\kappa_{S_0}/\kappa_0 = 2$.

is generally difficult to calculate the opacity associated with CIA, as it depends on the nature[6] of the collidants and the ambient pressure [1].

4.7.3   Anti-greenhouse cooling and atmospheric inversions

What happens when we increase the shortwave opacity to the point where it becomes larger than its longwave counterpart? Physically, this is akin to inserting extra absorbers of shortwave radiation in the atmosphere. On Earth, this role is played by ozone in the ultraviolet range of wavelengths. On exoplanets, a favored guess of the astronomers is titanium monoxide (TiO), due to its prevalence in the observed spectra of brown dwarfs [116], although it remains to be seen if the analogy between brown dwarfs and gas-giant exoplanets carries this far.

The thick, dashed and dotted curves in Figure 4.4 illustrate the effect of increasing the shortwave opacity to the point where it dominates the longwave opacity, which leads to the *anti-greenhouse effect*—the upper atmosphere warms and the lower atmosphere cools. The onset of the isothermal component shifts to lower pressures, since starlight is now deposited at higher altitudes. The

---

[6]More specifically, it depends on the stoichiometry of the collidants. For example, CIA may occur between a pair of hydrogen molecules or between a hydrogen molecule and a helium atom.

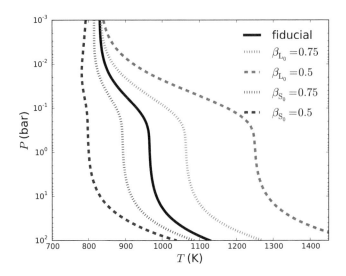

Figure 4.5: Elucidating the effects of shortwave/optical and longwave/infrared scattering on the temperature-pressure profiles. As the scattering parameter decreases, the strength of scattering increases. Shortwave scattering enhances the anti-greenhouse effect, while longwave scattering generally warms the entire atmosphere. In this set of models, we have also set $T_{int} = 150$ K.

thick, dotted curve develops a *temperature inversion*, where temperature now increases with altitude.

There is a long and rich debate in the astrophysical literature on the possible existence of temperature inversions in exoplanetary atmospheres [27, 28, 63, 64, 124, 152, 153, 155, 171, 229], although the shortwave absorber has yet to be clearly identified. If temperature inversions are present, they may negate the effects of disequilibrium chemistry by warming the upper atmosphere [175]—chemical time scales tend to be shorter at higher temperatures. This has consequences for interpreting chemical abundances.

4.7.4 Albedo variations

An important and measurable quantity is the Bond albedo of the atmosphere, which is the fraction of incident starlight reflected away from it across all wavelengths. The Bond albedo essentially controls the energy budget of an atmosphere. It is generally mediated by the presence of aerosols, condensates or dust grains, whose chemistry and compositions are not always easy to decipher. Nevertheless, it is of interest to understand the effects of a varying albedo on the thermal structure of an atmosphere.

One each of the thick, dotted and dashed curves in Figure 4.5 are for model atmospheres with shortwave scattering parameters of 0.75 and 0.5, respectively. In terms of the Bond albedo, these correspond to $A_B = 1/7 \approx 14\%$ and $A_B = 1/3 \approx 33\%$. A finite albedo exerts an effect similar to that of anti-greenhouse cooling, but with a key difference—not only do the upper and lower parts of the atmosphere warm and cool, respectively, the entire temperature-profile shifts to lower temperatures because of the $(1 - A_B)^{1/4}$ factor present in the irradiation temperature. In other words, the total energy content of the atmosphere is reduced by a factor of $(1 - A_B)$.

4.7.5 Scattering greenhouse effect

When scattering in the longwave or infrared is present, the opposite effect occurs—the atmosphere is warmer everywhere, as shown by the other set of thick, dotted and dashed curves in Figure 4.5. Thermal emission from the exoplanetary atmosphere attempts to escape, but some of it is scattered back into the exoplanet. This is known as the *scattering greenhouse effect*. If the thermal emission occurs predominantly in the infrared, then large particles (micron-sized or larger) are required for it to be noticeable.

Overall, the reason why aerosols or clouds, when they are present in an atmosphere, complicate any analysis or interpretation is that they are capable of exerting some combination of all of these effects: greenhouse warming and anti-greenhouse cooling in the form of both absorption and scattering. As we have seen, each of these effects has a significant influence on the thermal structure of an atmosphere. Sometimes, they reinforce each other; at other times, they cancel each other out. It is often difficult to accurately account for all of these effects in a real atmosphere, even in the presence of observational constraints.

4.8 MILNE'S SOLUTION AND THE CONVECTIVE ADIABAT

If an exoplanet possesses an internal source of heat, then it is possible that convection is active deep within its atmosphere. In our temperature-pressure profiles, this deep component is Milne's solution. It is worth investigating whether it is unstable to convection—and if so, to clarify the conditions under which convection is triggered.

If you work through Chapter 12 and Problem 12.9.1, then you will convince yourself that the atmosphere is stable against convection if *Schwarzschild's criterion* is fulfilled,

$$\frac{\partial \left(\ln \bar{T} \right)}{\partial \left(\ln P \right)} \leq \kappa_{\mathrm{ad}} = \frac{2}{2 + n_{\mathrm{dof}}}, \tag{4.42}$$

where κ_{ad} is the adiabatic gradient. The criterion depends on the number of degrees of freedom (n_{dof}) of the dominant atmospheric gas, which is generally a function of both chemical composition and temperature. If we include translation, rotation and vibration, then a diatomic molecule would have $n_{\mathrm{dof}} = 6$ and $\kappa_{\mathrm{ad}} = 1/4$. If we exclude vibration, then $\kappa_{\mathrm{ad}} = 2/7$. In what follows, we will assume that the buffer gas of the atmosphere is diatomic, which is not an unreasonable assumption—molecular nitrogen is the dominant gas, by mass, for the atmospheres of Earth and Titan, while molecular hydrogen seems to be the buffer gas of choice for Jupiter, Saturn and many detected exoplanets.

By differentiating Milne's solution with respect to pressure, we obtain

$$
\frac{\partial \left(\ln \bar{T} \right)}{\partial \left(\ln P \right)} = \frac{P}{4 \epsilon_{L_3} \beta_{L_0}^2 g} \left(\kappa_0 + \frac{\kappa_{CIA} P}{P_0} \right) \left(\frac{1}{\epsilon_L} + \frac{\kappa_0 P}{\epsilon_{L_3} \beta_{L_0}^2 g} + \frac{\kappa_{CIA} P^2}{2 \epsilon_{L_3} \beta_{L_0}^2 P_0 g} \right)^{-1}.
$$

(4.43)

At high pressures, this dimensionless temperature gradient becomes

$$
\lim_{P \to \infty} \frac{\partial \left(\ln \bar{T} \right)}{\partial \left(\ln P \right)} = \begin{cases} \frac{1}{4}, & \kappa_{CIA} = 0, \\ \frac{1}{2}, & \kappa_0 = 0. \end{cases}
$$

(4.44)

These asymptotic values lead us to a remarkable conclusion, independent of the strength of absorption or scattering in the infrared and the surface gravity of the exoplanet—if collision-induced absorption is operating deep in the atmosphere, then Milne's solution is unconditionally unstable to convection. In this situation, it describes the fully-convective interior of an exoplanet (e.g., a gas giant). If collision-induced absorption is inactive, then it is unconditionally stable to convection.

These considerations highlight an equally remarkable aspect of atmospheres: opacities, determined by the laws of quantum physics, are conspiring with thermodynamics and fluid dynamics, via Schwarzschild's criterion, to produce convective mixing, a distinctively macroscopic phenomenon. It is a stark reminder that to understand exoplanetary atmospheres, a holistic approach is needed.

4.9   PROBLEM SETS

4.9.1   The longwave Eddington coefficients

To use the formula for the temperature-pressure profile, one needs to fix the values of the first and third longwave Eddington coefficients. The latter is easy and follows directly from the third Eddington coefficient: $\epsilon_{L_3} = \epsilon_3 = 1/3$. For the former, show, using the two expressions for ϵ_2, that

$$
F_- = \frac{F_+^2}{2K_-}.
$$

(4.45)

Hence, show that

$$
\epsilon_L \equiv \frac{F_-}{J} = \frac{\epsilon^2}{2\epsilon_3} = \frac{3}{8}.
$$

(4.46)

Here, we have not used the subscript "L" in order to distinguish clearly between the net and total fluxes.

4.9.2 Detached convective regions

Higher up in the atmosphere, stellar irradiation generally acts to reduce the lapse rate and stabilize it against convection. However, if the shortwave opacity is sufficiently smaller than its longwave counterpart, such that the greenhouse effect is strong, it is possible to have *detached convective regions* straddled by zones of stability.

Consider a temperature-pressure profile, in the purely absorbing limit, with a constant shortwave opacity, no internal heat and a longwave opacity only due to collision-induced absorption,

$$
\bar{T}^4 = \frac{T_{\mathrm{irr}}^4}{8} \left[\frac{1}{2\epsilon_{\mathrm{L}}} + \mathcal{E}_2 \left(\frac{\kappa_{\mathrm{S}}}{\kappa_{\mathrm{L}}} - \frac{\kappa_{\mathrm{CIA}}\tilde{m}}{\epsilon_{\mathrm{L}^3}\kappa_{\mathrm{S}}\tilde{m}_0} \right) + \frac{\kappa_{\mathrm{CIA}}}{\epsilon_{\mathrm{L}^3}\kappa_{\mathrm{S}}^2\tilde{m}_0} \left(\frac{1}{2} - \mathcal{E}_3 \right) \right]. \qquad (4.47)
$$

(a) Derive the expression for $\frac{\partial(\ln \bar{T})}{\partial(\ln P)}$.

(b) What is the value of $\frac{\partial(\ln \bar{T})}{\partial(\ln P)}$ as $P \to \infty$?

(c) Compute the graphical solution of $\frac{\partial(\ln \bar{T})}{\partial(\ln P)}$ for $P_0 = 100$ bar, $g = 10^3$ cm s^{-2}, $\kappa_{\mathrm{S}} = 0.001$ cm^2 g^{-1} and $\kappa_{\mathrm{CIA}} = 0.01, 0.1$ and 1 cm^2 g^{-1}. Are there regions of convective instability within each model atmosphere?

4.9.3 Non-constant shortwave opacity

Write a Python program to compute temperature-pressure profiles for $n_{\mathrm{S}} = 0, 0.5$ and 1. What is the main qualitative difference between these profiles?

Chapter Five

Atmospheric Opacities: How to Use a Line List

5.1 FROM SPECTROSCOPIC LINE LISTS TO SYNTHETIC SPECTRA

In Chapter 3, we learned how to solve the radiative transfer equation using the two-stream approximation, which enabled us to calculate the fluxes passing through the layers of an atmosphere. In Chapter 4, we developed an intuition for the temperature-pressure profiles using idealized analytical solutions. A missing and key ingredient we have not yet discussed is the atmospheric opacity function. It is the cross section per unit mass, across all wavelengths, of all of the spectral lines of the atoms and molecules present in the atmosphere, which allows us to construct transmission functions, compute fluxes and ultimately predict the spectrum of an exoplanetary atmosphere as seen by an astronomer. If one desires to solve for radiative equilibrium,[1] then the temperature-, pressure- and wavelength-dependent opacity function is iterated with the temperature-pressure profile until a converged solution obtains. In this chapter, it is my intention to guide you through the calculation of the opacity function, so that you will finally have all of the ingredients required to either construct your own radiative transfer solver or at least be aware of the inputs that go into a solver when using one.

Figure 5.1 shows the opacity function of water calculated using the HITEMP database, which provides spectroscopic line list data needed to compute the strengths and shapes of these spectral lines. (In principle, any database may be used.) Quantum mechanics tells us that atoms and molecules may be described by harmonic oscillators with frequencies that are regularly spaced in frequency or wavenumber—for this entire chapter, we will express quantities in terms of wavenumber. The shapes of lines are described by the *Voigt profile*, except when one is far away from line center and the interactions between atoms or molecules alter the shape of the line wings. We will discuss the details involved in going from the line list data to the strengths and shapes of spectral lines. In the process, we will elucidate the assumptions and pitfalls involved.

A daunting challenge associated with calculating the atmospheric opacity function is the enormous number of spectral lines involved. In the example shown in Figure 5.1, there are $\sim 10^8$ lines of water—and this is merely at *one*

---

[1] As previously discussed, this is equivalent to enforcing local energy conservation between atmospheric layers.

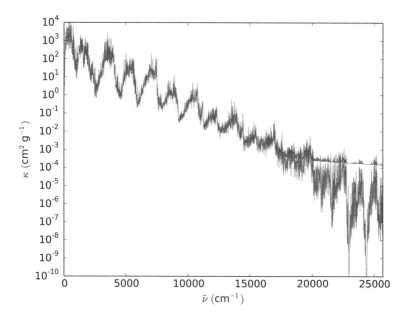

Figure 5.1: Sample opacity function constructed using the spectroscopic line list database of HITEMP [203]. As an example, we consider only water at $T = 1500$ K and $P = 1$ atm. For illustration, we include two calculations: one with every spectral line represented by the full Voigt profile and another with an ad hoc cutoff of 100 cm^{-1} imposed on the line wings. You will see that using the full Voigt profile drowns out the weak lines at larger wavenumbers (shorter wavelengths). Courtesy of Simon Grimm using the HELIOS-K software package [77].

specific pairing of the temperature and pressure! Calculations of radiative transfer typically require dozens, if not hundreds, of combinations of temperatures and pressures. To complete these calculations within any reasonable period of time, we need techniques to reduce the number of opacity points being sampled. We will describe a technique known as the *k-distribution method*, which is exact only under certain restricted conditions—we will elucidate what these conditions are.

By the end of the chapter, you will understand how to construct your own atmospheric opacity function for any temperature, pressure and mixture of atoms and molecules. I have focused on dealing with spectral lines; in practice, one also has to include the effects of collision-induced absorption, which introduces a continuum to the opacity function (see Chapter 4).

5.2   THE VOIGT PROFILE

Spectral lines are generally described by the Voigt profile, which is constructed from a combination of the Lorentz and Doppler profiles. This is standard fare for Astrophysics 101; we will review and discuss these line profiles and their properties.

In the zero-temperature limit, the shape of spectral lines associated with atoms or molecules is described by the *Lorentz profile* [55]

$$\Phi_{\mathrm{L}} = \frac{\Gamma_{\mathrm{L}}/\pi}{(\tilde{\nu} - \tilde{\nu}_0)^2 + \Gamma_{\mathrm{L}}^2}, \tag{5.1}$$

where

$$\tilde{\nu} \equiv \frac{1}{\lambda} \tag{5.2}$$

is the wavenumber, λ is the wavelength, $\tilde{\nu}_0$ is the wavenumber at line center and $2\Gamma_{\mathrm{L}}$ is the natural line width. One may show that the Lorentz profile normalizes to unity when integrated over all wavenumbers and that Γ_{L} is its *half-width at half-maximum* (HWHM) (Problem 5.6.1). Spectral lines have an intrinsic width because of the uncertainty principle of quantum mechanics.

When the gas has a finite temperature, atoms and molecules within it have a distribution of speeds. The mean speed is given by the *thermal speed*,

$$v_{\mathrm{th}} \equiv \left(\frac{2k_{\mathrm{B}}T}{m}\right)^{1/2}, \tag{5.3}$$

where k_{B} is the Boltzmann constant, T is the temperature and m is the mass of the atom or molecule. When a mixture of molecules is involved, m is the mean molecular mass. If we denote the wavenumber shift associated with the thermal speed by

$$\sigma_{\mathrm{th}} = \frac{\tilde{\nu}_0 v_{\mathrm{th}}}{c}, \tag{5.4}$$

where c is the speed of light, then the line shape is now described by the *Doppler profile*,

$$\Phi_{\mathrm{D}} = \frac{1}{\Gamma_{\mathrm{D}}} \left(\frac{\ln 2}{\pi}\right)^{1/2} e^{-(\tilde{\nu}-\tilde{\nu}_0)^2/\sigma_{\mathrm{th}}^2}. \tag{5.5}$$

There is more than one way to express the Doppler profile.[2] I have chosen this definition such that

$$\Gamma_{\mathrm{D}} \equiv \frac{\tilde{\nu}_0}{c} \left(\frac{2\ln 2 \; k_{\mathrm{B}}T}{m}\right)^{1/2} = (\ln 2)^{1/2} \sigma_{\mathrm{th}} \tag{5.6}$$

is its HWHM and $2\Gamma_{\mathrm{D}}$ is the Doppler line width.

---

[2]For example, Mihalas [172] and Draine [55] essentially do not distinguish between Γ_{D} and v_{th}. I prefer this distinction so as to emphasize the freedom one has in defining Γ_{D}, which is important for how one writes the Lorentz and Voigt profiles.

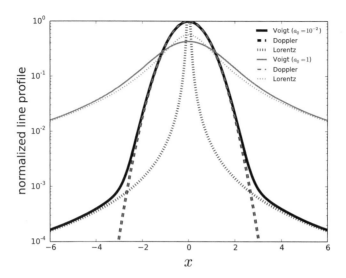

Figure 5.2: Examples of Lorentz, Doppler and Voigt profiles for two values of the damping parameter. When the damping parameter is small ($a_0 = 10^{-2}$), the Voigt profile is well-approximated by a Doppler core and Lorentz wings. This approximation breaks down when the damping parameter approaches unity. I have computed the Voigt profile using its alternative form (Problem 5.6.2) and normalized the profiles appropriately (Problem 5.6.3).

Physically, the shape of a spectral line involves some "mixing" of its natural and Doppler profiles. Mathematically, if we write $X \equiv \tilde{\nu} - \tilde{\nu}_0$, then the *Voigt profile* is a convolution of both profiles,

$$\Phi_V \equiv \int_{-\infty}^{\infty} \Phi_L\left(X - X'\right)\, \Phi_D\left(X'\right)\, dX'$$
$$= \left(\frac{\ln 2}{\pi}\right)^{1/2} \frac{H_V}{\Gamma_D}, \tag{5.7}$$

where the *Voigt H-function*, which is dimensionless, takes the form,

$$H_V \equiv \frac{a_0}{\pi} \int_{-\infty}^{+\infty} \frac{e^{-x'^2}}{\left(x - x'\right)^2 + a_0^2}\, dx'. \tag{5.8}$$

The *damping parameter* is almost the ratio of the Lorentz to the Doppler HWHMs,

$$a_0 \equiv \frac{(\ln 2)^{1/2}\,\Gamma_L}{\Gamma_D} = \frac{\Gamma_L}{\sigma_{\text{th}}}. \tag{5.9}$$

We have also defined $x \equiv X/\sigma_{\text{th}}$ and $x' \equiv X'/\sigma_{\text{th}}$. It is important not to confuse the Voigt H-function with the Voigt profile.

The damping parameter also happens to be the only parameter that controls the shape of the Voigt profile. Figure 5.2 shows examples of the Voigt profile with $a_0 = 10^{-2}$ and 1. For $a_0 = 10^{-2}$, we see that the Voigt profile is well-approximated by a Doppler core and Lorentz wings. When the damping parameter is unity, this approximation breaks down.

There is a serious shortcoming of the Voigt profile that is often ignored, even in the published literature. It was derived for an atom or molecule in isolation and is only accurate near the core of the spectral line. For an ensemble of atoms or molecules, an effect known as *pressure broadening* comes into play. Qualitatively, it is the phenomenon that atoms or molecules may make up for deficits or surpluses of energy, when activating a spectral line, by transferring them from or to other atoms or molecules via collisions, an effect that is pressure-dependent and already significant at moderate pressures ($\sim$ 0.1–1 bar). Pressure broadening is often mimicked by adding an extra term to the natural line width and inserting this generalized width into the Lorentz profile [172]. Such an approach is incorrect in the far line wings, but it remains unclear even what "far" means as the theory of pressure broadening is incomplete. While this may sound like a negligible correction to the spectral line shape, the errors accumulated over millions to billions of lines may be non-negligible [77]. This fundamental, unsolved physics problem awaits a general solution.

5.3 THE QUANTUM PHYSICS OF SPECTRAL LINES

5.3.1 Relating opacities and line profiles

The opacity associated with a spectral line is given by

$$\kappa = \mathcal{S}\Phi_{\rm V}, \tag{5.10}$$

where $\mathcal{S}$ is the *integrated line strength*. Exactly as its name suggests, it is the strength of the line integrated over all wavenumbers. It may be defined as

$$\mathcal{S} \equiv \int \kappa \, d\tilde{\nu}. \tag{5.11}$$

Depending on whether one integrates over wavenumber, frequency or wavelength, the integrated line strength takes on different physical units. In the definition given above, it has units of cm g^{-1}. We will see shortly that this choice of $\mathcal{S}$ ensures that it only depends on temperature and not pressure.

If one confuses κ to be the cross section or extinction coefficient [74], then $\mathcal{S}$ has 9 possibilities for its physical units (Problem 5.6.4). There is really no reason to be confused on this point.

5.3.2 Calculating the integrated line strength

The exact expression for $\mathcal{S}$ originates from quantum mechanics and statistical mechanics. In principle, one would solve the Schrödinger equation to obtain the various quantities needed to construct the integrated line strength. In practice, one looks up a spectroscopic database, such as HITEMP [203], and reads off the necessary quantities at reference values of the temperature and pressure. All that is left to do is to use an analytical scaling relation to scale S to other temperatures and pressures. We will now derive this scaling relation. I emphasize that the derivation here applies generally and not just to a specific spectroscopic database. Again, we are interested in the physical principles associated with the integrated line strength, rather than the quirks associated with each database itself.

Quantum mechanics tells us that a molecule may be visualized as having a set of discrete energy levels. Consider a pair of energy levels labeled by 1 and 2, with the latter having more energy. Next, consider an ensemble of the same molecule. Some fraction of this ensemble will be in the lower energy level; let their number density be n_1. The rest of the ensemble will be in the higher energy level and have a number density of n_2. The molecules in the higher energy level may spontaneously decay to the lower energy one with the decay rate being the *Einstein A-coefficient*, A_{21} (with physical units of s^{-1}). Photons may be absorbed by molecules in the lower energy level, thus promoting them to the higher energy one, a process described by the *Einstein B-coefficient* (B_{12}). *Stimulated emission* may also occur, in which a photon of the appropriate energy (corresponding to exactly the energy difference between the two levels) induces decay, described by another Einstein B-coefficient (B_{21}). The Einstein A- and B-coefficients are not independent quantities and may be related via the oscillator strength and statistical weights of the pair of energy levels [232].

The *principle of detailed balance* relates the rates of population and decay of the energy levels [232],

$$(n_1 B_{12} - n_2 B_{21}) \, \mathcal{E} = n_2 A_{21}, \tag{5.12}$$

where $\mathcal{E}$ is the energy density per unit wavelength, frequency or wavenumber. Since there are three different ways of expressing the physical units of $\mathcal{E}$ (erg cm^{-4}, erg cm^{-3} Hz^{-1} or erg cm^{-2}), the Einstein B-coefficients have three sets of physical units. By contrast, the Einstein A-coefficient always has physical units of s^{-1}.

If we choose the physical units of S to be cm g^{-1}, then the integrated line strength is [187]

$$S = \frac{h \tilde{\nu} k_{\mathrm{B}} T}{mP} \left(n_1 B_{12} - n_2 B_{21} \right), \tag{5.13}$$

where h denotes Planck's constant and P is the pressure.

We invoke the assumption of *local thermodynamic equilibrium* (LTE), which is actually a *set* of assumptions. Among other things, it asserts that the radiation field is Planckian and the population of the energy levels may be described by a Boltzmann distribution. The latter states that

$$\frac{n_2}{n_1} = \frac{g_2}{g_1} e^{-\Delta E / k_{\mathrm{B}} T} \left(1 + \frac{n_{\mathrm{cr}}}{n_e} \right)^{-1}, \tag{5.14}$$

where g_1 and g_2 are the statistical weights of the respective energy levels. The energy difference between the levels is represented by ΔE. LTE further assumes that the electron density (n_e) greatly exceeds the critical density for collisional de-excitation (n_{cr}).

We choose $\mathcal{E}$ to be the energy density per unit wavenumber and relate it to the Planck function,

$$\mathcal{E} = \frac{4\pi B}{c}. \tag{5.15}$$

The Planck function in units per wavenumber is

$$B = 2hc^2\tilde{\nu}^3 \left(e^{h\tilde{\nu}c/k_B T} - 1 \right)^{-1}. \tag{5.16}$$

The last ingredient we need is to relate n_1 to the total number density, n. This requires a *partition function* $Q_p(T)$ [187, 232],

$$\frac{n_1}{n} = \frac{g_1}{Q_p} e^{-\Delta E/k_B T}. \tag{5.17}$$

The functional form of $Q_p(T)$ is usually tabulated by the spectroscopic databases.

Putting it all together, we get

$$S = \frac{g_2 A_{21}}{8\pi c\tilde{\nu}^2 m Q_p} e^{-\Delta E/k_B T} \left(1 - e^{-h\tilde{\nu}c/k_B T} \right). \tag{5.18}$$

It is important to note that this form of S, with physical units of cm g^{-1}, has no dependence on n and thus P, which are related by the ideal gas law. The integrated line strength depends solely on T.

Often, spectroscopic databases provide S at a reference temperature (T_0). To compute S at another temperature, we use

$$\frac{S}{S_0} = \frac{Q_{p0}}{Q_p} e^{-\Delta E/k_B T + \Delta E/k_B T_0} \frac{1 - e^{-h\tilde{\nu}c/k_B T}}{1 - e^{-h\tilde{\nu}c/k_B T_0}}, \tag{5.19}$$

where $S_0 \equiv S(T_0)$ and $Q_{p0} \equiv Q_p(T_0)$.

5.3.3   Pressure broadening

An approximate way of describing pressure broadening is to include it in the form of a generalized HWHM for the Lorentz profile [172, 202],

$$\Gamma_L = \frac{A_{21}}{4\pi c} + \left(\frac{T}{T_0} \right)^{-n_{coll}} \left[\frac{\alpha_{air}(P - P_{self})}{P_0} + \frac{\alpha_{self} P_{self}}{P_0} \right], \tag{5.20}$$

where the first term after the equality is the natural HWHM of a spectral line in wavenumber units. Pressure broadening is included via the second term, which is an empirical fit with fitting parameters $(n_{coll}, \alpha_{air}$ and $\alpha_{self})$ given by, for example, the `HITRAN` database [202], which uses a reference pressure[3] of $P_0 = 1$ atm.

The subscript "air" refers to *air-broadening*, which is pressure broadening induced by the buffer gas in the atmosphere. In the case of Earth, this is molecular nitrogen with some contribution from molecular oxygen. One should not assume that this configuration applies generally to exoplanets, since air-broadening does depend on the stoichiometry of the buffer gas. The subscript "self" refers to *self-broadening*, which is pressure broadening induced by mutual collisions between molecules of the same species. Pressure broadening should not be confused with collision-induced absorption, which is a continuum effect.

Pressure broadening also introduces a shift (δ_{shift}) in the central wavenumber of the spectral line, as given by [202]

$$\nu_0 \to \nu_0 + \frac{\delta_{\text{shift}} P}{P_0}, \tag{5.21}$$

where δ_{shift} is again a tabulated quantity in spectroscopic databases.

I emphasize that this empirical way of including pressure broadening does not obviate the issues associated with the unsolved physics problem of the unknown shape of the far wings of spectral lines [77], especially since it does not afford us the generality we need to explore the broad parameter space describing exoplanetary atmospheres. It is a situation where precision should not be mistaken for accuracy—just because this prescription allows you to compute the pressure-broadening contributions to many decimal places does not necessarily mean it is correct.

5.4 THE MILLION- TO BILLION-LINE RADIATIVE TRANSFER CHALLENGE

The number of spectral lines that contributes to the opacity function of the atmosphere is generally a steep function of the temperature. Under Earth-like conditions, there are $\sim 10^5$ lines of the water molecule—not a small number, but still manageable on modern-day computers. As the temperature approaches that of the hot exoplanets (~ 1000–2000 K), the number of water lines balloons to $\sim 10^8$ or more. This is daunting given that the opacity function itself needs to be sampled at even more points than there are number of lines! If the radiative transfer calculation needs to be performed at dozens to hundreds of combinations of temperatures and pressures for different mixtures of atoms and molecules, then the problem quickly becomes computationally intractable. It is clear that brute-force computation is infeasible. Fortunately, methods exist to greatly reduce the number of points one needs to effectively sample the opacity function.

---

[3]In Grimm & Heng [77], we stated that 1 atm = 0.98692 bar in the caption of Figure 1, which is erroneous. Rather, we have 1 bar = 0.98692 atm.

If one somehow manages to solve the problem by brute force and sample the opacity function very finely—at generally more points than there are number of lines[4]—then one is performing a *line-by-line* calculation. If one wishes to sample the opacity function using a much smaller number of points—typically orders of magnitude smaller—then one is using *opacity sampling*. Generally, opacity sampling needs to be validated against line-by-line calculations in order to demonstrate that one's calculation has adopted a sufficient numerical resolution.

Another approach—and one we will discuss extensively—is the *k-distribution method*, a technique that has been honed in the Earth and planetary sciences [65, 74, 135]. It is a clever strategy—instead of integrating over the opacity function itself, one constructs its cumulative counterpart. This has the advantage that the somewhat erratic opacity function—which varies by many orders of magnitude in value and is often sparsely populated between lines—is now transformed into a smooth and monotonically increasing function, which is pleasing for computation. The opacity function itself needs first to be finely sampled, but its *k-distribution function* may be resampled using a much smaller number of points, which allows it to be pre-computed and tabulated efficiently. By constructing grids of k-distribution functions over temperature, pressure and chemical mixtures, one may interpolate between the grid points to perform radiative transfer calculations.

The catch is that the k-distribution function may only be used with certain classes of functions. Fortunately, we will see that the transmission function is one of these functions. Another catch is that the k-distribution method may only be used for homogeneous atmospheric paths and when the atmosphere contains only a single chemical species, a situation that is never encountered in practice. To apply the k-distribution method to more general situations, we have to invoke the *correlated-k approximation*, which we will elucidate later in the chapter. The correlated-k approximation is exact only in the limit of low temperature and Lorentzian lines.

5.4.1   The k-distribution method

Consider an arbitrary function $f(x)$, where x is the wavenumber normalized by the entire range considered. We wish to evaluate the integral over the range $x_{min} \leq x \leq x_{max}$,

$$\mathcal{I} = \int_{x_{min}}^{x_{max}} f(x) \ dx. \tag{5.22}$$

Imagine that $f(x)$ may be recast as $f(y)$ such that the quantity y is the fractional area under the curve that satisfies $f(x) \leq f_0$, where f_0 is an arbitrary value of

---

[4]In special circumstances, such as when pressure broadening is dominant or when lines are heavily blended, it may be possible to reduce the number of sampling points needed.

the function. The same integral may be evaluated as

$$\mathcal{I} = \int_0^1 f(y) \; dy. \tag{5.23}$$

Practically all of the functions we encounter in astrophysics may be integrated using this alternative expression. The "regular" way of evaluating the integral is *Riemann integration*, while the alternative one is *Lebesgue integration*.

A less obvious situation occurs when we have

$$\mathcal{I} = \int F \; dx = \int \mathcal{H} \; dF, \tag{5.24}$$

where $\mathcal{H}$ is the *fractional cumulative distribution function* of another arbitrary function, $F(f(x))$, that satisfies $F \leq F_0$ and F_0 is an arbitrary value of F. In other words, $\mathcal{H}$ gives the fractional area under the curve corresponding to $F \leq F_0$.

These concepts may be made less abstract by applying them to an atmosphere. Consider the simplest case of

$$F = f = \kappa, \tag{5.25}$$

which leads us to

$$\int \mathcal{H} \; dF = \int \mathcal{H} \, \kappa(x) \; dx = \int \kappa(y) \; dy. \tag{5.26}$$

The quantity

$$y \equiv \int_0^x \mathcal{H} \; dx \tag{5.27}$$

is the cumulative sum of intervals. It is comforting that one gets the same answer whether one evaluates $\int \kappa(x) \; dx$ or $\int \kappa(y) \; dy$. We shall refer to $\kappa(x)$ and $\kappa(y)$ as the *opacity function* and *k-distribution function*, respectively.

A less trivial case that is of greater interest to us is

$$f = \kappa, \; F = e^{-\kappa \tilde{m}}, \tag{5.28}$$

where $\tilde{m}$ is the column mass, since the transmission function,

$$\mathcal{T} = \int_0^\infty e^{-\kappa(x) \; \tilde{m}} \; dx = \int_0^1 e^{-\kappa(y) \; \tilde{m}} \; dy, \tag{5.29}$$

is a quantity that is indispensable for computing synthetic spectra. In this case, the cumulative sum of intervals is

$$y \equiv \int_0^\kappa \mathcal{H}\tilde{m} \; d\kappa. \tag{5.30}$$

Generally, this procedure does not work for all functional forms of F.

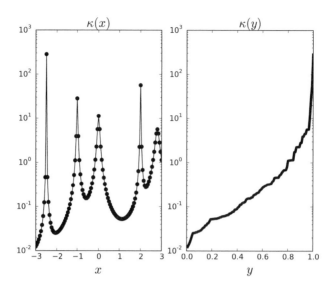

Figure 5.3: Idealized example of an atmospheric opacity function with spectral lines represented by Lorentz profiles with arbitrary units (left panel), which is then transformed to its k-distribution function (right panel). Notice how a uniform sampling in x generally corresponds to a non-uniform sampling in $\kappa(x)$. Spectral lines may be under-sampled near their cores if insufficient numerical resolution is used to capture the sudden jumps in value.

5.4.2   Implementing the k-distribution method

Generally, the opacity function may be divided into multiple bins. Within each bin, we may construct the k-distribution function, which allows us to compute the transmission function. With the transmission function of every bin in hand, we may compute the flux emerging from the exoplanetary atmosphere.

Consider a wavenumber bin with a width of $\Delta x \equiv x_{\mathrm{max}} - x_{\mathrm{min}}$. Within the bin, we consider equal intervals in x and let the interval be denoted by δx. Such a uniform grid in x generally leads to a non-uniform grid in $\kappa(x)$. Its virtue is that it reduces our problem to one of sorting and ordering, since every value of $\kappa(x)$ is associated with δx (and we do not have to keep track of changing values of the interval). For a fixed value of the opacity (κ_0), we count the number of points that satisfy $\kappa(x) \leq \kappa_0$. If N_x points are counted, then we have

$$y = \frac{N_x \, \delta x}{\Delta x}. \tag{5.31}$$

We also have $\delta x = \Delta x / N$, where N is the total number of intervals in x. It implies that the interval in y is also equal,

$$\delta y = \frac{\delta x}{\Delta x} = \frac{1}{N}. \tag{5.32}$$

By running through all possible values of κ_0, one constructs $\kappa(y)$. Since $\kappa(y)$ is a monotonic function that is typically smoother than $\kappa(x)$, it may be resampled and defined over a much smaller number of points, $N_y \ll N$. It is then used to calculate $\mathcal{T}$ for any value of $\tilde{m}$.

5.4.3 Correlated-k approximation

The k-distribution method is exact for a homogeneous atmosphere, meaning that it is isothermal and isobaric. This never happens in practice. Even for calculations involving pairs of atmospheric layers, we expect each layer to have its own temperature and pressure values. The path between these atmospheric layers is then inhomogeneous as it traverses a gradient in temperature and pressure. Formally, this implies that the transmission function is integrated over a range of column masses,

$$\mathcal{T} = \int_0^\infty e^{-\int \kappa(x)\, d\tilde{m}}\, dx \neq \int_0^1 e^{-\int \kappa(y)\, d\tilde{m}}\, dy, \tag{5.33}$$

and the k-distribution method no longer applies (as indicated by the non-equality).

We may demonstrate the inapplicability of the k-distribution method using the simple example of an atmosphere with two layers. Each layer has its own opacity function and column mass, subscripted by 1 and 2. Layer 1 sits above Layer 2 in altitude, as is our convention, such that the column mass of the former is smaller than that of the latter. If we write the integrand associated with the transmission function as

$$F = e^{-\kappa_1 \tilde{m}_1 - \kappa_2 \tilde{m}_2}, \tag{5.34}$$

then we obtain

$$dF = -F\left(\tilde{m}_1 d\kappa_1 + \tilde{m}_2 d\kappa_2\right). \tag{5.35}$$

The presence of the two terms in dF is important, because it implies that the transmission function also has two terms,

$$\begin{aligned} \mathcal{T} &= \int_0^1 e^{-\kappa_1(y_1)\tilde{m}_1 - \kappa_2(y_1)\tilde{m}_2}\, dy_1 + \int_0^1 e^{-\kappa_1(y_2)\tilde{m}_1 - \kappa_2(y_2)\tilde{m}_2}\, dy_2 \\ &\neq 2\int_0^1 e^{-\kappa_1(y)\tilde{m}_1 - \kappa_2(y)\tilde{m}_2}\, dy, \end{aligned} \tag{5.36}$$

where we have defined the cumulative sums of intervals as

$$dy_1 \equiv \mathcal{H}\tilde{m}_1\, d\kappa_1\ , dy_2 \equiv \mathcal{H}\tilde{m}_2\, d\kappa_2. \tag{5.37}$$

The non-equality in equation (5.36) derives from the fact that *even identical ranges of values in y_1 and y_2 generally correspond to different ranges of* *wavenumbers.* For example, $\kappa_1(y_1)$ and $\kappa_2(y_2)$ are cumulative functions constructed from their own cumulative sum of intervals. By contrast, $\kappa_1(y_2)$ and $\kappa_2(y_1)$ are cumulative functions constructed from the cumulative sum of intervals of their counterparts, meaning that the contributions are drawn from different wavenumber intervals even at the same value of the cumulative sum of intervals. Generally, we expect these four cumulative functions to have different functional forms. This peculiar property is an unavoidable consequence of working with cumulative functions.

Physically, in employing the k-distribution method, the price being paid is that the wavenumber information has been scrambled. If one *assumes* that $y = y_1 = y_2$, then one is making the *correlated-k approximation* and the transmission function may then be computed as a single integral across y. It is the assumption that each value of the cumulative opacity function is always drawn from the same wavenumber interval.

Curiously, the mathematics behind the reasoning is identical in the case of applying the correlated-k approximation to a homogeneous atmosphere with multiple atoms or molecules. Consider an atmosphere with only two molecules and a single value of the column mass ($\tilde{m}$). Let the mass mixing ratios[5] (relative abundance by mass) of the molecules be X_1 and X_2. We then have

$$
\mathcal{T} = \int_0^1 e^{-X_1\kappa_1(y_1)\tilde{m} - X_2\kappa_2(y_1)\tilde{m}} \; dy_1 + \int_0^1 e^{-X_1\kappa_1(y_2)\tilde{m} - X_2\kappa_2(y_2)\tilde{m}} \; dy_2
$$
$$
\neq 2 \int_0^1 e^{-X_1\kappa_1(y)\tilde{m} - X_2\kappa_2(y)\tilde{m}} \; dy.
$$
(5.38)

The two integrals originate from having

$$
F = e^{-(X_1\kappa_1 + X_2\kappa_2)\tilde{m}}
$$
(5.39)

and

$$
dF = -F\tilde{m} \left(X_1 d\kappa_1 + X_2 d\kappa_2 \right).
$$
(5.40)

Also, we have

$$
dy_1 = \mathcal{H}\tilde{m}X_1 \; d\kappa_1 \; , dy_2 = \mathcal{H}\tilde{m}X_2 \; d\kappa_2.
$$
(5.41)

I have intentionally written things out explicitly to illustrate the fact that one can avoid dealing with two integrals if a single, total opacity function is constructed first,

$$
\kappa = X_1\kappa_1 + X_2\kappa_2,
$$
(5.42)

before its cumulative function is computed.

Again, unless $y_1 = y_2$, the two integrals cannot be combined. Since this reasoning holds for multiple molecules in a homogeneous atmosphere, it must

---

[5]In Grimm & Heng [77], we stated that it is the mixing ratio (relative abundance by number), but this is only true if the two molecules have roughly the same mass, i.e., they are isotopologues.

also hold for multiple molecules in an inhomogeneous atmosphere. We conclude that one needs to first add the opacities of the various molecules in an atmosphere, weighted by their relative abundances, prior to constructing the cumulative function of the opacity. If one adds the cumulative opacity functions of different molecules, then one is effectively employing the correlated-k approximation. Both lines of reasoning can be straightforwardly generalized to an inhomogeneous atmosphere containing a single atom or molecule and with an arbitrary number of layers, a homogenous atmosphere with an arbitrary number of atomic or molecular species, or an inhomogeneous atmosphere with an arbitrary number of layers and species (Problem 5.6.5).

The correlated-k approximation has a deep connection to how spectral lines of different molecules are distributed across wavenumber. Consider two sets of lines from two molecular species. If the line positions of one molecule are randomly distributed with respect to the other, then we say that they are *perfectly uncorrelated* or *randomly overlapping.* If the line positions associated with each molecule completely avoid each other, then they are *disjoint.* If the lines of each molecule always populate the same wavenumber regions, then we say they are *perfectly correlated.* Problem 5.6.6 examines a toy model that works through these three possibilities. The correlated-k approximation does not directly invoke any of these limits, as it integrates over the wavenumber-dependent transmission functions of both molecules. But it implicitly assumes that the opacity contributions of each molecule are drawn from the same wavenumber interval, which is equivalent to assuming that the lines of the two molecules are correlated.

A common source of confusion in the published literature is the failure to distinguish the method (k-distribution) from the approximation (correlated-k). The "correlated-k method" is a misnomer.

5.4.4 The exactness of the correlated-k approach in the low-temperature, low-pressure limit

Besides the restrictive and unuseful limit of a homogeneous atmosphere with a single chemical species, is there a situation where the correlated-k approach is exact? Mathematically, it requires that the opacity function may be expressed as the product of two functions,

$$\kappa\left(T, P, \tilde{\nu}\right) = \mathcal{K}\left(\tilde{\nu}\right) \, \mathcal{K}'\left(T, P\right), \tag{5.43}$$

which allows us to separate out the wavenumber dependence from the temperature and pressure dependences. If the functions $\mathcal{K}$ and $\mathcal{K}'$ exist, then only the latter varies between atmospheric layers, but the former is the same for all layers. This property would allow us to construct k-distribution tables for $\mathcal{K}$ and use them for any temperature and pressure, provided we know the functional form of $\mathcal{K}'$.

Unfortunately, the profiles of real spectral lines do not behave in this way, as we have seen earlier in the chapter, because the wavenumber, temperature and pressure dependences of the Voigt profile cannot be easily separated. However, we can seek $\mathcal{K}$ and $\mathcal{K}'$ in the limit of low temperature. In this case, we have $h\tilde{\nu}c \gg k_{\mathrm{B}}T$ and the wavenumber and temperature dependences of $\mathcal{S}$ may be easily separated out. If we also approximate the line shape by a Lorentz profile and ignore pressure broadening, we end up with

$$\mathcal{K} \approx \frac{\Phi_{\mathrm{L}}}{\tilde{\nu}^2}, \ \mathcal{K}' \approx \frac{g_2 A_{21}}{8\pi mcQ_p}e^{-\Delta E/k_{\mathrm{B}}T}. \tag{5.44}$$

This is quaint, but not broadly applicable to atmospheres because of the severe approximations taken.

5.5   DIFFERENT TYPES OF MEAN OPACITIES

In certain situations, one may wish to condense the complexity of the opacity function into a mean opacity integrated over all wavenumbers, frequencies or wavelengths. For example, if one was interested in computing the temperature-pressure profiles described in Chapter 4, one would need mean opacities. Be aware that, depending on the averaging method being used, the value of the mean opacity may differ at the order-of-magnitude level.

5.5.1   The arithmetic mean (Planck)

A commonly used quantity is the *Planck mean opacity*, which is an arithmetic mean weighted by the Planck function,

$$\kappa_{\mathrm{P}} \equiv \frac{\int \kappa B \ d\tilde{\nu}}{\int B \ d\tilde{\nu}}. \tag{5.45}$$

Across a wavenumber range where the Planck function is approximately constant, it has the property that it lends equal weight to the smallest and largest opacities. The Planck mean opacity applies to a situation where the optical depth is on the order of unity—radiative transfer is controlled by the strongest lines.

Other arithmetic mean opacities include the absorption and flux mean opacities, which are weighted by the total intensity and flux, respectively [172]. Since these mean opacities depend on quantities that are themselves the outcome of a radiative transfer calculation, they cannot be pre-computed and tabulated, unlike for the Planck mean opacity. The eponym of the Planck mean opacity needs no introduction.

5.5.2   The harmonic mean (Rosseland)

Another commonly used and pre-tabulated quantity is the *Rosseland mean opacity*, which is a harmonic mean weighted by the gradient of the Planck function,

$$\kappa_{\mathrm{R}} \equiv \left(\frac{\int \frac{1}{\kappa} \frac{\partial B}{\partial T} \ d\tilde{\nu}}{\int \frac{\partial B}{\partial T} \ d\tilde{\nu}} \right)^{-1}. \tag{5.46}$$

Unlike the Planck mean opacity, it lends the bulk of the weight to the smallest opacities. Physically, this is appropriate for situations where the optical depth is very large, which means that the strong lines have long been saturated. The radiative transfer is instead controlled by the weak lines. The Rosseland mean opacity is named after the Norwegian astrophysicist Svein Rosseland.

5.5.3 The geometric mean

Formally, none of the mean opacities are geometric means, which would give equal weight to the opacities in a logarithmic sense.

5.6 PROBLEM SETS

5.6.1 Properties of line profiles

(a) Show that the Lorentz profile normalizes to unity when integrated over all wavenumbers.
(b) Prove that Γ_L is the HWHM of the Lorentz profile. Hint: equate Φ_L to *half* its value, when $\tilde{\nu} = \tilde{\nu}_0$, and solve for $(\tilde{\nu} - \tilde{\nu}_0)$.
(c) Next, show that the Doppler profile normalizes to unity as well.
(d) Prove that Γ_D is the HWHM of the Doppler profile.
(e) Starting from the definition that the Voigt profile is the convolution of the Lorentz and Doppler profiles, derive its expression.
(f) Prove that the Voigt profile normalizes to unity.

5.6.2 An alternative expression for the Voigt H-function

In its traditional form, the expression for H_V is challenging to evaluate numerically, because it is an indefinite integral. Prove that it can be recast in the form [201, 259],

$$
\begin{aligned}
H_V = {}& e^{a_0^2 - x^2}\ \texttt{erfc}\,(a_0)\ \cos\,(2a_0 x) \\
& + \frac{2}{\sqrt{\pi}} \int_0^x e^{x'^2 - x^2}\ \sin\,[2a_0\,(x - x')]\ dx',
\end{aligned}
\tag{5.47}
$$

where $\texttt{erfc}(a_0)$ is the *complementary error function* (with the argument a_0) [2, 7]. Numerically, this form of H_V is easier to handle because it consists of the sum of a term involving a special function (the complementary error function in this case) and a proper/finite integral.

5.6.3 Placing the Lorentz, Doppler and Voigt profiles on the same plot

We wish to place the Lorentz, Doppler and Voigt profiles on the same plot, as we have done in Figure 5.2. Rather than the Voigt profile, it is more natural to plot the dimensionless Voigt H-function.

(a) What normalization factor do we need to apply to the Doppler profile in order to render it dimensionless and place it on the same plot as the Voigt H-function?

(b) Rewrite the Lorentz profile in terms of u and a_0. What normalization factor do we need to apply to it to also render it dimensionless and place it on the same plot?

(c) Suppose we adopt a different definition of Γ_L such that it is the full width at half-maximum of the Lorentz profile. Suppose also that we now have $\Gamma_D = \sigma_{th}$ in the Doppler profile. Do these changes affect our results in (a) and (b)?

5.6.4   A variety of physical units for the integrated line strength and Einstein B-coefficient

(a) Suppose that, instead of using κ, we use the cross section to relate the integrated line strength and line profile,

$$\sigma = \mathcal{S}\Phi_V. \tag{5.48}$$

Derive the three different ways in which the physical units of $\mathcal{S}$ may be expressed.

(b) Now, consider

$$\alpha_e = \mathcal{S}\Phi_V. \tag{5.49}$$

Derive the other three ways in which the physical units of $\mathcal{S}$ may be expressed.

(c) Derive the three different ways in which the physical units of the Einstein B-coefficient may be expressed.

5.6.5   The correlated-k approximation for multiple molecules

Consider a pair of points associated with a homogeneous path across an atmosphere, which consists of N molecules, each with its own opacity function, such that

$$F = e^{-\sum_{i=1}^{N} \kappa_i \tilde{m}}. \tag{5.50}$$

(a) Show that evaluating $\int \mathcal{H} \, dF$ involves calculating N integrals, each with its own y_i.

(b) Show that assuming $y = y_i$ is mathematically equivalent to applying the correlated-k approximation across an inhomogeneous path consisting of N points.

5.6.6   Atmospheres with two molecules: uncorrelated, correlated and disjoint spectral lines

Pierrehumbert [191] has discussed in detail the different regimes in which the transmission functions associated with two atmospheric gases may be combined, but I feel that this subject is so important that I will recast his arguments as a problem set. Consider an atmosphere with two species of molecules. Each molecule is associated with a forest of spectral lines. To simplify matters, we assume that every line associated with a molecule has the same transmissivity (p_i). Across the same wavenumber range $(\Delta\tilde{\nu})$, the transmission function is

$$T_i = \begin{cases} p_i, & \text{for any spectral line,} \\ 1, & \text{not a spectral line,} \end{cases} \tag{5.51}$$

where $i = 1$ and 2 refer to the first and second molecular species, respectively. The spectral lines do not have to be lined up consecutively and may generally be distributed in some fashion across $\Delta\tilde{\nu}$. Collectively, they populate a fraction $f_i \leq 1$ of the wavenumber range. A fraction $(1 - f_i)$ of the range is transparent to radiation.

(a) Show that the mean transmission function associated with each molecule is

$$\bar{T}_i = \frac{1}{\Delta\tilde{\nu}} \int T_i \, d\tilde{\nu} = f_i (p_i - 1) + 1. \tag{5.52}$$

(b) Show that the product of the mean transmission functions is

$$\bar{T}_1 \bar{T}_2 = f_1 f_2 (1 - p_1)(1 - p_2) - f_1(1 - p_1) - f_2(1 - p_2) + 1. \tag{5.53}$$

(c) *Perfectly-uncorrelated* lines have the property that the wavenumber positions of the lines of the two molecules are randomly distributed. The fraction f_i may now be visualized as the probability that the ith molecule is absorbing radiation. For $i \neq j$, $f_i f_j$ is the probability that both molecules are absorbing at the same time. Using the four different permutations of probabilities, prove that

$$\bar{T} = \frac{1}{\Delta\tilde{\nu}} \int T_1 T_2 \, d\tilde{\nu}$$
$$= f_1(1 - f_2) p_1 + f_2(1 - f_1) p_2 + f_1 f_2 p_1 p_2 + (1 - f_1)(1 - f_2). \tag{5.54}$$

Next, prove that

$$\bar{T} = \bar{T}_1 \bar{T}_2. \tag{5.55}$$

Therefore, perfectly-uncorrelated lines have the special property that their mean transmission functions multiply each other without incurring a correction term.

(d) The spectral lines of the two molecules are *perfectly correlated* when the line positions always sync up in wavenumber and the transmission functions act in concert. In this case, $f = f_i$. Prove that the mean transmission function is

$$\bar{T} = \bar{T}_1 \bar{T}_2 + f(1 - f)(1 - p_1)(1 - p_2). \tag{5.56}$$

Therefore, perfectly-correlated lines acting in concert always absorb more radiation than if each set of lines was acting independently.

(e) *Disjoint* spectral lines have the property that the two sets of lines always act in anti-concert. The wavenumber locations of spectral lines for the first molecule completely avoid those of the second one. Prove that

$$\bar{T} = \bar{T}_1 \bar{T}_2 - f_1 f_2 (1 - p_1)(1 - p_2). \tag{5.57}$$

It is apparent that sets of lines that are disjoint absorb less radiation than if they were randomly overlapping.

Chapter Six

Introduction to Atmospheric Chemistry

6.1 WHY IS ATMOSPHERIC CHEMISTRY IMPORTANT?

In deciphering the astronomical observations of exoplanetary atmospheres, we wish to know the relative abundances of atoms and molecules contained within them. But we wish to do even better—we aim to know if the abundances we are measuring are surprising to us. Do we expect an abundance of methane in a cold atmosphere? What does "cold" exactly mean? Are carbon-rich atmospheres also water-poor? More specifically, given a specified set of atomic or elemental abundances (e.g., solar abundance), how do we predict the *molecular* abundances? Which are the molecules that are the dominant carriers of a particular atom? Understanding atmospheric chemistry teaches us *how* to be surprised. For example, atmospheres with temperatures below about 1000 K are expected to be rich in methane, with carbon monoxide being a secondary carrier of the carbon atom. For hotter exoplanets, carbon monoxide takes over as the dominant carbon carrier, unless the atmosphere is carbon-rich (i.e., the carbon-to-oxygen ratio is greater than unity).

Many books have been written on the subject of atmospheric chemistry. It is not my intention to provide an exhaustive course on it, especially since I do not consider myself to be a chemist. Rather, the intention of the current chapter is to provide a concise summary of what an exoplanet scientist would need to know to conduct research that involves atmospheric chemistry, including understanding the governing equations and rate coefficients involved, as well as the principles behind the computational techniques used to construct chemical kinetics networks or perform calculations of chemical equilibria. One of the main thrusts of this chapter is to demonstrate that the Arrhenius equation (which describes the chemical rate coefficients), the van't Hoff equation (which describes the equilibrium constant or coefficient) and various procedures associated with the Gibbs free energy (minimization, rescaling) all originate from the first law of thermodynamics, which in turn derives from statistical mechanics. Thus, the theoretical foundations of atmospheric chemistry are ultimately derived from statistical mechanics.

In the current chapter, I will focus on the basic concepts and principles behind atmospheric chemistry, rather than the chemistry one can learn from specific case studies in the Solar System, which are either data-based statements that have an expiry date or may not be broadly applicable to exoplanets. In Chapter 7, I will focus on building a hierarchy of systems in chemical equilibrium.

6.2   BASIC QUANTITIES: GIBBS FREE ENERGY, EQUILIBRIUM CONSTANT, RATE COEFFICIENTS

6.2.1   Basic setup

To cast the problem in general terms, consider a chemical reaction involving a pair of reactants (X_1 and X_2), which produces a pair of products (Z_1 and Z_2),

$$a_1 X_1 + a_2 X_2 \leftrightarrows b_1 Z_1 + b_2 Z_2. \tag{6.1}$$

In general, the reactants and products may be atoms or molecules. To put it in less abstract terms, consider the following example of the conversion of methane to carbon monoxide [24],

$$CH_4 + H_2O \leftrightarrows CO + 3H_2. \tag{6.2}$$

That we need one carbon atom (C) and four hydrogen atoms (H) to make methane (CH_4) is known as the *stoichiometry* of the molecule. The relative proportions of reactants and products needed are known as the *stoichiometric coefficients*. In this example, a single molecule of methane reacts with a single molecule of water to form a single molecule of carbon monoxide and three molecules of hydrogen—the stoichiometric coefficients are 1, 1, 1 and 3. The coefficients a_1, a_2, b_1 and b_2 are the stoichiometric coefficients for the general chemical reaction described in equation (6.1).

In both chemical reactions described above, the arrow pointing right indicates the *forward reaction*; the arrow pointing left is for the *reverse reaction*. Some chemists define the forward reaction as being exothermic—in proceeding with the chemical reaction, energy is released. Endothermic reactions require an input of energy. One of the things we seek, in this chapter, is the mathematical and physical underpinning of what "exothermic" and "endothermic" exactly mean.

Two questions immediately come to mind: how *likely* is it for the chemical reaction to proceed? And how fast or slow is the reaction? These are mutually exclusive considerations. It may be energetically favorable for a certain reaction to occur, but it may take the age of the Universe for it to complete. Conversely, a fast chemical reaction may encounter an energy barrier it cannot surmount.

6.2.2   Gibbs free energy and solving for chemical equilibrium

6.2.2.1   *General definitions: A myriad of energies*

What does "energy" actually mean? It turns out that there are various types of energy one needs to consider when it comes to chemical reactions. As a first guess, one may think that the excess energy associated with a reaction is the difference between its internal energy (E_{int}) and any losses associated with heat,

$$F = E_{int} - TS, \tag{6.3}$$

where T is the temperature and S is the specific entropy. This quantity is known as the *Helmholtz free energy*. However, it is not general enough, because one also has to account for the work done on the system. A more general quantity is the *Gibbs free energy*,

$$G = E_{\text{int}} + PV - TS, \qquad (6.4)$$

where P is the pressure, $V = 1/\rho$ is the volume per unit mass and ρ is the mass density. Recall that the specific enthalpy is $E_{\text{int}} + PV$, the sum of the internal energy and the work done on the system.

By combining the definition of the Gibbs free energy with the first law of thermodynamics, one gets (Problem 6.5.1a)

$$dG = V\,dP - S\,dT + \sum_j C_j dN_j, \qquad (6.5)$$

where C_j is the *chemical potential* and N_j is the number of the jth chemical species, respectively. In a chemically-active system, the number of particles of each species is generally not a conserved quantity, which explains the presence of the dN_j term. The summation is performed over all of the chemical species in the system. The preceding equation is the starting point for several useful expressions we will prove and may be regarded as a master equation of sorts.

6.2.2.2 Gibbs free energy minimization

The chemical potential is generally a function of both the pressure and temperature. At a fixed pressure and temperature, the expression for dG simplifies greatly,

$$dG = \sum_j C_j dN_j. \qquad (6.6)$$

Since we are considering an isobaric and isothermal system, the integration can be performed trivially to obtain [57, 72, 226, 245]

$$G = \sum_j C_j N_j. \qquad (6.7)$$

The preceding expression is one of the reasons why the chemical potential is sometimes referred to as the Gibbs free energy per particle, but it is often forgotten that this expression is only valid at a fixed pressure and temperature.

N_j refers to the molecules in a system with the index j labeling the species of molecule. Going back to the original motivation stated at the beginning of the chapter, it is of interest to us to relate N_j to the number of atoms present in the system, which we denote by N_i'. It is useful to visualize N_i' as the input and N_j as the outcome. All that is left to do is simply book-keeping: for a given atomic element, we count up the number of atoms used by every molecule and ensure that it matches what we put into the system. In other words, we need to enforce the stoichiometry of the system via the set of equations [226],

$$\sum_j A_{ij} N_j = N_i'. \tag{6.8}$$

The matrix A_{ij} counts the number of atoms of species i in each species of molecule indexed by j. If we return to the example of the chemical reaction in equation (6.2), which involves methane, water, carbon monoxide and molecular hydrogen, then the equations in (6.8) become

$$\begin{aligned}
N_{\text{CH}_4} + N_{\text{CO}} &= N_{\text{C}}, \\
4N_{\text{CH}_4} + 2N_{\text{H}_2\text{O}} + 2N_{\text{H}_2} &= N_{\text{H}}, \\
N_{\text{H}_2\text{O}} + N_{\text{CO}} &= N_{\text{O}}.
\end{aligned} \tag{6.9}$$

Simply put, the elemental building blocks of molecules cannot be created or destroyed.[1]

The constraints in equations (6.7) and (6.8) form a closed set of mathematical equations we can solve to obtain the equilibrium state of the system. Specifically, the values of N_j are varied until the total Gibbs free energy (G) of the system is minimized. This computational procedure is known as *Gibbs free energy minimization* and the solution obtained describes a system in *chemical equilibrium*. The Gibbs free energy is analogous to the Lagrangian of classical mechanics [245], which is the difference between the kinetic and potential energies of a system—one seeks the solution that minimizes the Lagrangian, known as the *principle of least action*. My motivation is to have you understand the principles behind Gibbs free energy minimization without actually laying down the exact recipe for how to implement it in a computer. In practice, one finds tabulated values of the chemical potential as a function of temperature and at a reference pressure. One then rescales the Gibbs free energy to different pressures using an analytical expression (Problem 6.5.1b).

The *change* in energy resulting from using the reactants to create the products is a measure of whether a chemical reaction is exothermic or endothermic. Strictly speaking, "exothermic" and "endothermic" are associated with negative and positive changes in the enthalpy, respectively. The proper terms associated with negative and positive changes in the Gibbs free energy are *exergonic* and *endergonic*, respectively, although these terms are seldom used in practice. In an exergonic reaction, the products end up with *less* Gibbs free energy than the reactants. Whether a reaction is exothermic/exergonic or endothermic/endergonic depends on the temperature and pressure—sometimes, we simply have to choose the forward reaction in order to proceed (especially when performing calculations of chemical kinetics), while being aware that it may not always be exothermic/exergonic.

---

[1] Unless nuclear reactions come into play.

6.2.3   Equilibrium constant or coefficient

A useful quantity in chemistry is the *equilibrium constant/coefficient*, which allows us to relate the forward and reverse rate coefficients of a reaction. There is also a direct relationship between the equilibrium constant and the change in the Gibbs free energy (between the reactants and products). The problem is that there is more than one definition of the equilibrium constant and it is worth the time, from the perspective of a physicist, to figure out these various definitions, so as to attain clarity of thought.

If you work through Problem 6.5.1b, you will convince yourself that when the temperature and number of particles associated with each chemical species are invariant, we obtain a simple expression for the Gibbs free energy [57],

$$G = G_0 + \mathcal{R}T \ln\left(\frac{P}{P_0}\right), \tag{6.10}$$

where $\mathcal{R}$ is the specific gas constant, P_0 is some reference value of the pressure and $G_0 \equiv G(P_0)$.

Now, consider again the chemical reaction in equation (6.1). By collecting the Gibbs free energies of the reactants and products, weighted by their stoichiometric coefficients, we obtain [57, 72, 118, 225, 226]

$$\Delta G_0 - \Delta G = -\mathcal{R}T \ln K_{\text{eq}}, \tag{6.11}$$

where the equilibrium constant is

$$K_{\text{eq}} \equiv \frac{(P_{Z_1}/P_0)^{b_1}\,(P_{Z_2}/P_0)^{b_2}}{(P_{X_1}/P_0)^{a_1}\,(P_{X_2}/P_0)^{a_2}}. \tag{6.12}$$

It is important to note that the equilibrium constant, in this definition, is a dimensionless quantity. It is not uncommon to see the factors associated with P_0 being omitted,[2] but this formally renders K_{eq} dimensional—such an approach cannot be correct,[3] because it introduces an inconsistency of physical units in the expression between ΔG and K_{eq}. P_{Z_1}, P_{Z_2}, P_{X_1} and P_{X_2} are the *partial pressures* of Z_1, Z_2, X_1 and X_2, respectively.

The differences in Gibbs free energy going from the reactants to the products are

$$\begin{aligned}\Delta G &\equiv b_1 G_{Z_1} + b_2 G_{Z_2} - a_1 G_{X_1} - a_2 G_{X_2}, \\ \Delta G_0 &\equiv b_1 G_{Z_1,0} + b_2 G_{Z_2,0} - a_1 G_{X_1,0} - a_2 G_{X_2,0}.\end{aligned} \tag{6.13}$$

The Gibbs free energy associated with each reactant or product is denoted in a self-explanatory way. If the subscript "0" is involved, then it is defined at the reference pressure.

---

[2]For example, in the paper by Visscher & Moses [247], their equation (7) lists an exponential function, which must be dimensionless, and equates it to a ratio of several pressures raised to generally different powers, which cannot be formally dimensionless.

[3]It has been mentioned to me that this omission is "understood," but I maintain the stand that physics should be mathematically self-evident and self-consistent.

To proceed, we need to provide a physical interpretation of the changes in Gibbs free energy at an arbitrary pressure and at the reference pressure. Physically, the system adjusts itself until $\Delta G = 0$, at which point it attains chemical equilibrium. By contrast, let the reference state, characterized by P_0, *not* be in equilibrium, such that $\Delta G_0 \neq 0$. If one is referring to a molecule, then ΔG_0 is the energy needed to construct it from its constituent atoms—it is the Gibbs free energy of formation. If one is referring to mixtures of molecules, then ΔG_0 is the difference in the Gibbs free energies of formation between the reactants and products. With these assertions, what follows is a useful relationship between the change in Gibbs free energy, at an arbitrary pressure, and the equilibrium constant [9, 57, 72, 113, 118, 174, 225, 226, 233, 245],

$$K_{\text{eq}} = e^{-\Delta G_0 / \mathcal{R}T}. \tag{6.14}$$

If we differentiate this expression with respect to the temperature, we end up with the *van't Hoff equation* [111],

$$\frac{\partial (\ln K_{\text{eq}})}{\partial T} = \frac{\Delta G_0}{\mathcal{R}T^2}. \tag{6.15}$$

Note that we are allowed to go from equation (6.14) to (6.15) only because ΔG_0 was constructed for a constant temperature. In most incarnations of the van't Hoff equation—named after the Dutch chemist Jacobus van't Hoff—it is the change in enthalpy, rather than the Gibbs free energy, which is stated. If the system is isothermic and adiabatic, then these two statements are equivalent.

A common source of confusion stems from the fact that there is another definition of the equilibrium constant, which is generally *not* dimensionless. It arises from the need to relate the forward and reverse rate coefficients. We again use the chemical reaction in (6.1) as the basis for our discussion. In chemical equilibrium, we expect the forward (k_{f}) and reverse (k_{r}) rate coefficients to be related in the following manner,

$$k_{\text{f}} n_{\text{X}_1}^{a_1} n_{\text{X}_2}^{a_2} = k_{\text{r}} n_{\text{Z}_1}^{b_1} n_{\text{Z}_2}^{b_2}. \tag{6.16}$$

These combinations of terms yield the forward and reverse *reaction rates*. The quantity n is the number density; its subscript denotes its association with each reactant or product. Each term in the preceding equation needs to ultimately have physical units of $\text{cm}^{-3}\ \text{s}^{-1}$ (reactions per unit volume per unit time). Given the values of the stoichiometric coefficients, the forward and reverse rate coefficients then adopt the appropriate physical units in order to make this happen. Generally, there is no reason for k_{f} and k_{r} to have the same physical units. An alternative definition of the equilibrium constant is [9]

$$K_{\text{eq}}' \equiv \frac{k_{\text{f}}}{k_{\text{r}}} = K_{\text{eq}} \left(k_{\text{B}}T \right)^{a_1 + a_2 - b_1 - b_2} P_0^{b_1 + b_2 - a_1 - a_2}, \tag{6.17}$$

with k_{B} being the Boltzmann constant. Notice the additional correction terms associated with the temperature and reference pressure. Collectively, these terms may be combined into a single correction term involving a reference number density, $n_0 \equiv P_0/k_B T$.

If the sums of stoichiometric coefficients of the reactants versus the products are identical (i.e., $a_1 + a_2 = b_1 + b_2$), then these correction terms vanish and we have $K_{eq} = K'_{eq}$. This is often overlooked in the literature[4] and is a crucial technique to master if one wishes to use K_{eq} or K'_{eq} to "reverse" the forward rate coefficients. In other words, K'_{eq} is used to "reverse" the forward rate coefficients, but it is K_{eq} that relates them to the Gibbs free energies. Equation (6.17) relates them properly.

6.2.4 Chemical rate coefficients and the activation energy

We may derive the expression for the rate coefficients, named the *Arrhenius equation* [111] after the Swedish physical chemist Svante Arrhenius, from the expressions for K_{eq} and K'_{eq}. It is useful to visualize the reactants and products as being two stable states residing at different energies. To transition from one to the other requires that an energy barrier, known as the *activation energy*, is surmounted (Figure 6.1). The activation energy itself has a quantum mechanical origin—it comes from having to overcome bond strengths. Furthermore, how likely or unlikely it is for a chemical reaction to proceed depends on the relative orientation of the reactants. Typically, the change in Gibbs free energy is used to refer to an entire network of reactions, whereas the activation energies refer to individual reactions in this network. For a single reaction, we may relate them,

$$\Delta G_0 = E_f - E_r - T\Delta S_0, \tag{6.18}$$

where E_f and E_r are the activation energies associated with the forward and reverse reactions, respectively, and ΔS_0 is the change in specific entropy at the reference pressure. The preceding expression allows us to cast the adjectives "exothermic" and "endothermic" in more precise, mathematical terms. If the activation energy of the forward reaction exceeds that of the reverse one, then one needs to inject energy into the system for it to proceed, i.e., $E_f - E_r > 0$. One refers to this as an endothermic reaction. Reactions with $E_f - E_r < 0$ are exothermic.

By combining the expressions for K_{eq} and K'_{eq}, we obtain

$$\ln k_f - \ln k_r = -\frac{E_f - E_r}{\mathcal{R}T} + \frac{\Delta S_0}{\mathcal{R}} + (a_1 + a_2 - b_1 - b_2)\ln\left(\frac{k_B T}{P_0}\right). \tag{6.19}$$

At first glance, it is very tempting to imagine that the preceding expression was constructed from two separate equations for k_f and k_r [243]. The problem is that it is mathematically suspect to "split" this expression—to do so, we need to appeal to physics. For starters, we expect that each rate coefficient is associated with its own activation energy. We also expect the pair of governing

---

[4]As has been pointed out by Visscher & Moses [247].

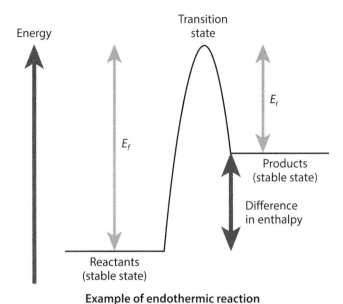

Example of endothermic reaction

Figure 6.1: Schematic relating the activation energies of a chemical reaction and the difference in enthalpy between the products and reactants.

equations for the rate coefficients to be symmetric. Thus, a plausible guess is that the preceding equation originated from the difference between these two equations,

$$
\begin{aligned}
\ln k_{\mathrm{f}} &= -\frac{E_{\mathrm{f}}}{\mathcal{R}T} + c_{\mathrm{f}} \ln T + \frac{c_{\mathrm{f}}' \Delta S_0}{\mathcal{R}} + c_{\mathrm{f}}'', \\
\ln k_{\mathrm{r}} &= -\frac{E_{\mathrm{r}}}{\mathcal{R}T} + c_{\mathrm{r}} \ln T + \frac{c_{\mathrm{r}}' \Delta S_0}{\mathcal{R}} + c_{\mathrm{r}}''.
\end{aligned}
\tag{6.20}
$$

The coefficients c_{f} and c_{r} cannot be stated uniquely. For example, we can have $c_{\mathrm{f}} = a_1 + a_2$ and $c_{\mathrm{r}} = b_1 + b_2$; we may also have $c_{\mathrm{f}} = -b_1 - b_2$ and $c_{\mathrm{r}} = -a_1 - a_2$. This mathematical freedom implies that c_{f} and c_{r} may take on a range of values and be positive or negative.

Finally, we end up with the *Arrhenius equations*, [99]

$$
\begin{aligned}
k_{\mathrm{f}} &= A_{\mathrm{f}} \; T^{c_{\mathrm{f}}} \; e^{-E_{\mathrm{f}}/\mathcal{R}T}, \\
k_{\mathrm{r}} &= A_{\mathrm{r}} \; T^{c_{\mathrm{r}}} \; e^{-E_{\mathrm{r}}/\mathcal{R}T},
\end{aligned}
\tag{6.21}
$$

where we necessarily have

$$
\begin{aligned}
c_{\mathrm{f}} - c_{\mathrm{r}} &= a_1 + a_2 - b_1 - b_2, \\
c_{\mathrm{f}}' - c_{\mathrm{r}}' &= 1, \\
c_{\mathrm{f}}'' - c_{\mathrm{r}}'' &= (a_1 + a_2 - b_1 - b_2) \ \ln\left(\frac{k_{\mathrm{B}}}{P_0}\right).
\end{aligned}
\tag{6.22}
$$

The pre-exponential factor A_{f} absorbs terms associated with c_{f}', c_{f}'' and ΔS_0; its counterpart, A_{r}, does the same for c_{r}', c_{r}'' and ΔS_0.

Traditionally, the Arrhenius equations are written as

$$
\begin{aligned}
k_{\mathrm{f}} &= A_{\mathrm{f}} e^{-E_{\mathrm{f}}/\mathcal{R}T}, \\
k_{\mathrm{r}} &= A_{\mathrm{r}} e^{-E_{\mathrm{r}}/\mathcal{R}T}.
\end{aligned}
\tag{6.23}
$$

Previous formulations have typically stated derivations for these simple exponential forms and tagged on the power-law terms after the fact [111], usually to justify the need for more general functional forms of the Arrhenius equation in order to fit experimental data. Our derivation leading to equation (6.21) demonstrates that such ad hoc "corrections" to the Arrhenius equation are unnecessary. In fact, the presence of a power-law component is a natural outcome of our derivation. Typically, $c_{\mathrm{f}} = 0$ and $c_{\mathrm{r}} = 0$ suffice for low temperatures; "non-Arrhenius" behavior, where $c_{\mathrm{f}} \neq 0$ and $c_{\mathrm{r}} \neq 0$, is important at high temperatures [72].

We are now in a position to quantify the terms "energetically favorable" and "fast/slow." An exergonic reaction ($\Delta G_0 < 0$) is energetically favorable, but if the activation energy is large compared to the thermal energy, then the rate coefficient becomes small. The reaction will proceed, but it will be slow. Generally, chemical reactions proceed faster when the temperature is high.

There remains a flaw to the Arrhenius equation, because it does not account for additional effects associated with three-body reactions. Some chemical reactions require a very specific amount of energy to proceed, and if only two particles were present the chances of them having exactly this energy are slim. If a third body is present to act as a "lender" or "borrower" of energy, then it is much more likely for the reaction to proceed. Let the number density of the third body be denoted by n_{third}. In the limit of low n_{third}, the reaction rate depends on n_{third} itself. Physically, this is because there are not enough third bodies to aid the reaction and thus how fast or slow the reaction proceeds depends on the number of third bodies present. At some point, this effect saturates and the reaction rate becomes independent of n_{third} and is described by the rate coefficient k_{∞}. Therefore, we need a fitting function[5] that transitions smoothly between the two limits. Typically, what is used in the literature is the

---

[5] Meaning that this equation is not derived from a conservation law of Nature, but is rather "pulled out of a hat."

modified rate coefficient [111, 247],

$$k'_f = \frac{k_f}{1 + n_{\text{third}} k_f / k_\infty},$$ (6.24)

such that the reaction rate, given by $k'_f n_{X_1}^{a_1} n_{X_2}^{a_2} n_{\text{third}}$, correctly produces the two limits.

Determining these rate coefficients is a non-trivial and thankless task, which requires either a first-principles calculation (i.e., solving the Schrödinger equation) or a laboratory experiment. In practice, one looks up extensive databases and use them as inputs for one's calculations. For our purposes, we will regard the inputs from these databases as a given, while being aware that there are systematic uncertainties and considerable heterogeneity associated with them, and leave the discussion of these caveats to the chemists.

6.3 CHEMICAL KINETICS: TREATING CHEMISTRY AS A SET OF MASS CONSERVATION EQUATIONS

6.3.1 Basics

Gibbs free energy minimization is a way of solving for chemical equilibrium, but it is not general enough for our purposes because we do not expect atmospheres to always be in chemical equilibrium. By contrast, chemical kinetics is the treatment of chemistry as a set of mass continuity/conservation equations, with each pathway being its own chemical reaction characterized by a rate coefficient. When a system is not in chemical equilibrium, the forward and reverse reaction rates are not necessarily equal. Chemical kinetics is a powerful method, because it allows for disequilibrium chemistry due to either atmospheric mixing or photochemistry to be considered and is able to solve for chemical equilibrium as a limiting case.[6] It is an essential tool if one desires to study the coupling between atmospheric chemistry, radiation and dynamics. To make the analogy to classical mechanics, chemical kinetics mirrors the direct solution of Newton's equation, while Gibbs free energy minimization is akin to minimizing the Lagrangian.

For the chemical reaction described in equation (6.1), we may describe the evolution of the number density of the reactant X_1 by the partial differential equation,

$$\frac{1}{a_1} \left(\frac{\partial n_{X_1}}{\partial t} - K_{zz} \frac{\partial^2 n_{X_1}}{\partial x^2} \right) = \mathcal{P} - \mathcal{L} n_{X_1}^{a_1} - \mathcal{J}_{X_1}.$$ (6.25)

The preceding equation is essentially a statement of mass conservation and records the evolution of X_1 across time and space, mediated by diffusion and various source and sink terms. Dividing the reaction rates by the stoichiometric

---

[6]In fact, a crucial test of one's chemical kinetics code is for it to reproduce the calculations of chemical equilibrium using another computer code that performs Gibbs free energy minimization.

coefficient allows us to place reactions involving different species on the same footing. For example, if $a_1 = 2$ and $a_2 = b_1 = b_2 = 1$, then X_1 is being consumed twice as fast as the other species. The importance of the $1/a_1$ factor cannot be over-emphasized, because without it one would get the book-keeping of elemental abundances wrong.

The quantities $\mathcal{P}$ and $\mathcal{L}$ are the chemical production and loss rates, respectively, and are given by

$$\begin{aligned} \mathcal{P} &= n_{Z_1}^{b_1} n_{Z_2}^{b_2} k_{\mathrm{r}}, \\ \mathcal{L} &= n_{X_2}^{a_2} k_{\mathrm{f}}. \end{aligned} \tag{6.26}$$

The time scale associated with the loss of X_1 is $t_{\mathrm{chem}} \sim n_{X_1}^{1-a_1}/\mathcal{L}$. The quantity $\mathcal{J}_{X_1}$ is the reaction rate associated with *photochemistry* and is generally a function of the number densities. It is an intrinsically disequilibrium effect that we will describe shortly.

The diffusion coefficient (K_{zz}) is usually used to *mimic* advection, convection and turbulence and (crudely or even incorrectly) subsume their collective influence into a single free parameter. As we will discuss in Chapter 12, one may relate K_{zz} to convection using mixing length theory. One may also use K_{zz} to describe turbulence in some rough sense, but since we lack a first-principles theory of turbulence this is a dubious exercise. Generally, advection, convection and turbulence hardly resemble diffusion in any rigorous way—the qualitative justification for such an approach is to argue that these processes operate on scales that are so small, compared to the characteristic atmospheric length scale of interest, that it "looks" like diffusion. Only in the limit where molecular diffusion is the dominant effect is the use of K_{zz} rigorous and exact. Notwithstanding, the inclusion of a diffusion coefficient allows us to treat situations with disequilibrium chemistry induced by atmospheric motion or mixing without resorting to a full-blown, three-dimensional calculation.

For the product Z_1, the evolution equation is

$$\frac{1}{b_1}\left(\frac{\partial n_{Z_1}}{\partial t} - K_{zz}\frac{\partial^2 n_{Z_1}}{\partial x^2}\right) = \mathcal{P}' - \mathcal{L}' n_{Z_1}^{b_1} - \mathcal{J}_{Z_1}, \tag{6.27}$$

where the production and loss rates are

$$\begin{aligned} \mathcal{P}' &= n_{X_1}^{a_1} n_{X_2}^{a_2} k_{\mathrm{f}}, \\ \mathcal{L}' &= n_{Z_2}^{b_2} k_{\mathrm{r}}. \end{aligned} \tag{6.28}$$

The reaction rate associated with photochemistry is represented by $\mathcal{J}_{Z_1}$. You may have already noticed the presence of a pleasing symmetry between the evolution equations for X_1 and Z_1.

Even without explicitly solving these evolution equations, we may already learn a lot by manipulating them. In the absence of atmospheric mixing ($K_{zz} = 0$), we may add them to obtain

$$\frac{1}{a_1}\frac{\partial n_{X_1}}{\partial t} + \frac{1}{b_1}\frac{\partial n_{Z_1}}{\partial t} = -\mathcal{J}_{X_1} - \mathcal{J}_{Z_1}. \tag{6.29}$$

If we integrate this expression, we obtain

$$\frac{n_{X_1}}{a_1} + \frac{n_{Z_1}}{b_1} = -\int \left(\mathcal{J}_{X_1} + \mathcal{J}_{Z_1} \right) dt + \text{constant}. \tag{6.30}$$

In principle, there is nothing special about the species X_1 and Z_1. We could easily have considered different pairs of reactants and products: X_1 and Z_2, X_2 and Z_1, X_2 and Z_2. Each pairing would yield an expression analogous to the preceding one. If we sum up all of these pairings, then we would obtain

$$\frac{n_{X_1}}{a_1} + \frac{n_{X_2}}{a_2} + \frac{n_{Z_1}}{b_1} + \frac{n_{Z_2}}{b_2} = -\int \left(\mathcal{J}_{X_1} + \mathcal{J}_{X_2} + \mathcal{J}_{Z_1} + \mathcal{J}_{Z_2} \right) dt + \text{constant}. \tag{6.31}$$

This result immediately informs us that photochemistry is a disequilibrium effect, because it allows the total number density to vary in time. Without it, the total number density would be invariant.

Next, we return to the evolution equations and immediately consider their steady-state limit. We neglect atmospheric mixing ($K_{zz} = 0$) and photochemistry. In this limit, we obtain

$$n_{X_1}^{a_1} n_{X_2}^{a_2} k_f = n_{Z_1}^{b_1} n_{Z_2}^{b_2} k_r, \tag{6.32}$$

which is identical to our starting point for defining the equilibrium constant K'_{eq}. Thus, we have verified that the steady-state limit of our governing equations of chemical kinetics produces chemical equilibrium—as it should.

6.3.2   Why the equations of chemical kinetics are stiff

In a chemical network, one has to solve a large, coupled set of differential equations with a considerable number of chemical species—both transient and stable. Each differential equation is associated with its own chemical time scale and the entire ensemble of time scales may span many orders of magnitude. Mathematically, this is a *stiff* set of differential equations. One also has to diffuse these chemical species across space, which requires an efficient method of computation.

To appreciate why a network of chemical reactions constitutes a stiff system of equations, consider the following pair of dimensionless[7] evolution equations,

$$\begin{aligned}
\frac{\partial n_{X_1}}{\partial t} &= \mathcal{A}_1 n_{X_1} + \mathcal{A}_2 n_{Z_1} - n_{X_1}, \\
\frac{\partial n_{Z_1}}{\partial t} &= -\mathcal{A}_1 n_{X_1} - \mathcal{A}_2 n_{Z_1} - n_{Z_1}.
\end{aligned} \tag{6.33}$$

They are again based on the chemical reaction depicted in equation (6.1), but we have ignored atmospheric mixing and subsumed factors associated with the

---

[7]We have used the same notation as the rest of the chapter, but it should be understood that, in this subsection only, the equations have been normalized by characteristic time and length scales. The exact expressions for these time and length scales are unimportant. The non-dimensionalization is merely one of convenience.

stoichiometric coefficients into the coefficients $\mathcal{A}_1$ and $\mathcal{A}_2$, which in turn contain the quantities associated with the chemical source/sink terms. To render the problem analytically tractable, we assume that $\mathcal{A}_1$ and $\mathcal{A}_2$ are constant. We have also assumed a very simple model of photochemistry, namely that the reaction rates are linearly proportional to the number density of the reactant or product (with coefficients that are exactly unity).

Problem 6.5.2 shows you how to work through the pair of evolution equations in (6.33) and obtain the solutions,

$$
\begin{aligned}
n_{X_1} &= 2\mathcal{A}_3 e^{-t} - \mathcal{A}_4 e^{-(\mathcal{A}_1+1)t}, \\
n_{Z_1} &= \mathcal{A}_4 e^{-(\mathcal{A}_1+1)t} - \mathcal{A}_3 e^{-t},
\end{aligned}
\tag{6.34}
$$

where $\mathcal{A}_3$ and $\mathcal{A}_4$ are constants of integration that are unimportant for our purposes. Photochemistry depletes the abundances of X_1 and Z_1 over time,

$$
n_{X_1} + n_{Z_1} = \mathcal{A}_3 e^{-t}.
\tag{6.35}
$$

What is of interest is the magnitude of $\mathcal{A}_1$. If $\mathcal{A}_1 \sim 1$, then both terms in the expressions for the number densities contribute equally. If $\mathcal{A}_1 \gg 1$, then the exponential term associated with it is mathematically unimportant, but yet introduces a severe computational limitation, because it corresponds to a characteristic time scale that is very small.

There are two time scales in this particular problem: one that is unity (in these dimensionless units) and another that is $\sim 1/\mathcal{A}_1 \ll 1$. If we solve the evolution equations numerically, instead of seeking their analytical solution, then the computational time step is bottlenecked by the shorter time scale, even though the physically interesting behavior may occur on much longer time scales. This is exactly what we mean when we say that the pair of evolution equations is "stiff." In general, one deals with a much larger set of coupled equations that are not amenable to analytical solution, so stiffness is a very real challenge.[8]

6.3.3   Casting the evolution across space as a matrix problem

To sharpen our discussion of how to evolve a system across space, we focus on a single chemical species with a number density denoted by n_j. We necessarily describe the number density on a grid with a set of discrete points marked by the index j. To simplify our notation, we write

$$
\dot{n}_j \equiv \frac{\partial n_j}{\partial t}.
\tag{6.36}
$$

It follows that the discretized version of the evolution equation is

$$
\dot{n}_j + K_{zz} \left[\frac{n_{j+1} - 2n_j + n_{j-1}}{(\Delta x)^2} \right] = \mathcal{P}_j - \mathcal{L}_j n_j - \mathcal{J}_j,
\tag{6.37}
$$

---

[8]The solution to this computational dilemma is to implement implicit or semi-implicit methods to step the system forward in time. Reliable workhorses include the *backward Euler* and *Rosenbrock* methods [195].

where the spatial separation between grid points is Δx. For simplicity and illustration, we have discretized the spatial derivative of the number density using a second-order, central-difference scheme. In the same spirit, we have assumed that all of the physical quantities "live" on the same grid, i.e., we do not consider a staggered grid, where half-steps in space are allowed.

If we re-arrange the preceding expression, we obtain

$$n_{j-1}\left[\frac{K_{zz}}{(\Delta x)^2}\right] + n_j\left[\mathcal{L}_j - \frac{2K_{zz}}{(\Delta x)^2}\right] + n_{j+1}\left[\frac{K_{zz}}{(\Delta x)^2}\right] = \mathcal{P}_j - \mathcal{J}_j - \dot{n}_j. \quad (6.38)$$

In this formulation, we are assuming that $\mathcal{J}_j$ does not depend on the number density. You may recognize that if we run the index j through its paces (from 1 to N_x), we end up with a set of equations that may be arranged into a matrix,

$$\mathcal{X}_j \tilde{n}_{j-1} + \mathcal{Y}_j \tilde{n}_j + \mathcal{Z}_j \tilde{n}_{j+1} = \mathcal{W}_j, \quad (6.39)$$

where the number densities have necessarily been non-dimensionalized. Specifically, it is a tridiagonal matrix that may be inverted using *Thomas's algorithm* [195], which states that

$$\mathcal{Z}_j' = \begin{cases} \mathcal{Z}_j/\mathcal{Y}_j, & j = 1, \\ \mathcal{Z}_j/\left(\mathcal{Y}_j - \mathcal{X}_j \mathcal{Z}_{j-1}'\right), & j = 2, 3, ..., N_x - 1. \end{cases} \quad (6.40)$$

and

$$\mathcal{W}_j' = \begin{cases} \mathcal{W}_j/\mathcal{Y}_j, & j = 1, \\ \left(\mathcal{W}_j - \mathcal{X}_j \mathcal{W}_{j-1}'\right)/\left(\mathcal{Y}_j - \mathcal{X}_j \mathcal{Z}_{j-1}'\right), & j = 2, 3, ..., N_x. \end{cases} \quad (6.41)$$

The solution for the number densities are then obtained via

$$\tilde{n}_j = \begin{cases} \mathcal{W}_j', & j = N_x, \\ \mathcal{W}_j' - \mathcal{Z}_j' \tilde{n}_{j+1}, & j = N_x, N_x - 1, ..., 1. \end{cases} \quad (6.42)$$

However, there is a catch: Thomas's algorithm only works if the matrix is diagonally dominant [195], namely

$$|\mathcal{Y}_j| \geq |\mathcal{X}_j| + |\mathcal{Z}_j|, \quad (6.43)$$

which implies that

$$\frac{(\Delta x)^2}{K_{zz}} \geq \frac{4}{\mathcal{L}_j}. \quad (6.44)$$

The quantity on the left is the computational time step, while the one on the right is essentially the chemical loss time scale. Physically, this requirement is equivalent to enforcing causality: the system cannot be diffused faster than the time it takes for each grid point to communicate via chemical reactions.

6.3.4   The quenching approximation

The *quenching approximation* is a way of avoiding having to perform the full-blown chemical kinetics calculation by arguing that the abundance of a chemical species becomes "frozen in" when the time scale associated with chemistry (t_{chem}) becomes longer than the dynamical time scale. Besides the issues associated with describing all forms of atmospheric motion by diffusion, the dynamical time scale,

$$t_{dyn} = \frac{l^2}{K_{zz}}, \tag{6.45}$$

suffers from the ambiguity that the expression for the characteristic length scale (l) is uncertain, ranging from $0.1H$ to H, where H is the pressure scale height [247]. Thus, t_{dyn} is uncertain by two orders of magnitude.

In practice, one performs an equilibrium chemistry calculation throughout the atmosphere and identifies the altitude or pressure at which $t_{dyn} = t_{chem}$ [153]. This is known as the *quench point*. The abundance of the chemical species under consideration is then assumed to take on its value at the quench point for all locations in the atmosphere where $t_{dyn} < t_{chem}$. The challenge with the quenching approximation is to identify the main, intermediate chemical reaction that drives the net reaction. For example, if one is interested in the net reaction involving the conversion of methane to carbon monoxide, one has to sort through dozens, if not hundreds, of intermediate reactions and figure out which one of them is the *rate-determining reaction*. A key challenge of the quenching approximation is to identify the correct rate-determining reaction and use the associated rate coefficient to compute the chemical time scale. Calculations that employ the quenching approximation always need to be checked by a chemical kinetics calculation.

Overall, the quenching approximation is a crude approach for obtaining rough answers and should be used with caution.

6.4   SELF-CONSISTENT ATMOSPHERIC CHEMISTRY, RADIATION AND DYNAMICS: A FORMIDABLE COMPUTATIONAL CHALLENGE

Constructing a rigorous model of an exoplanetary atmosphere involves taking into account the following couplings between chemistry, radiation and dynamics.

- The starlight incident upon an atmosphere heats it and sets up its temperature-pressure profile, provided we have knowledge of its opacity function (Chapter 4).

- The opacity function itself depends on the types of atoms and molecules present in the atmosphere (Chapter 5), and also on their relative abundances.

- But the relative abundances of the atoms and molecules is an outcome of the atmospheric chemistry. We will explore simple models for this in Chapter 7.

- In turn, whether parts of an atmosphere are in chemical equilibrium or disequilibrium depends on the dynamical versus the chemical time scales. Chemical disequilibrium occurs when the dynamical time scale is shorter than the chemical time scale. Atmospheric dynamics has a significant effect on this outcome. It may also drive the atmosphere away from radiative equilibrium.

A full-blown calculation would solve the equations of fluid dynamics and radiative transfer self-consistently with a chemical network involving all of the relevant reactions (which may number in the hundreds, if not thousands). And all of this complexity comes without even considering magnetic fields! This is a formidable, if not intractable, problem. Unsurprisingly, researchers have adopted approximations in order to move forward. Atmospheric dynamicists usually employ simplified treatments of the chemistry by drastically cutting down on the number of reactions involved [45]. Atmospheric chemists adopt fixed temperature-pressure profiles as a background upon which to compute the chemistry without self-consistently accounting for how the chemistry and opacities would alter the temperature-pressure profile [175]. Photochemistry adds another layer of complexity, as the incident ultraviolet light directly alters the abundances of certain molecules and thus has a direct influence on the thermal structure of the atmosphere. Radiative transfer specialists typically include thermochemistry and/or photochemistry in their calculations, but adopt a crude treatment of the atmospheric dynamics via the prescription of K_{zz} [229].

6.5   PROBLEM SETS

6.5.1   Rescaling the Gibbs free energy for different pressures

At any given temperature, one may rescale the Gibbs free energy, at a reference pressure, to other pressures. To do this, we first invoke a more general form of the first law of thermodynamics [238],

$$T dS = dE_{\text{int}} + P \, dV - \sum_j C_j dN_j. \tag{6.46}$$

In addition to the internal energy and work done on the system, we need to account for the change in the number of particles associated with each chemical species.

(a) By using the product rule, the definition of the Gibbs free energy and this generalized form of the first law, derive equation (6.5).

(b) Consider a situation in which the system is held at a fixed temperature $(dT = 0)$ and the number of particles associated with each species is invariant.

By invoking the ideal gas law and integrating the expression for dG from a reference pressure (P_0) to an arbitrary pressure (P), show that

$$G = G_0 + \mathcal{R}T \ln \left(\frac{P}{P_0} \right). \tag{6.47}$$

The preceding expression may be used to rescale the Gibbs free energy only when the system is isothermal.

6.5.2   Time-dependent system with chemical source/sink terms and photochemistry

Consider the pair of evolution equations in (6.33) depicting a system without atmospheric mixing, but with simplified chemical source/sink terms and photochemistry. For this problem, we will work in dimensionless units as originally described in the text surrounding equation (6.33).

(a) By using the variable transformations,

$$\begin{aligned} n_{X_1} &= 2n_1 - n_2, \\ n_{Z_1} &= -n_1 + n_2, \end{aligned} \tag{6.48}$$

show that

$$\begin{aligned} \frac{\partial n_1}{\partial t} &= -n_1, \\ \frac{\partial n_2}{\partial t} &= n_1 \left(\mathcal{A}_2 - 2\mathcal{A}_1 \right) - \left(\mathcal{A}_2 - \mathcal{A}_1 + 1 \right) n_2. \end{aligned} \tag{6.49}$$

(b) Obtain the solutions for n_1 and n_2, and hence n_{X_1} and n_{Z_1}, by setting $\mathcal{A}_2 = 2\mathcal{A}_1$.

(c) Seek general solutions for n_{X_1} and n_{Z_1} that do not require $\mathcal{A}_2 = 2\mathcal{A}_1$.

6.5.3   Time-dependent system dominated by atmospheric mixing

Consider a system with only a single chemical species and no chemical source or sink terms. Photochemistry is absent. In this limit, the evolution equation is mathematically identical to the heat equation,

$$\frac{\partial n}{\partial t} - K_{zz} \frac{\partial^2 n}{\partial x^2} = 0. \tag{6.50}$$

The diffusion coefficient (K_{zz}) is assumed to be constant.

(a) Proceed by assuming that n may be written as the product of two functions

$$n = n_t n_x, \tag{6.51}$$

where the functions $n_t = n_t(t)$ and $n_x = n_x(x)$ depend only on time and space, respectively. This method is known as the *separation of variables*. Show that one ends up with

$$\frac{1}{n_t}\frac{\partial n_t}{\partial t} = \frac{K_{zz}}{n_x}\frac{\partial^2 n_x}{\partial x^2}. \tag{6.52}$$

(b) Since each term in the preceding equation depends only on time or space, it must independently vanish and be equal to some constant. We interpret this constant to be $1/t_{\mathrm{diff}}$, where t_{diff} is the diffusion time scale. Derive the solutions for n_t and n_x and hence for n.

(c) What is the diffusion length scale?

Chapter Seven

A Hierarchy of Atmospheric Chemistries

7.1 A HIERARCHY OF MODELS FOR UNDERSTANDING ATMOSPHERIC CHEMISTRY

It is difficult to obtain analytical solutions for systems that are in chemical disequilibrium, as it involves solving a set of partial differential equations across time and space. Systems in chemical equilibrium are much easier to deal with, as it essentially involves stoichiometric book-keeping—counting atoms and molecules. The transformation of one molecule to another is accounted for via the equilibrium constant. If we deal only with pure-gas chemistry,[1] then analytical solutions to simple systems may be obtained provided that the net reactions are identified.

In this chapter, our goal is to present a *hierarchy* of atmospheric chemistries, starting from a system with only hydrogen to ones with carbon (C), oxygen (O) and hydrogen (H). By incrementally increasing the sophistication of our chemical system, we may develop intuition for trends with temperature, pressure and carbon-to-oxygen ratio (C/O). Our most sophisticated model contains acetylene (C_2H_2), carbon dioxide (CO_2), carbon monoxide (CO), ethylene (C_2H_4), methane (CH_4), water (H_2O) and molecular hydrogen (H_2).

7.2 EQUILIBRIUM CHEMISTRY WITH ONLY HYDROGEN

The simplest system we may consider contains only hydrogen,

$$2H + M \leftrightarrows H_2 + M, \tag{7.1}$$

which involves a third body (of arbitrary stoichiometry M). Using what we learned in Chapter 6, we may write down the pair of evolution equations for atomic and molecular hydrogen in the absence of atmospheric mixing or photochemistry,

$$\frac{1}{2}\frac{\partial n_H}{\partial t} = -n_H^2 n_M k_f + n_{H_2} n_M k_r,$$
$$\frac{\partial n_{H_2}}{\partial t} = n_H^2 n_M k_f - n_{H_2} n_M k_r, \tag{7.2}$$

---

[1] More realistic calculations will account for the formation of condensates within the atmosphere.

Note the factor of $1/2$—it is present because atomic hydrogen is being consumed at twice the rate of its molecular counterpart. If we add the preceding pair of equations, we obtain

$$n_{\mathrm{H}} + 2n_{\mathrm{H}_2} = n_{\mathrm{total}}, \tag{7.3}$$

where n_{total} is the total number density of hydrogen atoms in the system, regardless of whether a hydrogen atom leads a solitary existence or is sequestered in molecular hydrogen. If we had neglected the factor of $1/2$, we would essentially get the book-keeping between atomic and molecular hydrogen wrong.

In chemical equilibrium, we may write down the (dimensional) equilibrium constant of the reaction,

$$K'_{\mathrm{eq}} = \frac{n_{\mathrm{H}_2}}{n_{\mathrm{H}}^2}. \tag{7.4}$$

Notice how n_{M} appears in both the numerator and denominator and thus cancels out. If we plug the expression for the equilibrium constant into that for n_{total}, we obtain a quadratic equation for n_{H},

$$2K'_{\mathrm{eq}}n_{\mathrm{H}}^2 + n_{\mathrm{H}} - n_{\mathrm{total}} = 0, \tag{7.5}$$

which we may solve to obtain the solution for the number density of atomic hydrogen [66],

$$n_{\mathrm{H}} = \frac{-1 + \left(1 + 8K'_{\mathrm{eq}}n_{\mathrm{total}}\right)^{1/2}}{4K'_{\mathrm{eq}}}. \tag{7.6}$$

Before visualizing the solution for arbitrary values of the equilibrium constant and total number density, it is instructive to examine its asymptotes. Recall that the equilibrium constant is a measure of the relative strengths of the forward versus the reverse reaction. When K'_{eq} is small, we have $n_{\mathrm{H}} \approx n_{\mathrm{total}}$. In this limit, the forward reaction is endothermic/endergonic and dominated by what we started with: atomic hydrogen. When K'_{eq} is large, the reaction becomes exothermic/exergonic and we get $n_{\mathrm{H}} \approx (n_{\mathrm{total}}/K'_{\mathrm{eq}})^{1/2} \sim 0$. This usually happens at low temperatures, as molecular hydrogen is broken up into its atomic counterparts when the temperature is too high (above about 3000 K). Almost all of the exoplanets that have been discovered exist in the regime where their buffer or inert gaseous component is molecular hydrogen, if hydrogen is the dominant element in their atmospheres.

To plot the number densities in the form of a figure, it helps to render them dimensionless. By dividing n_{H} and n_{H_2} by n_{total}, we obtain

$$\begin{aligned}
\tilde{n}_{\mathrm{H}} &= \frac{-1 + (1 + 8K')^{1/2}}{4K'}, \\
\tilde{n}_{\mathrm{H}_2} &= \frac{1 - \tilde{n}_{\mathrm{H}}}{2},
\end{aligned} \tag{7.7}$$

where $\tilde{n}_{\mathrm{H}} \equiv n_{\mathrm{H}}/n_{\mathrm{total}}$, $\tilde{n}_{\mathrm{H}_2} \equiv n_{\mathrm{H}_2}/n_{\mathrm{total}}$ and $K' \equiv K'_{\mathrm{eq}}n_{\mathrm{total}}$. In Figure 7.1, we plot the normalized number densities as a function of K'. We see that the trends

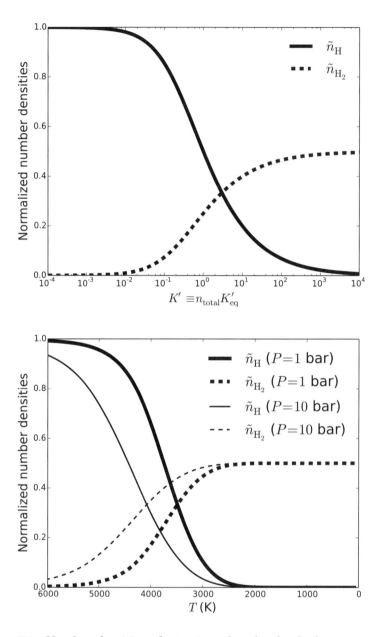

Figure 7.1: Number densities of atomic and molecular hydrogen, normalized by the total density, as a function of the product of the equilibrium constant and the total density (top panel). An increasing K' corresponds to decreasing temperature. In the bottom panel, the number densities are shown as functions of the temperature for two values of the pressure.

shown in the figure are consistent with the asymptotic behavior we derived earlier. We also compute the number densities as functions of temperature and for two values of the pressure (Exercise 7.6.1).

7.3   EQUILIBRIUM C-H-O CHEMISTRY: FORMING METHANE, WATER, CARBON MONOXIDE AND ACETYLENE

Of great interest to us is understanding the relative abundances of molecules formed from hydrogen, oxygen and carbon. Typically, the major molecules formed are methane, water and carbon monoxide—the chemical net reaction that describes the balance between them is listed in equation (6.2), which we reproduce here for convenience and label as our first net reaction,

$$CH_4 + H_2O \leftrightarrows CO + 3H_2. \tag{7.8}$$

However, when the temperature and ratio of carbon to oxygen atoms are high enough, hydrogen cyanide (HCN) and hydrocarbons such as acetylene start to appear [155]. To account for this effect, we assume acetylene to be the dominant hydrocarbon and include it into our chemical network via the following, second net reaction [147],

$$2CH_4 \leftrightarrows C_2H_2 + 3H_2. \tag{7.9}$$

We wish to derive analytical expressions for the abundances of methane, water, carbon monoxide and acetylene, based on an approach that is mathematically identical to the one we used to study the pure-hydrogen system. It allows us to cast the molecular abundances in terms of the atomic ones and the equilibrium constants.

The dimensional equilibrium constants for the two net reactions are

$$
\begin{aligned}
K'_{\text{eq}} &= \frac{n_{CO} n_{H_2}^3}{n_{CH_4} n_{H_2O}} = \frac{\tilde{n}_{CO} n_{H_2}^2}{\tilde{n}_{CH_4} \tilde{n}_{H_2O}}, \\
K'_{\text{eq},2} &= \frac{n_{C_2H_2} n_{H_2}^3}{n_{CH_4}^2} = \frac{\tilde{n}_{C_2H_2} n_{H_2}^2}{\tilde{n}_{CH_4}^2}.
\end{aligned}
\tag{7.10}
$$

The number densities associated with a tilde have been normalized by n_{H_2}. Written in this form, they are usually termed *mixing ratios* by atmospheric scientists. We are assuming that molecular hydrogen is the dominant buffer gas in the atmosphere and that $P = n_{H_2} k_B T$ is essentially the total pressure of the atmosphere, where k_B denotes the Boltzmann constant and T the temperature. Atomic hydrogen is assumed to be present in amounts small enough that it introduces only a small correction to P—this approximation allows us to avoid specifying an additional equilibrium constant to describe the atomic to molecular transformation (and vice versa) of hydrogen. Figure 7.1 demonstrates that this is a reasonable assumption as long as $T \lesssim 3000$ K, which is expected to be the case in most, if not all, exoplanetary atmospheres—at least in the vicinity of the infrared photosphere.

Conservation of atoms states that

$$n_{CH_4} + n_{CO} + 2n_{C_2H_2} = n_C,$$
$$n_{H_2O} + n_{CO} = n_O, \tag{7.11}$$
$$4n_{CH_4} + 2n_{H_2O} + 2n_{C_2H_2} + 2n_{H_2} = n_H,$$

where the atomic abundances (n_C, n_O and n_H) will be used as input parameters. By dividing the equations involving n_C and n_O by that involving n_H, we obtain

$$\tilde{n}_{CH_4} + \tilde{n}_{CO} + 2\tilde{n}_{C_2H_2} = \tilde{n}_C \left(4\tilde{n}_{CH_4} + 2\tilde{n}_{H_2O} + 2\tilde{n}_{C_2H_2} + 2\right),$$
$$\tilde{n}_{H_2O} + \tilde{n}_{CO} = \tilde{n}_O \left(4\tilde{n}_{CH_4} + 2\tilde{n}_{H_2O} + 2\tilde{n}_{C_2H_2} + 2\right). \tag{7.12}$$

It is important to note that the atomic abundances marked by tildes have been normalized by n_H and not n_{H_2}. This is because elemental (atomic) abundances are typically listed relative to atomic hydrogen.

All that remains to be done is to express the normalized molecular abundances in terms of the atomic abundances, K'_{eq} and $K'_{eq,2}$. After some algebra, we obtain a cubic equation for the mixing ratio of methane [99],

$$\mathcal{C}_3 \tilde{n}_{CH_4}^3 + \mathcal{C}_2 \tilde{n}_{CH_4}^2 + \mathcal{C}_1 \tilde{n}_{CH_4} + \mathcal{C}_0 = 0, \tag{7.13}$$

which has the coefficients,

$$\begin{aligned}
\mathcal{C}_3 &= 2K'K'_2 \left(\tilde{n}_O - \tilde{n}_C + 1\right), \\
\mathcal{C}_2 &= K' \left(4\tilde{n}_O - 4\tilde{n}_C + 1\right) \\
&\quad - K'_2 \left[4\tilde{n}_O\tilde{n}_C + 2\left(1 - 2\tilde{n}_O\right)\left(\tilde{n}_C - 1\right)\right], \\
\mathcal{C}_1 &= 2K' \left(\tilde{n}_O - \tilde{n}_C\right) - 4\tilde{n}_C - 2\tilde{n}_O + 1, \\
\mathcal{C}_0 &= -2\tilde{n}_C.
\end{aligned} \tag{7.14}$$

Analogous to the case study of pure hydrogen, we have defined

$$K' \equiv \frac{K'_{eq}}{n_{H_2}^2}, \quad K'_2 \equiv \frac{K'_{eq,2}}{n_{H_2}^2}. \tag{7.15}$$

The other mixing ratios may be obtained via

$$\begin{aligned}
\tilde{n}_{H_2O} &= \frac{2\tilde{n}_O \left(K'_2\tilde{n}_{CH_4}^2 + 2\tilde{n}_{CH_4} + 1\right)}{1 + K'\tilde{n}_{CH_4} - 2\tilde{n}_O}, \\
\tilde{n}_{CO} &= K'\tilde{n}_{CH_4}\tilde{n}_{H_2O}, \\
\tilde{n}_{C_2H_2} &= K'_2\tilde{n}_{CH_4}^2.
\end{aligned} \tag{7.16}$$

When acetylene is absent ($K'_2 = 0$), the solution for the mixing ratio of methane can be easily written down,

$$\tilde{n}_{CH_4} = \frac{-\mathcal{C}_1 + \left(\mathcal{C}_1^2 - 4\mathcal{C}_0\mathcal{C}_2\right)^{1/2}}{2\mathcal{C}_2}. \tag{7.17}$$

Notice how the coefficient $\mathcal{C}_3$, and thus K_2', controls the extent to which the mixing ratio of methane is described by a quadratic versus cubic equation. Physically, we expect that at low temperatures, the mixing ratio of acetylene is negligible. In this limit $(K', K_2' \ll 1)$, we have $\tilde{n}_{CH_4} \approx 2\tilde{n}_C$ and $\tilde{n}_{H_2O} \approx 2\tilde{n}_O$. These asymptotic solutions predict the ratio of methane to carbon monoxide to be simply

$$\frac{\tilde{n}_{CH_4}}{\tilde{n}_{H_2O}} \approx \frac{\tilde{n}_C}{\tilde{n}_O}. \tag{7.18}$$

In other words, measuring the relative abundances of these molecules, at low temperatures, directly yields the carbon-to-oxygen ratio.

Instead of pursuing graphical solutions for our current C-H-O system, we defer this endeavor and next add carbon dioxide to the mix.

7.4 EQUILIBRIUM C-H-O CHEMISTRY: ADDING CARBON DIOXIDE

7.4.1 Deriving the quintic equation for the mixing ratio of methane

Carbon dioxide is known to participate in the C-H-O system via the (third) net reaction [175],

$$CO_2 + H_2 \leftrightarrows CO + H_2O. \tag{7.19}$$

When the elemental abundance of carbon greatly exceeds its solar value, carbon dioxide may even be the dominant gas by mass [176]. It is therefore of interest to include it in our chemical network. The equilibrium constant associated with the preceding reaction is

$$K_{eq,3}' \equiv \frac{n_{CO}n_{H_2O}}{n_{H_2}n_{CO_2}} = \frac{\tilde{n}_{CO}\tilde{n}_{H_2O}}{\tilde{n}_{CO_2}} \equiv K_3'. \tag{7.20}$$

Stoichiometric book-keeping is generalized to

$$\begin{aligned}
n_{CH_4} + n_{CO} + n_{CO_2} + 2n_{C_2H_2} &= n_C, \\
n_{H_2O} + n_{CO} + 2n_{CO_2} &= n_O, \\
4n_{CH_4} + 2n_{H_2O} + 2n_{H_2} + 2n_{C_2H_2} &= n_H.
\end{aligned} \tag{7.21}$$

Working through the algebra, we obtain a pair of coupled quadratic equations for the mixing ratios of water and methane,

$$\begin{aligned}
&\frac{K'\tilde{n}_{H_2O}^2\tilde{n}_{CH_4}}{K_3'} + 2K_2'\tilde{n}_{CH_4}^2\left(1 - \tilde{n}_C\right) - 2\tilde{n}_C\tilde{n}_{H_2O} \\
&+ \tilde{n}_{CH_4} + K'\tilde{n}_{H_2O}\tilde{n}_{CH_4} - 4\tilde{n}_C\tilde{n}_{CH_4} - 2\tilde{n}_C = 0, \\
&\frac{2K'\tilde{n}_{H_2O}^2\tilde{n}_{CH_4}}{K_3'} - 2K_2'\tilde{n}_O\tilde{n}_{CH_4}^2 + \tilde{n}_{H_2O}\left(1 - 2\tilde{n}_O\right) \\
&+ K'\tilde{n}_{H_2O}\tilde{n}_{CH_4} - 4\tilde{n}_O\tilde{n}_{CH_4} - 2\tilde{n}_O = 0.
\end{aligned} \tag{7.22}$$

This is an undesired outcome, as it is simpler to solve a single polynomial equation, rather than a pair of coupled polynomial equations. It turns out that we may employ an algebraic trick if we recast the equations in (7.22) in terms of $\tilde{n}_{CO}$, rather than $\tilde{n}_{H_2O}$, which allows us to combine them into a quadratic equation for the mixing ratio of carbon monoxide [100],

$$
\begin{aligned}
& \frac{\tilde{n}_{CO}^2}{K'K_3'} + \frac{\tilde{n}_{CO}}{K'}\left(1 + 2\tilde{n}_C - 2\tilde{n}_O\right) - 2K_2'\tilde{n}_{CH_4}^3\left(1 - \tilde{n}_C + \tilde{n}_O\right) \\
& - \tilde{n}_{CH_4}^2 - 2\tilde{n}_{CH_4}\left(\tilde{n}_O - \tilde{n}_C\right)\left(2\tilde{n}_{CH_4} + 1\right) = 0.
\end{aligned}
\tag{7.23}
$$

Notice how the mixing ratios of carbon monoxide and methane are no longer coupled to each other within the same equation—there are no "mixed" terms, unlike for each equation in (7.22) between water and methane. This property has the virtue that we may cleanly take the approximation $\tilde{n}_C, \tilde{n}_O \ll 1$ and end up with a relatively simple expression for the mixing ratio of carbon monoxide,

$$
\tilde{n}_{CO} \approx K'\tilde{n}_{CH_4}\left[2K_2'\tilde{n}_{CH_4}^2 + \tilde{n}_{CH_4} + 2\left(\tilde{n}_O - \tilde{n}_C\right)\right].
\tag{7.24}
$$

Despite not being exact, the preceding expression retains generality, because the only assumption we have made so far is that the elemental abundances are small compared to hydrogen. It is less obvious how to take the $\tilde{n}_C, \tilde{n}_O \ll 1$ approximation directly using the equations in (7.22).

The expression for $\tilde{n}_{CO}$ leads to an approximate expression for the mixing ratio of water,

$$
\tilde{n}_{H_2O} \approx 2K_2'\tilde{n}_{CH_4}^2 + \tilde{n}_{CH_4} + 2\left(\tilde{n}_O - \tilde{n}_C\right),
\tag{7.25}
$$

and a quintic equation for the mixing ratio of methane [100],

$$
\sum_{i=0}^{5} \mathcal{A}_i \tilde{n}_{CH_4}^i \approx 0,
\tag{7.26}
$$

where the coefficients are

$$
\begin{aligned}
\mathcal{A}_5 &= \frac{8K'K_2'^2}{K_3'}, \\
\mathcal{A}_4 &= \frac{8K'K_2'}{K_3'}, \\
\mathcal{A}_3 &= \frac{2K'}{K_3'}\left[1 + 8K_2'\left(\tilde{n}_O - \tilde{n}_C\right)\right] + 2K'K_2', \\
\mathcal{A}_2 &= \frac{8K'}{K_3'}\left(\tilde{n}_O - \tilde{n}_C\right) + 2K_2' + K', \\
\mathcal{A}_1 &= \frac{8K'}{K_3'}\left(\tilde{n}_O - \tilde{n}_C\right)^2 + 1 + 2K'\left(\tilde{n}_O - \tilde{n}_C\right), \\
\mathcal{A}_0 &= -2\tilde{n}_C.
\end{aligned}
\tag{7.27}
$$

The mixing ratios of carbon monoxide, carbon dioxide and acetylene may be obtained using

$$\tilde{n}_{\text{CO}} = K' \tilde{n}_{\text{CH}_4} \tilde{n}_{\text{H}_2\text{O}},$$

$$\tilde{n}_{\text{CO}_2} = \frac{\tilde{n}_{\text{CO}} \tilde{n}_{\text{H}_2\text{O}}}{K_3'}, \tag{7.28}$$

$$\tilde{n}_{\text{C}_2\text{H}_2} = K_2' \tilde{n}_{\text{CH}_4}^2.$$

An indication that the quintic equation is correct comes from the fact that it automatically yields $\tilde{n}_{\text{CH}_4} \approx 2\tilde{n}_{\text{C}}$ when all of the normalized equilibrium constants vanish (i.e., the low-temperature limit).

There are no general analytical solutions to the quintic equation in (7.26). Instead, we use canned routines in the Python programming language to solve it numerically. The Python script is provided in Appendix E. But before we can solve the quintic equation, we need to relate the normalized equilibrium constants to the Gibbs free energies of formation of each molecule. As described in Chapter 6, it is the dimensionless equilibrium constant that is directly related to the change in (molar) Gibbs free energy,

$$K_{\text{eq}} = e^{-\Delta \tilde{G}_{0,1}/\mathcal{R}_{\text{univ}} T}, \tag{7.29}$$

where $\Delta \tilde{G}_{0,1} \equiv \Delta G_{0,1} N_A / N$, $N_A = 6.02214129 \times 10^{23}$ mol^{-1} is Avogrado's constant, N is the number of molecules, $\Delta G_{0,1}$ is the change in Gibbs free energy going from the reactants to the products, at the reference pressure (P_0), associated with the first net reaction and $\mathcal{R}_{\text{univ}}$ is the universal gas constant.

It follows that the dimensionless (K_{eq}), dimensional (K_{eq}') and normalized (K') equilibrium constants are related by [99]

$$K_{\text{eq}}' = n_0^2 K_{\text{eq}} = n_{\text{H}_2}^2 K', \tag{7.30}$$

where $n_0 = P_0 / k_B T$ is the number density corresponding to the reference pressure. Thus, we obtain

$$K' = \left(\frac{P_0}{P} \right)^2 e^{-\Delta \tilde{G}_{0,1}/\mathcal{R}_{\text{univ}} T}. \tag{7.31}$$

For the second and third net reactions, we have

$$K_2' = \left(\frac{P_0}{P} \right)^2 e^{-\Delta \tilde{G}_{0,2}/\mathcal{R}_{\text{univ}} T},$$

$$K_3' = e^{-\Delta \tilde{G}_{0,3}/\mathcal{R}_{\text{univ}} T}. \tag{7.32}$$

In Appendix D, you will find tabulated values of $\Delta \tilde{G}_{0,1}$, $\Delta \tilde{G}_{0,2}$ and $\Delta \tilde{G}_{0,3}$, which are the changes in the molar Gibbs free energies of the net reactions, for a range of temperatures and $P_0 = 1$ bar. We are now ready to study graphical solutions of our C-H-O network.

7.4.2 Elucidating trends as functions of temperature, pressure and C/O

Figures 7.2 and 7.3 display graphical solutions for the mixing ratios of the various molecules we have considered so far. Figure 7.2 shows the mixing ratios as functions of the temperature and for two values of the carbon-to-oxygen ratio: carbon-poor (C/O = 0.1) and C/O = 1. For reference, the value of C/O at the solar photosphere is between 0.5 and 0.6. Figure 7.3 shows the mixing ratios as functions of the carbon-to-oxygen ratio and for low ($T = 800$ K) and high ($T = 1200$ K) temperatures. Taken together, these graphical solutions allow us to elucidate a series of fundamental trends.

- In a carbon-poor atmosphere, water is expected to be the carrier of oxygen [176]. In fact, it is the dominant molecule, being more abundant than methane or carbon monoxide. When C/O takes on the solar value or higher, carbon monoxide becomes the dominant carrier of oxygen at high temperatures [155].

- Methane dominates carbon monoxide as the carrier of carbon at low temperatures [196]. This trend reverses at high temperatures [24, 147]. What "low" and "high" mean depend on the pressure and C/O. In other words, we expect cold atmospheres to be methane-rich and hot atmospheres to be carbon monoxide-rich.

- When the elemental abundance of carbon is low ($\tilde{n}_C \ll 1$), carbon dioxide is less abundant than carbon monoxide. Therefore, we expect carbon dioxide to be a minor carrier of carbon in hydrogen-dominated atmospheres [100]. In exotic situations where $\tilde{n}_C \sim 0.1$, carbon dioxide could be the dominant gas by mass [176], although this trend is not elucidated by our Figures 7.2 and 7.3.

- Acetylene (and therefore the higher-order hydrocarbons) is typically sub-dominant in carbon-poor atmospheres. It may become the dominant carrier of carbon if the atmosphere is carbon-rich and hot.

- When the temperature is low ($T = 800$ K in our graphical example), the trends exhibited by the mixing ratios versus C/O is relatively simple: the mixing ratio of water is constant, while those of methane, carbon monoxide, carbon dioxide and acetylene have simple scalings. In Figure 7.3, we have held the elemental abundance of oxygen to be fixed, which explains why $\tilde{n}_{H_2O} \approx 2\tilde{n}_O$ is constant. At low temperatures, we have $\tilde{n}_{CH_4} \approx 2\tilde{n}_C$, which explains the other trends.

- Hot atmospheres ($T = 1200$ K in our graphical example) exhibit more complex behavior. They are methane-poor and water-rich when they are carbon-poor, but they are methane-rich and water-poor when they are carbon-rich [155]. A carbon-to-oxygen ratio of unity becomes an important

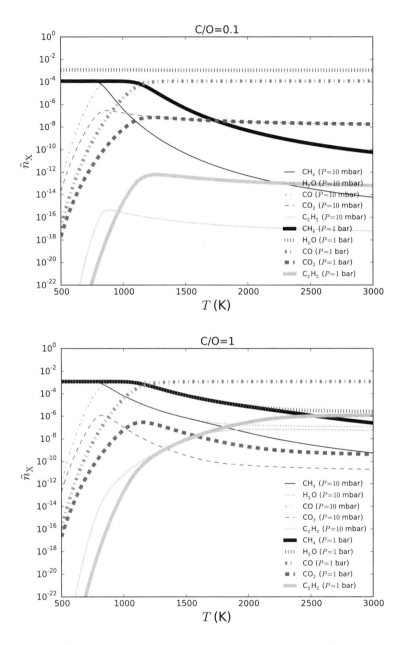

Figure 7.2: Mixing ratios for methane, water, carbon monoxide, carbon dioxide and acetylene as functions of the temperature and at two values of the pressure. The top and bottom panels are for C/O = 0.1 and 1, respectively. The elemental abundance of oxygen has been set to $\tilde{n}_O = 6 \times 10^{-4}$, which is roughly the solar value. (More specifically, it is the value at the solar photosphere.)

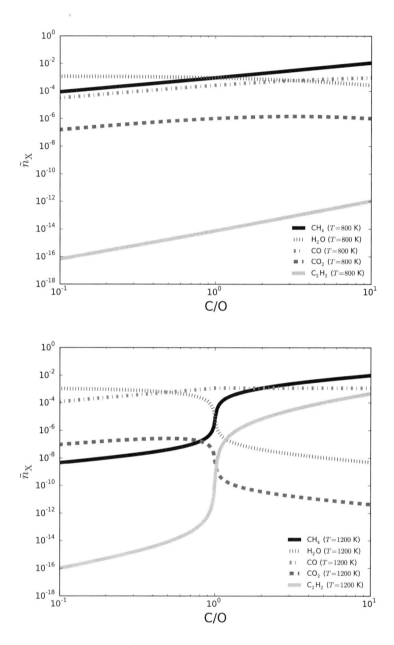

Figure 7.3: Mixing ratios for methane, water, carbon monoxide, carbon dioxide and acetylene as functions of the carbon-to-oxygen ratio. The top and bottom panels are for low ($T = 800$ K) and high ($T = 1200$ K) temperatures, respectively. The elemental abundance of oxygen has been set to $\tilde{n}_O = 6 \times 10^{-4}$, which is roughly the solar value. For illustration, we have set $P = 10$ mbar.

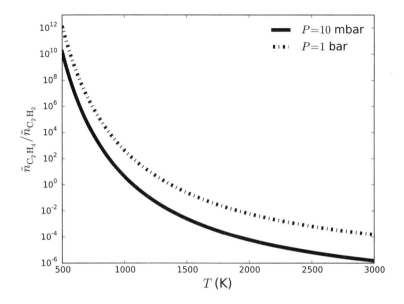

Figure 7.4: Relative abundance of ethylene to acetylene as a function of temperature and for two values of the pressure.

transition point. Carbon monoxide has a roughly constant abundance across C/O.

Another trend that is not elucidated by our analytical model is that if C/O is high enough, then graphite may condense out of the atmospheric gas [176].

7.5   EQUILIBRIUM C-H-O CHEMISTRY: ADDING ETHYLENE

Acetylene may be converted into higher-order hydrocarbons via a process known as *hydrogenation*, the lowest order of which produces ethylene (which is sometimes termed "ethene"),

$$C_2H_4 \leftrightarrows C_2H_2 + H_2. \tag{7.33}$$

The dimensional equilibrium constant of this fourth net reaction is

$$K'_{eq,4} = \frac{n_{C_2H_2} n_{H_2}}{n_{C_2H_4}}, \tag{7.34}$$

which implies that the normalized equilibrium constant is

$$K'_4 \equiv \frac{K'_{eq,4}}{n_{H_2}} = \frac{\tilde{n}_{C_2H_2}}{\tilde{n}_{C_2H_4}}. \tag{7.35}$$

Since we have $K_4' = K_{eq,4}P_0/P$, where $K_{eq,4}$ is the dimensionless equilibrium constant, it follows that

$$\frac{\tilde{n}_{C_2H_4}}{\tilde{n}_{C_2H_2}} = \frac{P}{P_0}e^{\Delta\tilde{G}_{0,4}/\mathcal{R}_{univ}T}. \tag{7.36}$$

This expression immediately informs us that ethylene is preferentially produced at high pressures over acetylene. Figure 7.4 shows that the production of ethylene is favored at low temperatures. Near the infrared photospheres of hot exoplanets with $T \gtrsim 1000$ K, ethylene is expected to be subdominant compared to acetylene. In this situation, acetylene dominates the contribution of the hydrocarbons.

7.6 PROBLEM SETS

7.6.1 Pure-hydrogen chemistry

Using the Gibbs free energies of formation for atomic hydrogen (as listed in Appendix D), reproduce the results in Figure 7.1.

7.6.2 C-H-O chemistry without temperature and pressure

Compute the graphical solution of equation (7.13) and plot the mixing ratios of methane, water, carbon monoxide and acetylene as functions of the carbon-to-oxygen ratio. Use the normalized equilibrium constants as input parameters. Set $\tilde{n}_O = 5 \times 10^{-4}$, which is roughly the solar abundance of atomic oxygen, and vary $\tilde{n}_C$. Are the trends in agreement with more rigorous calculations that express the equilibrium constants as functions of temperature and pressure (Figure 7.3)?

7.6.3 C-H-O chemistry with ethylene

Generalize the quintic equation in (7.26) by adding ethylene to the chemical network. With the aid of the `Python` scripts in Appendix E, generalize Figures 7.2 and 7.3 by including the mixing ratio of ethylene. Do the trends in Section 7.4.2 change with the addition of ethylene?

7.6.4 Nitrogen chemistry

Consider the net reaction,

$$N_2 + 3H_2 \leftrightarrows 2NH_3. \tag{7.37}$$

By collecting the Gibbs free energies of formation of molecular nitrogen (N_2) and ammonia (NH_3) from the JANAF database (see Appendix D for more information), set up an equilibrium chemical network to calculate their mixing ratios as functions of temperature and pressure. Is molecular nitrogen or ammonia the dominant nitrogen carrier at high temperatures?

Chapter Eight

Introduction to Fluid Dynamics

8.1 WHY IS THE STUDY OF FLUIDS RELEVANT TO EXOPLANETARY ATMOSPHERES?

To some physicists, the study of fluid dynamics is an intrinsically interesting endeavor. But since this is a textbook about exoplanetary atmospheres, it is reasonable to ask why we bother ourselves with this endeavor in the first place. The study of any atmosphere is a complex undertaking, as chemical and physical phenomena are occurring on a diverse—and possibly enormous—range of scales. If one magnifies a patch of atmospheric gas, one sees the constituent atoms and molecules. This might inspire some physicists to attempt to exhaustively compute the interactions between each and every atom or molecule, in order to understand the macroscopic behavior of the atmosphere.

Such a calculation is both infeasible—some would say impossible—and unnecessary, because it attempts to substitute intuitive understanding with brute-force computation. First, we need to understand that collisions between the atoms/molecules occur on a time scale that is typically the shortest in the system, meaning that different parts of it are causally connected. (Other time scales include those associated with advection, chemistry, convection, diffusion, friction, turbulence, radiative cooling, rotation, etc.) In other words, the *mean free path* for collisions,

$$l_{\mathrm{mfp}} = \frac{1}{n\sigma_{\mathrm{coll}}}, \qquad (8.1)$$

is short, where n is the number density and σ_{coll} is the collisional cross section. As long as one observes the system on a length scale (l) that is much longer than the mean free path ($l \gg l_{\mathrm{mfp}}$), it behaves like a fluid. One only needs to understand its behavior on scales $\sim l$. Its behavior on microscopic scales is irrelevant, except when information is needed on its chemistry, thermodynamics and sources of opacity (via quantum mechanics).

Certainly, the $l \gg l_{\mathrm{mfp}}$ approximation is not expected to hold everywhere in the atmosphere. At high enough altitudes, it breaks down. For Earth, this transition height is known as the *exopause* or *exobase*: it is the altitude where $l = l_{\mathrm{mfp}}$ (with l being interpreted as the pressure scale height). The visible and infrared radiation emanating from an exoplanetary atmosphere—and traveling across space to our telescopes and spectrographs—originates from lower altitudes, where the atmosphere may be reasonably described by a fluid.

Generally, it means that to understand the global structure of density, pressure, temperature and velocity in an atmosphere, one needs to solve a coupled set of fluid equations.

8.2 WHAT EXACTLY IS A FLUID?

Frequently, the answer to this question is, "Something that flows." But this answer is unsatisfactory in several regards. First, it is imprecise; does "to flow" refer to advection, convection or diffusion? Second, it suggests an element of human judgment being involved; or at least, the observation of the system on a time scale that allows a flow to manifest itself. From everyday experience, one would expect gases and liquids to be fluids. But here is a counter-example: bitumen appears to us as a solid, but it is a fluid that "flows" on time scales of years to decades; it changes its form at a glacial pace, due to its inherent viscosity.

More succinctly, a fluid is an entity that obeys a continuum, rather than a discrete, description. The behavior of its building blocks is less important than how those blocks "play together." It is governed by a set of equations describing the conservation of mass, momentum and energy in the continuum limit. These equations include the effects of (external) gravity,[1] rotation, molecular viscosity, radiative heating and even magnetic fields. Its solution allows for the existence of a diverse set of waves, which transport mass, energy and momentum throughout the fluid and determine its macroscopic structure. Waves for which the restoring force is gravity are aptly named *gravity waves*. Rotationally-modified gravity waves are known as *Poincaré waves*. *Rossby waves* arise from a gradient in the Coriolis force across latitude. In Chapter 10, we will study the behavior of these waves in detail.

8.3 THE GOVERNING EQUATIONS OF FLUID DYNAMICS

In this section, we will introduce and discuss the governing equations of fluid dynamics, as applied to exoplanetary atmospheres, in an intuitive way. We will defer their detailed derivation to Chapter 9.

8.3.1 The conservation of momentum

Many discussions of fluid dynamics begin with the Navier-Stokes equation,

$$\frac{D\vec{v}}{Dt} = \vec{g} - \frac{\nabla P}{\rho} - 2\vec{\Omega} \times \vec{v} + \nu\nabla^2\vec{v} + \frac{\nu}{3}\nabla\left(\nabla.\vec{v}\right) + \vec{F}_{\text{others}}. \qquad (8.2)$$

The velocity is represented by $\vec{v}$; its temporal derivative is the acceleration. Simply put, equation (8.2) is a restatement of Newton's second law for fluids, but

---

[1] In astrophysical situations, it is not uncommon to encounter *self-gravitating* fluids, which are systems that are so massive that the self-induced gravity plays a non-negligible role in their behavior.

in physical units of force per unit mass (acceleration). The quantities appearing after the equality are source and sink terms that seed this acceleration. The first term $(\vec{g})$ is the acceleration due to gravity or simply termed the "surface gravity" by exoplanet scientists. The second term is associated with the pressure gradient (∇P). The third term is due to the Coriolis force $(-2\vec{\Omega} \times \vec{v})$, which only manifests itself when one is in the co-rotating frame of reference. The angular frequency of the rotation of the exoplanet, about its own axis, is denoted by $\vec{\Omega}$. The fourth and fifth terms are associated with the molecular viscosity[2]— more precisely, it is the viscosity associated intrinsically with the microscopic properties of the fluid, rather than with convection or turbulence (see Chapter 12). Other sources/sinks associated with turbulence, magnetic fields, etc, are contained in $\vec{F}_{\text{others}}$. Equation (8.2) is written in a coordinate-free form, meaning that it holds regardless of whether one chooses to use Cartesian, cylindrical or spherical coordinates.

A quantity that is unfamiliar to non-practitioners of fluid dynamics is the *material derivative*,

$$\frac{D}{Dt} \equiv \frac{\partial}{\partial t} + \vec{v}.\nabla. \tag{8.3}$$

Physically, it is the temporal derivative of a quantity *following* a fluid element. In other words, we place ourselves in the reference frame of the moving fluid element and record the change of some quantity (mass, momentum, temperature, energy, etc) with time. Problem 8.6.1 works through the derivation and interpretation of the material derivative in detail.

A mathematically straightforward, but physically non-trivial, solution of equation (8.2) obtains from demanding that $\frac{D\vec{v}}{Dt} = 0$, which yields

$$\nabla P = \rho \vec{g}, \tag{8.4}$$

where ρ is the mass density of the atmosphere. The assumption being made is that gravity and pressure support are the dominant physical effects and cancel each other out. Equation (8.4) formally expresses the state of *hydrostatic equilibrium* (or balance). In one dimension (with the spatial coordinate represented by z), we may cast it in the form of integrals,

$$\int_{P_0}^{P} \frac{1}{P}\, dP = -\int_{z_0}^{z} \frac{mg}{k_{\mathrm{B}}T}\, dz, \tag{8.5}$$

via use of the ideal gas law. Note that the minus sign comes from the fact that gravity acts "inwards," towards the center of the exoplanet, such that $\vec{g} = -g\hat{z}$, where $\hat{z}$ is the unit vector in the lone spatial direction. Evaluating the integrals requires that a reference pressure (P_0) is defined corresponding to a reference height (z_0). The reference pressure is typically and conveniently chosen to be the surface pressure of the exoplanet.

---

[2]The quantity ν is referred to as the *kinematic viscosity*, while $\mu_\nu \equiv \rho\nu$ is the dynamic viscosity.

If the atmosphere is isothermal (constant temperature)—which never happens in practice—the solution is

$$P = P_0 \, e^{-z/H}. \tag{8.6}$$

This simple solution showcases an important characteristic length in atmospheres known as the *pressure scale height*,

$$H = \frac{k_B T}{mg}, \tag{8.7}$$

where k_B is the Boltzmann constant, T is the temperature and m is the mean molecular mass. On Earth, the atmosphere consists mostly of nitrogen, which yields $m \approx 29 m_H$ (with m_H being the mass of the hydrogen atom) and $H \approx 8$ km. In gas-giant exoplanets, which comprise mostly molecular hydrogen ($m \approx 2 m_H$), we have $H \sim 100$–1000 km (if $T \sim 10^3$ K).

The transit radius of an exoplanet varies by several pressure scale heights across the spectral features of atoms and molecules. It is for this reason that hot (high T), light (low m), and low-gravity (small g) atmospheres are easier to characterize.

8.3.2   The conservation of mass

Often solved in conjunction with the Navier-Stokes equation is the equation of mass continuity,

$$\frac{\partial \rho}{\partial t} + \nabla \cdot (\rho \vec{v}) = 0. \tag{8.8}$$

When rewritten in terms of the material derivative of the mass density, one obtains

$$\frac{1}{\rho} \frac{D\rho}{Dt} = -\nabla \cdot \vec{v}. \tag{8.9}$$

The quantity $\nabla \cdot \vec{v}$ is known as the *divergence* of the velocity. It quantifies how the velocity flows away from ($\nabla \cdot \vec{v} > 0$) or into ($\nabla \cdot \vec{v} < 0$) a point. A fluid is *incompressible* when $\nabla \cdot \vec{v} = 0$. From the preceding equation, we see that this implies a constant density: $\frac{D\rho}{Dt} = 0$.

Physically, the state of being incompressible is related to how fast density perturbations are communicated throughout a fluid. These perturbations are mediated by sound waves. Unlike its more famous cousin, the speed of light, the speed of sound depends on the local conditions of pressure and density, since it is

$$c_s = \left(\frac{\partial P}{\partial \rho} \right)^{1/2}. \tag{8.10}$$

Problem 8.6.2 shows you how to derive this expression in the limit of small perturbations. Incompressibility derives from having an infinite sound speed; or at least, sound waves are the fastest waves in the system.

There is a common misunderstanding about hydrostatic equilibrium that is worth elucidating. It is often believed that it implies a static atmosphere, which is generally incorrect. A static atmosphere appears only as a special, limiting case of hydrostatic equilibrium. In actuality, hydrostatic balance asserts that the time scale associated with sound waves is the second shortest in the system—bested only by the time scale of collisions between atoms/molecules. When an atmosphere is perturbed away from hydrostatic balance, it is returned to it essentially instantaneously. It is not that the atmosphere is static—rather, its motion and circulation are ponderous compared to the propagation of sound waves.

This physical view may be reconciled with the mathematics. Consider a two-dimensional system with the constant velocity $\vec{v} = (v_x, v_z)$, where x and z are the horizontal and vertical directions, respectively. The corresponding, characteristic length scales are the radius of the exoplanet (R) and the isothermal pressure scale height (H). We demand that $v_x \gg v_z$ and $R \gg H$, which is typical for an atmosphere. The constant velocity implies that hydrostatic balance obtains: $\frac{\partial P}{\partial z} = -\rho g$. For the mass continuity equation, one obtains

$$\frac{\partial \rho}{\partial t} + \frac{\partial}{\partial x}\left(\rho v_x\right) + \frac{\partial}{\partial z}\left(\rho v_z\right) = 0. \tag{8.11}$$

At the order-of-magnitude level, we have

$$\frac{\partial(\rho v_x)}{\partial x} \sim \frac{\rho v_x}{R} \quad \text{and} \quad \frac{\partial(\rho v_z)}{\partial z} \sim \frac{\rho v_z}{H}. \tag{8.12}$$

Both of these spatial derivatives need to be retained, since they are expected to be comparable in magnitude. In other words, even though $v_x \gg v_z$ and $R \gg H$, we may have $v_x/R \sim v_z/H$. The vertical component of the velocity may not appear in the momentum equation, but it is non-negligible in the equation of mass conservation. Numerical solutions of the fluid equations are able to produce v_z as an output, even when hydrostatic equilibrium is *assumed*, because of this property—it emerges from the mass continuity equation.

Finally, we note that whether hydrostatic equilibrium holds depends on the length scale (l) on which we are examining the atmosphere. On scales of $l \gg H$, hydrostatic balance is a fine approximation. On scales of $l \lesssim H$, it breaks down and non-hydrostatic effects have to be considered. The atmosphere of Earth provides an example of these scales at play: on global scales (~ 1000 km), it is hydrostatic to a very good approximation, while it is non-hydrostatic on meso-scales (~ 10–100 km), where cloud formation, convection and turbulence are important.

8.3.3 The conservation of energy

There are five unknown variables describing a fluid: its three components of velocity, its mass density and its temperature, necessitating a final equation to "close" the system of equations. Alongside mass and momentum conservation, we naturally have the conservation of energy,

$$\frac{DT}{Dt} = \frac{\kappa_{\mathrm{ad}} T}{P} \frac{DP}{Dt} + \frac{Q}{c_P},$$

(8.13)

where κ_{ad} is the adiabatic coefficient and c_P is the specific heat capacity at constant pressure. For the uninitiated, Appendix C provides some essential formulae of the thermodynamics needed to understand atmospheres. As we will see in Chapter 9, the preceding equation derives from the first law of thermodynamics. The first term is associated with work done due to changes in volume, while the second term derives from external heating.

There are several ways of expressing energy conservation. We have chosen to write down an expression for T and then relate it to P via the ideal gas law,

$$P = \frac{\rho k_{\mathrm{B}} T}{m},$$

(8.14)

which is an *equation of state*, but one may also write down a governing equation for the pressure, internal energy or total energy. This range of choices will be explored in Chapter 9.

8.4 POTENTIAL TEMPERATURE AND POTENTIAL VORTICITY

Intuitively, one would like to compare basic quantities or observables from different patches or regions of an atmosphere. From the perspective of fluid dynamics, such a comparison is not always fair. Two particular quantities, for which a direct comparison between parcels of atmosphere would be unfair, are the temperature and vorticity.

8.4.1 Potential temperature (and its relationship to specific entropy)

Consider the following situation: you are standing on the surface of Earth. At a specific point in time, you measure the temperature of a parcel of air next to you. Suppose you have some way of measuring, at the same point in time, the temperature of a parcel of air a kilometer off the ground. You compare the temperatures of the two parcels of air and conclude that the parcel of air located at the higher altitude is colder. What is wrong with this comparison?

The answer is that we have forgotten to take into account the adiabatic[3] contraction or expansion of the parcels of air. Since the Earth's atmosphere is approximately in hydrostatic equilibrium, the pressure decreases with increasing altitude. If a parcel of air near the surface is lifted a kilometer upwards, it will

---

[3]Meaning that no energy is injected into or extracted from a parcel of air.

expand simply because of the decrease in its ambient pressure. In doing so, it will cool adiabatically. Conversely, a parcel of air situated at a kilometer in altitude will contract when brought down to the surface of Earth, because of the higher ambient pressure. This simple thought experiment suggests that the temperature is an incomplete quantity to compare, since it needs to take into account adiabatic cooling and heating.

It turns out that the correct quantity to compare is the *potential temperature*. We will defer its derivation to Chapter 12 (when we discuss convection) and simply state its definition,

$$\Theta \equiv T \left(\frac{P}{P_0} \right)^{-\kappa_{\mathrm{ad}}} , \tag{8.15}$$

where P_0 is a reference pressure, which we nominally define to be the surface pressure. The potential temperature generally increases with decreasing pressure to compensate for adiabatic cooling, a property that allows two parcels of air at different altitudes to be fairly compared. Another way of understanding the potential temperature is to relate it to the specific entropy (entropy per unit mass),

$$S = \int c_P \, d \left(\ln \Theta \right) = c_P \ln \Theta. \tag{8.16}$$

The second equality is only valid if c_P is constant.

Here is another illuminating thought experiment: imagine an atmosphere with a constant mass density. Using the ideal gas law and hydrostatic balance, we conclude that its temperature decreases exponentially with altitude. In reality, the mass density is certainly not constant with altitude, but it serves to illustrate the point that the temperature generally decreases with altitude (unless some sort of special absorber, like the ozone layer in Earth's atmosphere, is at work). We know from everyday experience that having cold air sit above hot air leads to a situation that is unstable to convection. So is this atmosphere always unstable to convection?

As you may have anticipated, it turns out that the temperature is again the wrong quantity to inspect. We will see in Chapter 12 that the criterion for convective stability is for the gradient of the potential temperature to be zero or positive. This is known as *Schwarzschild's criterion*, after the German physicist and astronomer Karl Schwarzschild, who formulated it for the atmospheres of stars.

Thus, atmospheric motion may sometimes be understood as resulting from gradients in the potential temperature or specific entropy. By examining maps of the potential temperature, one may gain a deeper understanding for the dynamical structures present in an atmosphere.

8.4.2 Potential vorticity

Analogously, the vorticity of a parcel of atmosphere,

$$\omega \equiv \nabla \times \vec{v}, \tag{8.17}$$

is not the most general quantity to examine and compare. Intuitively, the vorticity measures the amount of "spin" of a fluid parcel, and we expect some quantity related to it to be conserved, in the same way that ice skaters spin faster when they retract their limbs.

In the 1940s, the German meteorologist and geophysicist Hans Ertel formulated what he called a new conservation law of hydrodynamics, first by himself [58] and later with Carl-Gustaf Rossby [59]. Ertel discovered that a quantity, now known as the *potential vorticity*,[4]

$$\Phi \equiv \frac{\vec{\omega} + 2\vec{\Omega}}{\rho} . \nabla \Theta, \tag{8.18}$$

is invariant in time in the reference frame of a fluid element. In other words, we have

$$\frac{D\Phi}{Dt} = 0. \tag{8.19}$$

The potential vorticity is a challenging quantity to even define, because it first requires that one understands what the potential temperature is, since it is defined only on surfaces of constant gradients of the potential temperature. Similar to the potential temperature, examining maps of the potential vorticity yields insight into the dynamical structures of an atmosphere.

A source of confusion is the use of the term *vortensity* in the astrophysical literature studying protoplanetary disks [179]. The vortensity is sometimes described as being $\hat{z}.(\nabla \times \vec{v})/\rho$, where ρ is the surface density (height-integrated mass density), rather than the mass density. This definition is a special case of the potential vorticity in the barotropic limit. To confuse matters further, some researchers use the terms "vortensity" and "potential vorticity" interchangeably [141, 179], something I do not recommend.

We will see in Chapter 9 that the conservation of potential vorticity follows directly from manipulating the Navier-Stokes equation, but with a series of assumptions and caveats, rather than being an independent conservation law on its own. It is of interest to us to elucidate what these assumptions and caveats are.

8.5 DIMENSIONLESS FLUID NUMBERS

A useful concept in fluid dynamics is to describe systems with a set of dimensionless numbers that capture some essential property. Different systems with

---

[4]In the absence of a universal notation for the potential vorticity, we have opted for using Φ and avoided the use of ambiguous notation like PV, which may be mistaken for being the product of the pressure and volume.

vastly different scales, but the same fluid numbers, are expected to exhibit the same macroscopic behavior. Here, we describe some of the more commonly-used fluid numbers related to atmospheres.

The first fluid number we will introduce is the *Knudsen number*,

$$\mathcal{K}_n \equiv \frac{l_{\mathrm{mfp}}}{l}, \tag{8.20}$$

where l is the characteristic length scale on which we are examining the system. If $\mathcal{K}_n \ll 1$, then it is reasonable to regard the system as being a fluid and study it using the governing equations of fluid dynamics. If $\mathcal{K}_n \gtrsim 1$, one either has to adopt a non-fluid approach or introduce correction factors expressed in terms of the Knudsen number [229].

An obvious and commonly-used fluid number to introduce is the *Reynolds number*. It derives naturally from non-dimensionlizing the Navier-Stokes equation. Consider a typical length (l) and velocity (U) scale, which in turn yields a typical time scale (l/U). By labeling all of the dimensionless quantities with a prime, we obtain

$$\frac{D\vec{v}'}{Dt'} = \vec{g}' - \frac{\nabla' P'}{\rho'} - 2\vec{\Omega}' \times \vec{v}' + \frac{1}{\mathcal{R}_e}\nabla'^2\vec{v}' + \frac{1}{3\mathcal{R}_e}\nabla'\left(\nabla'.\vec{v}'\right). \tag{8.21}$$

The Reynolds number is defined as

$$\mathcal{R}_e \equiv \frac{Ul}{\nu}. \tag{8.22}$$

Physically, it is a measure of the importance of advection versus molecular viscosity. Viscous fluids are typically described by $\mathcal{R}_e \ll 1$. Turbulent fluids have $\mathcal{R}_e \gg 1$. The Navier-Stokes equation with $\nu = 0$ or $\mathcal{R}_e \to \infty$ is sometimes termed the *Euler equation*.

The Mach number,

$$\mathcal{M} \equiv \frac{v}{c_s}, \tag{8.23}$$

measures if a flow is subsonic $(\mathcal{M} < 1)$ or supersonic $(\mathcal{M} > 1)$. As we will see in Chapter 11, supersonic flow is a necessary but insufficient condition for *shocks* to form.

The Rossby number quantifies the importance of advection versus rotation,

$$\mathcal{R}_o \equiv \frac{U}{\Omega l}, \tag{8.24}$$

where Ω is the angular rotational frequency. Rotation is an important factor in determining the structure of Earth's atmosphere $(\mathcal{R}_o \ll 1)$, while it is less important for tidally locked exoplanets $(\mathcal{R}_o \sim 1)$. The Rossby length scale (U/Ω) is the typical size of vortices present in an atmosphere.

Other commonly-used fluid numbers include the Prandtl number (molecular viscosity versus thermal conduction), the Richardson number (buoyancy versus advection and hence the susceptibility of the fluid to the Kelvin-Helmholtz instability) and the Stokes number (the strength of coupling of a particle or aerosol to the fluid flow).

8.6 PROBLEM SETS

8.6.1 Deriving the material derivative

We wish to derive the expression for the material derivative or operator, as stated in equation (8.3). At a time t, consider a fluid element with a velocity $\vec{v} = (v_x, v_y, v_z)$. After a time interval of δt, it moves from (x, y, z) to $(x + v_x \delta t, y + v_y \delta t, z + v_z \delta t)$. By tracking the movement of this fluid element, we may record any changes associated with it in its own reference frame.

(a) Let any property of the fluid be represented by a function $f(x, y, z, t)$. Let the change of this property be

$$\delta f = f\left(x + v_x \delta t, y + v_y \delta t, z + v_z \delta t, t + \delta t\right) - f\left(x, y, z, t\right). \tag{8.25}$$

Perform Taylor expansion, to linear order, on the first term after the equality to obtain a reduced expression for δf.

(b) Show that

$$\lim_{\delta t \to 0} \frac{\delta f}{\delta t} = \frac{Df}{Dt}. \tag{8.26}$$

8.6.2 Deriving the expression for the sound speed

Consider the mass continuity and momentum equations in one dimension,

$$\frac{\partial v}{\partial t} + v \frac{\partial v}{\partial z} = -\frac{1}{\rho} \frac{\partial P}{\partial z}, \quad \frac{\partial \rho}{\partial t} + \frac{\partial}{\partial z}(\rho v) = 0. \tag{8.27}$$

(a) Perform a perturbation analysis by considering the following state,

$$\rho = \rho_0 + \rho', \; P = P_0 + P', \; v = v'. \tag{8.28}$$

We will assume that the basic state, which is at rest and described by ρ_0 and P_0, experiences small perturbations, such that $\rho' \ll \rho_0$ and $P' \ll P_0$. The background density (ρ_0) and pressure (P_0) are assumed to be constant. Derive the governing equations describing ρ' and P'.

(b) By combining this pair of governing equations, show that

$$\frac{\partial^2 \rho'}{\partial t^2} = \frac{\partial P'}{\partial \rho'} \frac{\partial^2 \rho'}{\partial z^2}. \tag{8.29}$$

One recognizes that this is a wave equation describing ρ' with the wave speed being given by $c_s^2 = \partial P'/\partial \rho'$, where c_s is the sound speed.

(c) By assuming an adiabatic gas $(P \propto \rho^{\gamma_{ad}})$, show that

$$c_s = \left(\frac{\gamma_{ad} k_B T}{m}\right)^{1/2}, \tag{8.30}$$

where γ_{ad} is the adiabatic gas index.

8.6.3   Dimensionless fluid numbers

Estimating the values of dimensionless fluid numbers, even at the order-of-magnitude level, serves as a very useful guide to developing intuition about a given fluid.

(a) A cup of water is stirred with a spoon at room temperature and pressure. What is the value of the Reynolds number? How about a cup of honey? Which fluid do you expect to develop turbulence more easily?

(b) What is the value of the Reynolds number for the atmospheres of Earth, Venus, Jupiter and a hot Jupiter? Is it appropriate to use the Euler equation for modeling these atmospheres?

(c) What is the value of the Mach number for the atmospheres of Earth, Venus, Jupiter and a hot Jupiter? Do you expect shocks to develop in these atmospheres?

(d) What is the value of the Rossby number for the atmospheres of Earth, Venus, Jupiter and a hot Jupiter? In which of these atmospheres is rotation expected to be important?

8.6.4   Classic solutions for viscous flows

Viscous flow between a pair of infinite plates and through a cylindrical pipe are classic examples of toy models often used to teach fluid dynamics. They are also used to illustrate that much of the physics is contained within the boundary conditions.

(a) Consider two parallel plates that extend indefinitely in the x-direction. The y-axis is oriented perpendicular to the surfaces of these plates. We locate $y = 0$ at one of these surfaces. The two plates are separated by $y = d$. Fluid flows between these plates, but only in the x-direction. Assume that the mass density is constant in space. Consider this two-dimensional system in the steady-state, incompressible limit without gravity, rotation or magnetic fields. Using the Navier-Stokes equation, solve for the function form of v_x, which is the x-component of the velocity. Apply the boundary conditions,[5]

$$
\begin{aligned}
v_x\left(y = 0\right) &= 0, \\
v_x\left(y = d\right) &= v_0,
\end{aligned}
\tag{8.31}
$$

where v_0 is an arbitrary constant. If the pressure is constant everywhere between the plates, then where does the fluid flow the fastest?

(b) Consider fluid flowing through a pipe. We use cylindrical coordinates to describe this flow. The z-axis is oriented parallel to the pipe. We choose $r = 0$ to be at the center of the pipe. Assume that flow only occurs in the z-direction and is symmetric about ϕ. Assume again that the mass density is constant in space. Consider this three-dimensional system in the steady-state, incompressible limit

---

[5]Physically, one of the plates has the "no slip" boundary condition, where the speed transitions from some finite value to being zero at its surface, while the other plate has the "free slip" boundary condition, where its surface is frictionless.

without gravity, rotation or magnetic fields. Using the Navier-Stokes equation, solve for the functional form of v_z, which is the z-component of the velocity. Apply the "no slip" boundary condition,

$$v_z \left(r = r_0 \right) = 0, \tag{8.32}$$

where r_0 is the radius of the pipe. Also apply the secondary boundary condition that v_z needs to be finite at the center of the pipe. What is the speed of the flow when the pressure is constant everywhere?

8.6.5 A thought experiment

Imagine that you are observing a system at length scales much larger than the mean free path for collisions between atoms or molecules. However, you observe it between time intervals that are much shorter than the time scale associated with these collisions. Can the system still be regarded as a fluid?

8.6.6 The Boltzmann equation

Show that the mass continuity and Navier-Stokes equations are limiting cases of the Boltzmann equation, when it is meaningful to define ensemble averages of quantities.

Chapter Nine

Deriving the Governing Equations of Fluid Dynamics

9.1 PREAMBLE

The derivation of the governing equations of fluid dynamics has a rich history and has been treated in numerous textbooks [105, 134, 244]. In fact, the Navier-Stokes equation remains one of the unsolved Millennium Prize Problems (the solution of which comes with the reward of a million dollars). It is not my intention to repeat these detailed derivations. Rather, the spirit of this textbook is to opt for short, physically intuitive derivations to complement these existing ones. Intuitive descriptions and discussions of the governing equations were previously presented in Chapter 8.

9.2 THE MASS CONTINUITY EQUATION (MASS CONSERVATION)

Consider an infinitesimal element within the fluid. It does not even need to have a regular shape. Its volume is denoted by V. The mass enclosed by this fluid element is

$$M_{\text{element}} = \int \rho \, dV, \tag{9.1}$$

where ρ is the mass density of the enclosed fluid. The fluid element sits in the laboratory frame; fluid enters and exits it. The rate of change of mass is

$$\dot{M}_{\text{element}} = \int \frac{\partial \rho}{\partial t} \, dV. \tag{9.2}$$

The mass flux of fluid through all faces of the element is

$$F_M = \oint \rho \vec{v}.d\vec{A} = \int \nabla.(\rho \vec{v}) \, dV, \tag{9.3}$$

where the surface areas of the faces is represented by $\vec{A}$ and $\vec{v}$ is the velocity of the fluid. The second equality in the preceding equation is a consequence of using Gauss's theorem.

Since the rate of change of mass is related to the mass flux by $\dot{M}_{\text{element}} = -F_M$ (conservation of mass) and this relationship must hold for any volume, we have

$$\frac{\partial \rho}{\partial t} + \nabla.(\rho \vec{v}) = 0. \tag{9.4}$$

Among the conservation equations for a fluid, this is the easiest one to derive.

9.3   THE NAVIER-STOKES EQUATION (MOMENTUM CONSERVATION)

We precede our derivation of the Navier-Stokes equation by stating the *generalized Leibniz theorem*, sometimes known as the *Reynolds transport theorem* [134],

$$\frac{D}{Dt} \int \vec{X} \, dV = \int \frac{\partial \vec{X}}{\partial t} \, dV + \oint \vec{X} \, \vec{v}.d\vec{A}. \qquad (9.5)$$

The first term describes the change of any fluid property $\vec{X}$ within the element, while the second term describes the change in shape of the bounding surfaces.

Next, we specialize to $\vec{X} = \rho\vec{v}$. It follows that

$$\frac{D}{Dt} \int \rho\vec{v} \, dV = \int \rho\frac{D\vec{v}}{Dt} \, dV, \qquad (9.6)$$

via the application of Gauss's theorem on equation (9.5) and the use of the mass continuity equation.

Physically, we expect equation (9.6) to equate to two types of forces: body and surface forces. Body forces arise from "action at a distance" and do not require physical contact. They are generally attributed to force fields such as gravity and electromagnetism. Surface forces are established through direct contact with the fluid element and are usually expressed as a force per unit area. Using this reasoning, we have

$$\int \rho\frac{Dv_i}{Dt} \, dV = \int F_i \, dV + \oint S_{ij} \, dA_j. \qquad (9.7)$$

Notice that we now label each component of the velocity by v_i. For example, we have $i = x, y, z$ in Cartesian coordinates. Correspondingly, each component of the surface area of the fluid element is indexed by j. The reason to use this notation is that the quantity S_{ij} is the *stress tensor*, which depends on both indices i and j. (It is usually easier and more elegant to write tensors in terms of their indices.) Following the Einstein notation convention,[1] repeated indices imply a summation over them, i.e., we are summing over the j index. The body force per unit volume is denoted by F_i.

Using Gauss's theorem and the reasoning that equation (9.7) must hold for any volume, we obtain

$$\rho\frac{Dv_i}{Dt} = F_i + \frac{\partial S_{ij}}{\partial r_j}, \qquad (9.8)$$

where r_j is a component of the position vector. The preceding equation is sometimes known as *Cauchy's equation of motion* [134]. For now, the only body force we will consider is gravity,

$$F_i = \rho g_i. \qquad (9.9)$$

---

[1] Essentially, for the vector X_j and tensor Y_{ij}, we write $\sum_j X_j Y_{ij}$ simply as $X_j Y_{ij}$ and omit the summation sign out of laziness.

We shall state, without proof,[2] the expression for the stress tensor S_{ij} [134],

$$S_{ij} = -\delta_{ij}\left(P + \frac{2\mu_\nu}{3}\nabla.\vec{v}\right) + 2\mu_\nu W_{ij}, \qquad (9.10)$$

where $\mu_\nu \equiv \rho\nu$ is the *dynamic viscosity* and ν is the *kinematic viscosity*. As far as possible, I will use ν as I consider it to be a more fundamental quantity. The first and second terms after the equality are the isotropic components of the stress tensor: pressure and viscous compression, respectively. If S_{ij} is viewed as a matrix, they comprise its diagonal elements. The third term accounts for the shearing of adjacent fluid layers and makes up the off-diagonal matrix elements.

The *Kronecker delta* is a convenient quantity that is defined as

$$\delta_{ij} = \begin{cases} 0, & i \neq j, \\ 1, & i = j. \end{cases} \qquad (9.11)$$

The quantity W_{ij} is the *strain rate tensor* [134]. To express it in terms of other quantities requires that we understand the concept of Newtonian versus non-Newtonian fluids. Physically, a *Newtonian fluid* has no memory of its previous states and offers more resistance when sheared. Mathematically, the strain rate tensor for a Newtonian fluid takes on a simple form,

$$W_{ij} = \frac{1}{2}\left(\frac{\partial v_i}{\partial r_j} + \frac{\partial v_j}{\partial r_i}\right), \qquad (9.12)$$

where the velocity gradient $\frac{\partial v_i}{\partial r_j}$ is termed the *linear* or *normal strain rate* if $i = j$ [134]. It is the *shear strain rate* if $i \neq j$ [134]. Essentially, one is assuming that the strain rate tensor is linearly proportional to the strain rates.

Curiously, several common fluids are distinctively non-Newtonian. Any baker will tell you that dough has a memory of its previous states. Ketchup flows more readily when shear is applied to it. Blood, toothpaste and paint are other examples of non-Newtonian fluids. When using the Navier-Stokes equation to model exoplanetary atmospheres, one should always be mindful of these built-in assumptions. A theoretical model is only as good as its assumptions, and one should question whether they apply to the situation one is interested in representing. A good theorist knows when his or her model breaks.

---

[2]The stress tensor is first separated into its isotropic (diagonal) and non-isotropic (off-diagonal) components. The former is the pressure, while the latter is the *deviatoric stress tensor*. One assumes a linear relationship between the deviatoric stress tensor and the strain rate tensor, with the quantity of proportionality being a fourth-order tensor. By demanding that the stress-strain relationship is invariant to rotation, one may show that the fourth-order tensor is isotropic, which allows one to express it in terms of three pairs of Kronecker deltas. Each pair of Kronecker deltas is associated with a scalar. This set of three scalars may be reduced to one scalar (the dynamic viscosity) by asserting that the deviatoric stress tensor is symmetric and invoking *Stokes's assumption* or *hypothesis*. By following through on the contraction of indices demanded by these Kronecker deltas, one obtains the final expression for the stress tensor.

Combining equations (9.7) and (9.10) and reverting to regular vector notation, we obtain

$$\frac{D\vec{v}}{Dt} = \vec{g} - \frac{\nabla P}{\rho} + \nu \nabla^2 \vec{v} + \frac{\nu}{3} \nabla \left(\nabla . \vec{v} \right). \tag{9.13}$$

Additionally, the $-2\vec{\Omega} \times \vec{v}$ term comes from transforming the system into the co-rotating frame.

9.4 THE THERMODYNAMIC EQUATION (ENERGY CONSERVATION)

The thermodynamic equation is a statement of energy conservation and should not be confused with the equation of state. The latter expresses a relationship between the temperature (T), pressure (P) and mass density (ρ). Throughout the textbook, our assumption for the equation of state of an atmosphere is that of an ideal gas: $P = \rho k_{\mathrm{B}} T / m$, where k_{B} is the Boltzmann constant and m is the mean molecular mass.

The first law of thermodynamics states that the heat content of a system is the sum of its internal energy and work done due to changes in volume (often termed "PdV" work),

$$dq = c_V \ dT + P \ d \left(\frac{1}{\rho} \right), \tag{9.14}$$

where c_V is the specific heat capacity at constant volume and $V = 1/\rho$ is the *specific volume* (volume per unit mass).

Transforming into the fluid reference frame and invoking the standard thermodynamic relations (see Appendix C), we obtain

$$\frac{DT}{Dt} = \frac{\kappa_{\mathrm{ad}} T}{P} \frac{DP}{Dt} + \frac{Q}{c_P}, \tag{9.15}$$

where κ_{ad} is the adiabatic coefficient and

$$Q \equiv \frac{Dq}{Dt} \tag{9.16}$$

is the specific power (energy per unit time per unit mass).

A common source of confusion arises from the fact that equation (9.15) may be recast to solve for the pressure and internal energy (per unit volume),

$$E_{\mathrm{int}} = \frac{P}{\gamma_{\mathrm{ad}} - 1}, \tag{9.17}$$

where γ_{ad} is the adiabatic gas index, or the total energy, which is the sum of the internal, potential and kinetic energies (per unit volume),

$$E = E_{\mathrm{int}} + E_{\mathrm{g}} + E_{\mathrm{k}}, \tag{9.18}$$

where $E_{\mathrm{g}} \equiv \rho g r$ is the potential energy per unit volume, $E_{\mathrm{k}} \equiv \rho v^2 / 2$ is the kinetic energy per unit volume and r is the radial coordinate.

It is possible to show that the equation governing the pressure is (see Problem 9.8.2)

$$\frac{\partial P}{\partial t} + \nabla . \left(P \vec{v} \right) = \left(1 - \gamma_{\mathrm{ad}} \right) P \nabla . \vec{v} + \rho \left(\gamma_{\mathrm{ad}} - 1 \right) Q. \tag{9.19}$$

Obtaining the governing equation for the total energy is trickier and involves several intermediate steps. First, the mass continuity equation is re-expressed in terms of E_{g}, yielding

$$\frac{\partial E_{\mathrm{g}}}{\partial t} + \nabla . \left(E_{\mathrm{g}} \vec{v} \right) = \rho g v_r, \tag{9.20}$$

where v_r is the radial component of the velocity (assuming spherical coordinates). Next, the Navier-Stokes equation is recast in terms of E_{k} (see Problem 9.8.3),

$$\frac{\partial E_{\mathrm{k}}}{\partial t} + \nabla . \left(E_{\mathrm{k}} \vec{v} \right) = -\rho g v_r - \vec{v} . \nabla P. \tag{9.21}$$

Adding equations (9.19) (transformed into the equation for internal energy, rather than for pressure), (9.20) and (9.21) yields

$$\frac{\partial E}{\partial t} + \nabla . \left(E \vec{v} \right) = -\nabla . \left(P \vec{v} \right) + \rho Q. \tag{9.22}$$

In principle, all of these different governing equations are equivalent and arise from the conservation of energy. Generally, they are not amenable to analytical solution. In practice, numerical[3] methods are required to solve the coupled set of mass, momentum and thermodynamic equations.

9.5   THE CONSERVATION OF POTENTIAL VORTICITY

The vorticity of a fluid is generally not a conserved quantity. Rather, it may be shown that a quantity known as the *potential vorticity* is generally invariant across time. What is less appreciated is that the definition of the potential vorticity, as well as its general conservation, derives naturally from considering the Navier-Stokes equation. The conservation of potential vorticity is thus equivalent to the conservation of momentum.

9.5.1   Basic derivation: Euler equation with no rotation

By writing the vorticity as $\vec{\omega} \equiv \nabla \times \vec{v}$, one may show that [244]

$$\vec{v} . \nabla \vec{v} = \frac{\nabla v^2}{2} + \vec{\omega} \times \vec{v}. \tag{9.23}$$

---

[3]Whether to adopt the governing equation for T, P, E_{int} or E depends on which quantity we choose to conserve to machine precision.

We ignore the terms associated with the Coriolis force and molecular viscosity for now; we will consider them later. By taking the curl of the Navier-Stokes equation, we obtain

$$\frac{\partial \vec{\omega}}{\partial t} + \nabla \times (\vec{\omega} \times \vec{v}) = -\nabla \times \left(\frac{\nabla P}{\rho}\right) + \nabla \times \vec{g} - \cancel{\nabla \times \left(\frac{\nabla v^2}{2}\right)}. \qquad (9.24)$$

Since we are assuming the acceleration due to gravity to be constant throughout the atmosphere, we have $\nabla \times \vec{g} = 0$. Also, the last term in the preceding equation is zero by definition, since the curl of the gradient of a scalar is always zero. Similarly, the term associated with the pressure gradient may be expanded,

$$\nabla \times \left(\frac{\nabla P}{\rho}\right) = \cancel{\frac{\nabla \times \nabla P}{\rho}} - \frac{\nabla \rho \times \nabla P}{\rho^2}, \qquad (9.25)$$

and it may be recognized that the first term after the equality vanishes.

To proceed further, we need to expand the following term,

$$\nabla \times (\vec{\omega} \times \vec{v}) = \vec{v}.\nabla\vec{\omega} - \vec{\omega}.\nabla\vec{v} + \vec{\omega}(\nabla.\vec{v}) - \cancel{\vec{v}(\nabla.\vec{\omega})}. \qquad (9.26)$$

The last term in the preceding equation vanishes, because we always have $\nabla.(\nabla \times \vec{v}) = 0$ (for any vector, not just $\vec{v}$). It is important to note that we do not assume $\vec{\omega}(\nabla.\vec{v})$ to vanish, because we are generally interested in compressible flows.

By using the identity in equation (9.26) and invoking mass continuity, we obtain

$$\frac{D\vec{\omega}'}{Dt} = \vec{\omega}'.\nabla\vec{v} + \frac{\nabla \rho \times \nabla P}{\rho^3}. \qquad (9.27)$$

This is the equation governing $\vec{\omega}' \equiv \vec{\omega}/\rho$. By itself, it is not enough to demonstrate the conservation of potential vorticity, although it is a key, intermediate step towards this goal. It is, however, enough to reproduce a well-known theorem, often included as part of the standard fare of textbooks on fluid dynamics, known as *Kelvin's circulation theorem* [105, 134]. It is formulated in the limit of an incompressible ($\frac{D\rho}{Dt} = 0$) and barotropic ($\nabla \rho \times \nabla P = 0$) fluid. The former property allows us to shift factors of ρ around with impunity. Physically, the latter property describes a situation where surfaces of constant density and constant pressure coincide. Equation (9.27) is then integrated over a bounding surface $\vec{A}$. Using Stokes's theorem (see Appendix B), we obtain Kelvin's circulation theorem,

$$\frac{D}{Dt} \oint \vec{v}.d\vec{l} = \int [(\nabla \times \vec{v}).\nabla\vec{v}].d\vec{A} = 0, \qquad (9.28)$$

where $\vec{l}$ is the curve bounding the surface $\vec{A}$. The integral involving $\vec{A}$ vanishes, because for every component of the integrand one can always find an opposing component that cancels it out—in other words, it is an odd integral. Physically, the circulation of fluid within a closed curve remains invariant across time. It is a prelude to potential vorticity conservation, but is much less general because of the approximations taken.

The last step requires a series of somewhat subtle mathematical manipulations, but has profound physical implications. Consider some scalar quantity χ. One may show that [244]

$$(\vec{\omega}'.\nabla\vec{v}).\nabla\chi = (\vec{\omega}'.\nabla)\frac{D\chi}{Dt} - \vec{\omega}'.\frac{D(\nabla\chi)}{Dt}. \tag{9.29}$$

Next, we take the dot product of equation (9.27) with $\nabla\chi$,

$$\nabla\chi.\frac{D\vec{\omega}'}{Dt} = (\vec{\omega}'.\nabla\vec{v}).\nabla\chi + \frac{\nabla\chi}{\rho^3}.(\nabla\rho \times \nabla P). \tag{9.30}$$

By eliminating the first term after the equality via the use of equation (9.29), we obtain

$$\frac{D(\vec{\omega}'.\nabla\chi)}{Dt} = (\vec{\omega}'.\nabla)\frac{D\chi}{Dt} + \frac{\nabla\chi}{\rho^3}.(\nabla\rho \times \nabla P). \tag{9.31}$$

What exactly is the physical quantity χ? We may anticipate that it should be a quantity that makes the baroclinic[4] term $(\nabla\rho \times \nabla P)$ vanish. Taking $\chi = T$ does not do the job, because there is no *general* reason for surfaces of constant temperature to align with surfaces of constant density or pressure. Adopting $\chi = \rho$ allows the baroclinic term to vanish, but at the expense of effectively assuming incompressibility in order to set the other term to zero ($\frac{D\rho}{Dt} = 0$). This is clearly not a general choice for χ. It turns out that a more general choice is the potential temperature, $\chi = \Theta \propto TP^{-\kappa_{ad}}$. By assuming an ideal gas, we obtain

$$\nabla\chi = C_1\nabla\rho + C_2\nabla P, \tag{9.32}$$

where the exact functional forms of the coefficients C_1 and C_2 are inconsequential for our purposes—what matters is that we generically obtain $\nabla\chi.(\nabla\rho \times \nabla P) = 0$. By demanding that the potential temperature is invariant with time ($\frac{D\Theta}{Dt} = 0$), which is equivalent to assuming an adiabatic system, we obtain

$$\frac{D\Phi}{Dt} = 0 \text{ where } \Phi \equiv \vec{\omega}'.\nabla\Theta. \tag{9.33}$$

The conservation of potential vorticity is the fluid dynamical equivalent of inertia—the amount of "spin" a fluid element has remains constant. What we have just derived is the exact form of what "spin" means when gradients of temperature, pressure and density exist in a general manner. The freedom to choose the exact form of χ is also the reason why several definitions of the potential vorticity exist in the literature.

---

[4]In the present context, one may take "baroclinic" to mean "not barotropic."

9.5.2 Potential vorticity conservation and other physical effects

What happens when we consider other physical effects? The most important of these is rotation. If the Coriolis term is added back in, we obtain

$$\frac{\partial \vec{\omega}}{\partial t} + \nabla \times (\vec{\omega} \times \vec{v}) = -\nabla \times \left(\frac{\nabla P}{\rho}\right) - \nabla \times \left(2\vec{\Omega} \times \vec{v}\right). \tag{9.34}$$

Notice how the terms involving $\vec{\omega}$ and $2\vec{\Omega}$ have exactly the same mathematical form. It immediately implies that we may simply *add* them together to obtain a more general form of the potential vorticity,

$$\Phi \equiv \frac{\vec{\omega} + 2\vec{\Omega}}{\rho} \cdot \nabla\Theta, \tag{9.35}$$

as we previously stated in Chapter 8. However, this requires us to assume that

$$\frac{\partial \vec{\Omega}}{\partial t} = 0, \tag{9.36}$$

so that we may simply replace $\vec{\omega}$ with $\vec{\omega} + 2\vec{\Omega}$ in the subsequent steps of the derivation. This is not an unreasonable assumption, as we do not expect the rate of rotation of the exoplanet to change on dynamical time scales.

When the term involving molecular viscosity ($\nu\nabla^2\vec{v}$) is added, we obtain

$$\frac{D\vec{\omega}'}{Dt} = \vec{\omega}'.\nabla\vec{v} + \frac{\nabla\rho \times \nabla P}{\rho^3} + \frac{\nu\nabla^2\vec{\omega}}{\rho}. \tag{9.37}$$

Since there is generally no reason for $\nabla^2\vec{\omega}.\nabla\chi$ to vanish, we conclude that molecular viscosity breaks the conservation of potential vorticity. Physically, we expect any form of friction or dissipation to violate potential vorticity conservation.

We reach the same conclusion for magnetic tension. Generally, we expect

$$\frac{1}{\rho}\left[\nabla \times \left(\frac{\vec{B}.\nabla\vec{B}}{4\pi\rho}\right)\right].\nabla\chi \neq 0, \tag{9.38}$$

implying that magnetic tension would also generally violate potential vorticity conservation.

9.5.3 Putting it all together: When does potential vorticity conservation hold?

It is useful to summarize the conditions under which potential vorticity conservation holds (or does not).

- Gravity is assumed to have a constant acceleration throughout the atmosphere. It is also assumed that the atmosphere obeys the ideal gas law.

- The potential temperature is assumed to be invariant across time. If the specific heat capacity (at constant pressure) is constant throughout the atmosphere, then it implies that the specific entropy is invariant.

- The rotation of the exoplanet (expressed via the Coriolis term) may be added to the definition of the potential vorticity, but it requires the assumption that the angular frequency of rotation is constant with time.

- The fluid is generally *not* assumed to be incompressible or barotropic. These assumptions may be made, but they are not required.

- Molecular viscosity and magnetic tension generally cause potential vorticity conservation to be violated.

9.6 VARIOUS APPROXIMATE FORMS OF THE GOVERNING EQUATIONS OF FLUID DYNAMICS

It is very challenging to solve the set of mass continuity, Navier-Stokes and thermodynamic equations. Frequently, one has to adopt various approximations to render the set of equations amenable to analytical solution or reduce the difficulty of solving them numerically. The approximation is usually, but not always, taken on the Navier-Stokes equation. In this section, we will review the various approximate forms of the governing equations.

9.6.1 The Euler equation

As mentioned in Chapter 8, the Reynolds number plays a central role in the behavior of a fluid and determines if it is laminar ($\mathcal{R}_e \ll 1$) or turbulent ($\mathcal{R}_e \gg 1$). In atmospheres, the Reynolds number is typically enormous,

$$\mathcal{R}_e = \frac{Ul}{\nu} \sim 10^{11}\text{--}10^{14}, \qquad (9.39)$$

regardless of whether we have slow winds ($U \sim 1\text{--}10$ m s^{-1}) for Solar System bodies or fast winds ($U \sim 1$ km s^{-1}) for hot exoplanets. The typical length scale is the (exo)planetary radius, which is $l \sim 10^8\text{--}10^{10}$ cm. The molecular viscosity is $\sim 0.1\text{--}10$ cm^2 s^{-1}, since

$$\nu \sim l_{\mathrm{mfp}} c_s = \frac{1}{n\sigma_{\mathrm{coll}}} \left(\frac{\gamma_{\mathrm{ad}} k_B T}{m} \right)^{1/2}, \qquad (9.40)$$

where $l_{\mathrm{mfp}} = 1/n\sigma_{\mathrm{coll}}$ is the mean free path for collisions, n is the number density and $\sigma_{\mathrm{coll}} \sim 10^{-15}$ cm^2 is the cross section for atomic/molecular collisions. Air at room temperature has $\nu \sim 0.1$ cm^2 s^{-1}, whereas hot hydrogen gas at $T \sim 10^3$ K and $P \sim 1$ bar has $\nu \sim 10$ cm^2 s^{-1}. Even if the typical length scale is interpreted to be the pressure scale height, the Reynolds number is still much, much larger than unity.

When $\mathcal{R}_e \gg 1$, we obtain the *Euler equation* [134, 244],

$$\frac{D\vec{v}}{Dt} = \vec{g} - \frac{\nabla P}{\rho} - 2\vec{\Omega} \times \vec{v} + \vec{F}_{\text{others}}, \tag{9.41}$$

where $\vec{F}_{\text{others}}$ is again the force per unit mass associated with other physical effects. In Solar System atmospheres, it typically includes source/sink terms attributed to turbulence and the *planetary boundary layer*, which is the "no slip" interface between the atmosphere and the surface where friction is energetically important.

9.6.2   The primitive equations of meteorology

Originally developed for the atmosphere of Earth, the *primitive equations of meteorology* describe a reduced form of the mass and momentum conservation equations in spherical geometry [244],

$$\frac{\partial \rho}{\partial t} + \frac{\partial}{\partial r}(\rho v_r) + \frac{1}{R \sin \theta}\left[\frac{\partial}{\partial \theta}(\rho v_\theta \sin \theta) + \frac{\partial}{\partial \phi}(\rho v_\phi)\right] = 0,$$

$$\frac{Dv_\theta}{Dt} - \frac{v_\phi^2}{R \tan \theta} - 2\Omega v_\phi \cos \theta = -\frac{1}{\rho R}\frac{\partial P}{\partial \theta}, \tag{9.42}$$

$$\frac{Dv_\phi}{Dt} + \frac{v_\theta v_\phi}{R \tan \theta} + 2\Omega v_\theta \cos \theta = -\frac{1}{\rho R \sin \theta}\frac{\partial P}{\partial \phi},$$

where R is the radius of the exoplanet and the velocity is $\vec{v} = (v_r, v_\theta, v_\phi)$. The radial, polar and azimuthal coordinates are given by r, θ and ϕ, respectively. Problem 9.8.4 works through the derivation of the governing equations in spherical geometry, an intermediate and essential step towards deriving the primitive equations, and finally the derivation of the primitive equations themselves. We have not stated the thermodynamic equation, as it is identical to equation (8.13) and we do not have to deal with the subtleties of vector calculus in spherical geometry, unlike for the momentum equation.

Arriving at equation (9.42) involves invoking a set of assumptions, which we will now describe.

- **Hydrostatic balance:** Hydrostatic equilibrium describes a state in which gravity and pressure support are the dominant and opposing forces along the radial or vertical direction of the atmosphere. In the context of the primitive equations, it means that terms associated with the radial or vertical component of the velocity, the Coriolis effect and spherical geometry (so-called *metric terms*) are neglected.

- **Shallow atmosphere:** The radial spatial coordinate is expressed as

$$r = R + z, \tag{9.43}$$

which naturally means $\frac{\partial}{\partial r} = \frac{\partial}{\partial z}$, since the radius of the exoplanet is a constant. This specific approximation requires that in places where we have $1/r$, we replace it with $1/R$ on the premise that $R \gg z$.

- **Traditional approximation:** Coriolis and metric terms involving the radial component of the velocity are assumed to be small and may be neglected.

The primitive equations are an excellent approximation for the Earth's atmosphere and are the frequent starting point of *general circulation models* (GCMs), which are three-dimensional numerical simulations of the circulation of air in an atmosphere (atmospheric dynamics). Numerical solvers that focus exclusively on solving the primitive equations are often termed *dynamical cores*. GCMs use the dynamical core as a basis for adding physics and chemistry: radiation, precipitation, aerosol formation and distribution, etc.

9.6.3 The shallow water equations

As the name suggests, the *shallow water equations* assume that the horizontal dimensions of the atmosphere are much larger than its vertical one. They are amenable to analytical solution, which we will explore extensively in Chapter 10, because of a clever trick: instead of considering a thermodynamic equation, an equation of state (the ideal gas law, in our case) and the mass continuity equation, these considerations are combined in the form of a single equation for the shallow water height, denoted by h. One ends up with only a pair of equations to solve,

$$
\begin{aligned}
\frac{D\vec{v}}{Dt} &= -g\nabla h - 2\vec{\Omega} \times \vec{v} + \nu\nabla^2\vec{v} + \vec{F}_{\text{others}}, \\
\frac{\partial h}{\partial t} &+ \nabla . (h\vec{v}) = 0.
\end{aligned}
\tag{9.44}
$$

Note that the preceding equations are only meaningful in the two horizontal dimensions.

Notice that the pressure, temperature and mass density do not appear in the preceding equations. Instead, they have all been "absorbed" into h. Also notice how the governing equation for h is identical to the mass continuity equation, except that ρ is replaced by h. As we will see in Chapter 10, the equation for h derives from the condition of incompressibility ($\nabla.\vec{v} = 0$), which explains why the momentum equation does not have the term associated with viscous compression. Furthermore, the pressure gradient and gravity terms have been replaced by a single term ($-g\nabla h$), which arises from the assumption of hydrostatic equilibrium.

Despite these simplifications, the shallow water system remains a very useful *theoretical* laboratory for developing intuition for the global structure of an atmosphere and the waves contained within it. Surprisingly, it remains amenable to analytical solution even when a host of effects are considered: rotation, viscosity, radiative heating ("forcing") and even magnetic fields. This allows us to study the different types of waves produced as a consequence of the different physical effects coupling to one another.

9.6.4   The Boussinesq and anelastic approximations

The *Boussinesq approximation* applies to flows driven primarily by buoyancy. It is assumed that the background state has a constant density. Deviations in density contribute negligibly to inertia and are assumed to be small, unless they are mediated by gravity. Sound waves, which feed off density variations, are filtered out by such an approximation—essentially, the sound speed becomes infinite.

Consider a hydrostatic background state of constant mass density (ρ_0) and pressure (P_0). The total mass density and pressure are given by

$$\rho = \rho_0 + \rho', \ P = P_0 + P'. \tag{9.45}$$

The perturbations are assumed to be small ($\rho' \ll \rho_0$, $P' \ll P_0$). The Navier-Stokes equation becomes

$$\frac{D\vec{v}}{Dt} = \frac{\rho'\vec{g}}{\rho_0} - \nabla\left(\frac{P'}{\rho_0}\right) - 2\vec{\Omega} \times \vec{v}. \tag{9.46}$$

The preceding equation is the *Boussinesq equation for momentum conservation*. The quantity $-g\rho'/\rho_0$ is sometimes called the *buoyancy* [244]. Physically, the essence of the Boussinesq approximation is to ignore density perturbations except where they are associated with gravity.

The Boussinesq approximation may only be used to model systems where the background density (ρ_0) is approximately constant. Even for the atmosphere of Earth, this is a lousy approximation, unless one is interested in a patch of atmosphere smaller than a pressure scale height (which is ~ 10 km). It has been used to approximately describe the planetary boundary layer of Earth, which is the thin (~ 100 m) layer of atmosphere directly above its surface [105]. It is a good approximation for modeling the Earth's oceans, since the dynamically active layer (called the *oceanic mixed layer*) is only ~ 10–100 m thick and has a roughly constant density. The Boussinesq approximation has also been used to model convection in the Earth's mantle—the domain of geophysicists.

The *anelastic approximation* takes this one step farther by allowing the background state of density to be non-constant. Instead of solving for deviations in density, one solves for deviations in the specific entropy [244]. We will not pursue the anelastic approximation further, except to note that it has been used to understand fluid motions in the Sun and stars in general.

Both the Boussinesq and anelastic approximations are not broadly applicable to exoplanetary atmospheres, especially when wind speeds approach the speed of sound. Naturally, both approximations are incapable of accounting for the development of shocks in the atmospheric flow, since this requires that the sound speed be explicitly calculated. Supersonic (faster than sound) flows are a necessary, but insufficient, condition for shock formation, as we will see in Chapter 11.

9.6.5   The diffusion approximation

Purely diffusing fluids are never encountered in exoplanetary atmospheres as they require that the Reynolds number be less than unity ($\mathcal{R}_e < 1$), but it is worth listing this limit for completeness,

$$\frac{D\vec{v}}{Dt} = \nu\nabla^2\vec{v}. \tag{9.47}$$

In calculations of atmospheric chemistry, it is not uncommon to encounter crude treatments of atmospheric dynamics where all fluid motion is parametrized by a diffusion coefficient. Even though the fluid dynamics is poorly described by the diffusion approximation, the passage of radiation through the atmosphere (radiative transfer) sometimes resembles diffusion, as we have seen in Chapter 3.

9.7   MAGNETOHYDRODYNAMICS

9.7.1   Characteristic length scales

To understand magnetohydrodynamics (MHD), we first need to understand what a *plasma* is. Often described as being the fourth state of matter, it occurs when the temperature is high enough that the electrons are no longer bound to their nuclei. On small enough scales, charge separation occurs. A charged particle tends to attract other charged particles with the opposing charge, such that the effective charge of the ensemble, from a distance, is reduced overall, a phenomenon known as *charge shielding*. On appropriately-large scales, the plasma exhibits charge neutrality *and* is an electrical conductor. A plasma experiences all of the fluid dynamics of normal fluids and the additional collective behavior that comes from the ability to be coupled to a magnetic field, which requires that more characteristic length scales than a hydrodynamic description are needed to develop qualitative understanding.

What is the length scale on which charge neutrality occurs? In Gaussian or cgs units,[5] the current density is the flow of electric charge[6] per unit time. Now, imagine that there is an ensemble of positively and negatively charged particles (e.g., protons and electrons) that may be collectively described by a number density n. The net charge of the ensemble is zero. If we denote a unit of electric charge by e, then the energy per unit area associated with it is ne^2.

---

[5] Centimeters, grams and seconds.
[6] The magnetic field strength has the same physical units as the electric charge.

Consider an imaginary sphere around a collection of these particles, which are in thermodynamic equilibrium and may be described by a temperature T. If the radius of the sphere is denoted by l_D, then the total energy encompassed by it is $4\pi l_D^2 ne^2$. If we equate the energy associated with the electric charge with the thermal energy, then we obtain the *Debye length*,

$$l_D = \frac{1}{e}\left(\frac{k_B T}{4\pi n}\right)^{1/2}, \tag{9.48}$$

where k_B is Boltzmann's constant. Beyond the Debye length, the charges of the particles are electrically screened—it is essentially the characteristic length scale of electrostatic effects. We expect that the mean free path is much larger than the Debye length, such that collisions do not disrupt the collective behavior of the particles.

A related quantity is the *plasma parameter*, which is the number of particles contained within this sphere,

$$\Lambda \equiv \frac{4\pi n l_D^3}{3}. \tag{9.49}$$

We expect that $\Lambda \gg 1$, which immediately implies that the spacing between particles is much smaller than the Debye length.

On length scales that are much larger than the Debye length and mean free path for collisions, the plasma may be described by MHD. It is also assumed that the fluid is being examined on time scales that are much longer than those associated with the gyrofrequencies of the particles—the reciprocal of the gyrofrequency is roughly the time it takes for a charged particle to spiral around a magnetic field line. Furthermore, the fluid is also examined on length scales that are larger than the gyroradius.

When magnetic fields are present and being "felt" by the fluid, extra terms appear in the momentum equation. To close the set of governing equations, we need an additional one: the induction equation. Collectively, these equations describe MHD. There is a rich body of literature on MHD and it is not my intention to do it justice by reviewing it exhaustively. Rather, we will discuss how the fluid equations are generalized to MHD equations. In Chapter 10, we will use shallow water systems to study the effects of magnetic fields.

9.7.2   Magnetic tension and pressure

In hot exoplanets, we expect partial but widespread ionization of the atmospheric gas from collisional ionization, especially if alkali metals like sodium and potassium are present [14, 188, 189]. If the exoplanet has an intrinsic magnetic field, then partially ionized gas moving past it, in the form of fast winds, would induce an opposing force that works to slow down these winds (Lenz's law). We thus expect magnetic fields to be dynamically important in hot (~ 1000 K) exoplanets and for $\vec{F}_{\text{others}}$ to take the form

$$\vec{F}_{\text{others}} = \frac{\vec{J} \times \vec{B}}{\rho c} = \frac{1}{4\pi\rho} \left(\nabla \times \vec{B} \right) \times \vec{B} = \frac{\vec{B}.\nabla\vec{B}}{4\pi\rho} - \frac{\nabla P_B}{\rho}, \tag{9.50}$$

where the current density, as given by Ampère's law, is

$$\vec{J} = \frac{c}{4\pi}\nabla \times \vec{B}, \tag{9.51}$$

$\vec{B}$ is the magnetic field strength and c is the speed of light. We have used Gaussian units for the electromagnetic quantities.

The third equality in the preceding equation comes from using the identity,

$$\left(\nabla \times \vec{B} \right) \times \vec{B} = \vec{B}.\nabla\vec{B} - \frac{1}{2}\nabla \left(\vec{B}.\vec{B} \right). \tag{9.52}$$

I encourage you to work through Problem 9.8.5, as it will teach you a neat trick for working through vector algebra identities without having to memorize them.

The term associated with $\vec{B}.\nabla\vec{B}$ accounts for *magnetic tension*. It is best understood when the magnetic field is visualized as a taut string, which offers resistance when it is plucked. The *magnetic pressure* is

$$P_B = \frac{B^2}{8\pi}, \tag{9.53}$$

and is usually negligible compared to the gas pressure near the infrared photosphere. In subsequent discussions of the magnetized momentum equation, we will ignore the magnetic pressure term.

9.7.3 The induction equation

Faraday's law of induction, which is one of Maxwell's equations, describes how the magnetic field is evolving due to a number of different effects [10],

$$\frac{\partial\vec{B}}{\partial t} = \nabla \times \left[\vec{v} \times \vec{B} - \frac{4\pi\eta\vec{J}}{c} - \frac{\vec{J} \times \vec{B}}{en_e} + \frac{\left(\vec{J} \times \vec{B} \right) \times \vec{B}}{c\alpha_{\text{drag}}} \right], \tag{9.54}$$

where n_e is the number density of electrons. The exact form of α_{drag} is unimportant for our discussion; we will only note that it contains the mass densities of the neutrals and ions, as well as a drag coefficient [10]. If only the first term after the equality is present, then *ideal MHD* prevails. It describes the situation where the atmosphere is perfectly conducting and the magnetic field lines are perfectly coupled to it. In other words, the magnetic field lines are frozen into the fluid and move along with it; this is known as *Alfvén's theorem*, named after the Swedish Nobel laureate, Hannes Alfvén. Only atmospheric motions perpendicular to the field lines induce a change in the magnetic field. Ideal MHD is not expected to generally prevail in exoplanetary atmospheres.

The other terms after the equality exist when non-ideal MHD is present. The second term is the *Ohmic term*, which accounts for the fluid being an imperfect conductor. If the conductivity of the fluid, associated with electrons, is denoted by σ_e, then the *resistivity* is [110]

$$\eta = \frac{c^2}{4\pi\sigma_e}. \tag{9.55}$$

The third term is associated with the *Hall effect*—magnetic field lines threading an electric current feel a force, which causes them to drift. The last term accounts for *ambipolar diffusion*, which is the phenomenon that the electrons and ions are able to slip past the neutral species of the fluid, because of imperfect coupling. This "ion-neutral slip" produces an extra source of friction.

The Hall effect and ambipolar diffusion are usually subdominant in exoplanetary atmospheres, implying that the induction equation takes the reduced form,

$$\frac{\partial \vec{B}}{\partial t} = \nabla \times \left(\vec{v} \times \vec{B}\right) + \eta\nabla^2\vec{B} - \eta\nabla\left(\nabla.\vec{B}\right). \tag{9.56}$$

We can see the reason why η is sometimes also termed the *magnetic diffusivity*, since it behaves like molecular viscosity in the momentum equation: there is a term associated with the Laplacian of the magnetic field strength and another that resembles viscous compression. The last term after the equality in equation (9.56) is easy to ignore, since we expect that

$$\nabla.\vec{B} = 0, \tag{9.57}$$

unless magnetic monopoles exist. Magnetic monopoles have never been observed in Nature.

If we non-dimensionalize equation (9.56) by typical length (l) and velocity (U) scales, then it becomes

$$\frac{\partial \vec{B}'}{\partial t'} = \nabla' \times \left(\vec{v}' \times \vec{B}'\right) + \frac{1}{\mathcal{R}_B}\nabla'^2\vec{B}', \tag{9.58}$$

where the primed quantities are dimensionless and the *magnetic Reynolds number* is

$$\mathcal{R}_B = \frac{Ul}{\eta}. \tag{9.59}$$

The movement of the charged atmosphere past an ambient magnetic field, with a strength of B_0, induces a current, which in turn induces a secondary magnetic field that works to oppose the original motion. The secondary magnetic field has a strength $\sim \mathcal{R}_B B_0$, the tertiary field $\sim \mathcal{R}_B^2 B_0$ and so on [170]. In principle, this process proceeds ad infinitum, but if the magnetic Reynolds number is very small then these higher-order corrections are insignificant. When $\mathcal{R}_B \gtrsim 1$, a *magnetic dynamo* may operate. Physically, the induced magnetic fields due to advection are strong enough to overcome dissipation due to magnetic diffusivity.

9.8 PROBLEM SETS

9.8.1 Derivation of Navier-Stokes equation

(a) Derive equation (9.6) by applying Gauss's theorem and the conservation of mass on equation (9.5) with $\vec{X} = \rho\vec{v}$.
(b) Using Gauss's theorem, derive equation (9.8).
(c) Derive the stress tensor as stated in equation (9.10).
(d) Derive equation (9.13) by considering equations (9.7) and (9.10).
(e) Show that one obtains an additional $-2\vec{\Omega} \times \vec{v}$ term from transforming into the co-rotating frame.

9.8.2 Derivation of thermodynamic equation

(a) Derive equation (9.15) using the information in Appendix C.
(b) By using Appendix C, show that

$$\frac{1}{\mathcal{R}} - \frac{1}{c_P} = \frac{c_V}{\mathcal{R}c_P}, \quad \kappa_{\mathrm{ad}} = \frac{(\gamma_{\mathrm{ad}} - 1)}{\gamma_{\mathrm{ad}}}. \tag{9.60}$$

(c) Starting from equation (9.15) and using the ideal gas law and the expressions derived in (b), derive equation (9.19).
(d) Hence, write down the governing equation for the internal energy (E_{int}).

9.8.3 Derivation of governing equation for total energy

Recall the Navier-Stokes equation in the co-rotating frame,

$$\rho\frac{D\vec{v}}{Dt} = \rho\vec{g} - \nabla P - 2\rho\left(\vec{\Omega} \times \vec{v}\right). \tag{9.61}$$

(a) First, prove that
$$\rho\frac{D\vec{v}}{Dt} = \frac{D}{Dt}\left(\rho\vec{v}\right) + \rho\vec{v}\left(\nabla.\vec{v}\right). \tag{9.62}$$

(b) Take the dot product of the Navier-Stokes equation with $\vec{v}$. Show that

$$\rho\vec{v}.\frac{D\vec{v}}{Dt} = \frac{D}{Dt}\left(\rho v^2\right) + \rho v^2\left(\nabla.\vec{v}\right) + \rho g v_r + \vec{v}.\nabla P. \tag{9.63}$$

(c) The step before obtaining the governing equation for E_{k} requires us to prove that

$$\frac{\partial}{\partial t}\left(\rho v^2\right) + \rho v^2\left(\nabla.\vec{v}\right) + \vec{v}.\nabla\left(\rho v^2\right) = -2\rho g v_r - 2\vec{v}.\nabla P. \tag{9.64}$$

(d) Finally, show that equation (9.22) obtains from adding equations (9.19), (9.20) and (9.21).

9.8.4 Deriving the primitive equations

Deriving the primitive equations mainly involves casting the Navier-Stokes equation in spherical coordinates and applying the set of assumptions described earlier in the chapter.

(a) The challenge of transitioning from a coordinate-free, vectorial form of the Navier-Stokes equation to stating it in spherical coordinates comes from having to evaluate the term $\vec{v}.\nabla\vec{v}$, since the gradient operator acts not only on the scalar magnitude of the components of $\vec{v}$, but also its unit vectors. Using the identities in Appendix B and by collecting terms associated with $\hat{r}$, $\hat{\theta}$ and $\hat{\phi}$ (the unit vectors in spherical coordinates), show that

$$\vec{v}.\nabla\vec{v}\big|_r = \vec{v}.\nabla v_r - \frac{v_\theta^2 + v_\phi^2}{r},$$

$$\vec{v}.\nabla\vec{v}\big|_\theta = \vec{v}.\nabla v_\theta + \frac{v_r v_\theta}{r} - \frac{v_\phi^2}{r\tan\theta}, \qquad (9.65)$$

$$\vec{v}.\nabla\vec{v}\big|_\phi = \vec{v}.\nabla v_\phi + \frac{v_r v_\phi}{r} + \frac{v_\theta v_\phi}{r\tan\theta},$$

where v_r, v_θ and v_ϕ are the r-, θ- and ϕ-components of the velocity, respectively.

(b) Show that the Coriolis term $(-2\vec{\Omega}\times\vec{v}$, where $\vec{\Omega}$ is the angular frequency of rotation of the exoplanet on its own axis) has the components,

$$-2\vec{\Omega}\times\vec{v}\Big|_r = 2\Omega v_\phi \sin\theta,$$

$$-2\vec{\Omega}\times\vec{v}\Big|_\theta = 2\Omega v_\phi \cos\theta, \qquad (9.66)$$

$$-2\vec{\Omega}\times\vec{v}\Big|_\phi = -2\Omega v_\theta \cos\theta - 2\Omega v_r \sin\theta.$$

(c) Hence, show that

$$\frac{\partial v_r}{\partial t} + v_r\frac{\partial v_r}{\partial r} + \frac{v_\theta}{r}\frac{\partial v_r}{\partial\theta} + \frac{v_\phi}{r\sin\theta}\frac{\partial v_r}{\partial\phi} - \frac{v_\theta^2 + v_\phi^2}{r} = -g - \frac{1}{\rho}\frac{\partial P}{\partial r} + 2\Omega v_\phi \sin\theta. \quad (9.67)$$

Furthermore, show that hydrostatic equilibrium obtains from setting $v_r = 0$ and ignoring the metric and Coriolis terms.

(d) Finally, show that

$$\frac{\partial v_\theta}{\partial t} + v_r\frac{\partial v_\theta}{\partial r} + \frac{v_\theta}{r}\frac{\partial v_\theta}{\partial\theta} + \frac{v_\phi}{r\sin\theta}\frac{\partial v_\theta}{\partial\phi} + \frac{v_r v_\theta}{r} - \frac{v_\phi^2}{r\tan\theta}$$

$$= -\frac{1}{\rho r}\frac{\partial P}{\partial\theta} + 2\Omega v_\phi \cos\theta,$$

$$\frac{\partial v_\phi}{\partial t} + v_r\frac{\partial v_\phi}{\partial r} + \frac{v_\theta}{r}\frac{\partial v_\phi}{\partial\theta} + \frac{v_\phi}{r\sin\theta}\frac{\partial v_\phi}{\partial\phi} + \frac{v_r v_\phi}{r} + \frac{v_\theta v_\phi}{r\tan\theta} \qquad (9.68)$$

$$= -\frac{1}{\rho r\sin\theta}\frac{\partial P}{\partial\phi} - 2\Omega v_\theta \cos\theta - 2\Omega v_r \sin\theta.$$

(e) By taking the approximations associated with the primitive equations (as described earlier in the chapter), derive the equations in (9.42), including the governing equation for the mass density.

(f) Some textbooks prefer to list the primitive equations in terms of latitude and longitude, which requires that the following transformations be made,

$$\theta \to \frac{\pi}{2} - \theta, \ \hat{\theta} \to -\hat{\theta}, \ d\theta \to -d\theta, v_\theta \to -v_\theta. \tag{9.69}$$

Show that

$$\frac{Dv_\theta}{Dt} + \frac{v_\phi^2 \tan\theta}{R} + 2\Omega v_\phi \sin\theta = -\frac{1}{\rho R}\frac{\partial P}{\partial \theta},$$

$$\frac{Dv_\phi}{Dt} - \frac{v_\theta v_\phi \tan\theta}{R} - 2\Omega v_\theta \sin\theta = -\frac{1}{\rho R \cos\theta}\frac{\partial P}{\partial \phi}. \tag{9.70}$$

(Hint: the metric and Coriolis terms associated with $\hat{\theta}$ pick up a negative sign, whereas the θ-component of ∇P picks up *two* negative signs.)

9.8.5 The Levi-Civita symbol

The *Levi-Civita symbol* is a convenient way of expressing vector cross products in tensor notation. For the ith component of $\nabla \times \vec{B}$ (with $\vec{B}$ denoting the magnetic field strength), we have

$$\nabla \times \vec{B}\Big|_i = \epsilon_{ijk}\partial_j B_k, \tag{9.71}$$

where we have ignored the distinction between covariant and contravariant vectors (and written all of our indices as subscripts). Also, we have defined the operator

$$\partial_j \equiv \frac{\partial}{\partial r_j}, \tag{9.72}$$

where r_j is the jth component of the position vector. The Levi-Civita symbol is invariant to cyclic permutations of its indices [7],

$$\epsilon_{ijk} = \epsilon_{kij} = \epsilon_{jki}. \tag{9.73}$$

Anti-cyclic permutations incur a minus sign. A pair of Levi-Civita symbols contract to produce a collection of Kronecker deltas,

$$\epsilon_{imn}\epsilon_{ijk} = \delta_{jm}\delta_{kn} - \delta_{jn}\delta_{km}. \tag{9.74}$$

(a) Express $(\nabla \times \vec{B}) \times \vec{B}$ in terms of Levi-Civita symbols.

(b) By performing contraction of the indices using the Kronecker deltas, prove the identity in equation (9.52).

(c) Starting from equation (9.54) and retaining only the ideal MHD and Ohmic terms, derive equation (9.56) by evaluating the $\nabla \times (\nabla \times \vec{B})$ term using the Levi-Civita symbol.

9.8.6   A myriad of length scales

Consider a system with five length scales: l_D, the Debye length; l_{mfp}, the mean free path for collisions between the constituent particles; l, the typical length scale on which this system is being examined; l_{gyro}, the gyroradius of charged particles in the system; and Δl, the typical spacing between particles. For the system to be considered a fluid, what is the order in which these length scales need to be arranged (from the smallest to the largest)? If the system consists of more than one type of charged particle, what is the additional constraint(s) on the length scales such that it can be considered to be a one-fluid system?

9.8.7   A myriad of length scales II

Cast l_{mfp}/l_D and $\Delta l/l_D$ in terms of Λ. Are all plasmas (magnetized) fluids?

Chapter Ten

The Shallow Water System: A Fluid Dynamics Lab on Paper

10.1 A VERSATILE FLUID DYNAMICS LABORATORY ON PAPER

Beyond idealized, one-dimensional situations, it is generally difficult to obtain analytical (pen-and-paper) solutions of the equations of fluid dynamics. Analytical solutions are useful because it is easier to glean physical intuition from them and one does not have to deal with the subtleties of numerical methods (i.e., choice of grid, solution scheme, sources of numerical dissipation). The tradeoff is that one sacrifices some realism that may be afforded by numerical simulations.

Among the several options available (Boussinesq, anelastic, etc), I have chosen to emphasize the shallow water system of equations because they admit a versatile set of analytical solutions. This versatility allows for gravity, pressure support, rotation, sources of friction, radiative heating and cooling (often termed *forcing*[1]) and even magnetic fields to be studied.

Shallow water models have been used to understand the dynamics of the Earth's ocean [148, 149] and atmosphere [70, 165, 218], solar tachocline [71], atmospheres of stars [260] and neutron stars [85, 231], protoplanetary disks [242] and exoplanetary atmospheres [43, 89, 93, 146, 169, 219]. Despite its name, the shallow water system has more generality beyond studying water.

It is crucial to define what "shallow" actually means. If an exoplanet has a rocky core with a radius of R and the atmosphere has a thickness of ΔR, one may be tempted to think that $\Delta R \ll R$ implies that it is shallow. In fact, the real comparison is between the thickness of the atmosphere and the pressure scale height (H): a shallow atmosphere has $\Delta R \lesssim H$; in this case, it is reasonable to describe it using a two-dimensional model. If $\Delta R \gg H$, then a three-dimensional treatment is necessary. By definition, shallow water models only apply to atmospheres with $\Delta R \lesssim H$. One may also construct multi-layer shallow-water models [219].

In this chapter, our goal is to use the shallow water system to study the variety of waves produced when the atmosphere is perturbed, the properties of these waves and how they influence the global structure of an atmosphere.

---

[1]Radiative forcing is but one of several types of forcing, which may also be mechanical in nature.

10.2 DERIVING THE SHALLOW WATER EQUATIONS

10.2.1 The implications of incompressibility: The governing equation for the shallow water height

The starting point for deriving the shallow water equations is the assumption of incompressibility *in three dimensions,*

$$\nabla.\vec{v} = 0. \tag{10.1}$$

For clarity, we will work in Cartesian coordinates to avoid dealing with geometric terms inherent in spherical geometry, but our conclusions in this section hold regardless of the coordinate system chosen. The preceding expression becomes

$$v_z = -\int_0^h \frac{\partial v_x}{\partial x} + \frac{\partial v_y}{\partial y}\, dz. \tag{10.2}$$

Across the vertical direction, one integrates from the ground or surface ($z = 0$) to the shallow water height ($z = h$). The shallow water height generally varies across the horizontal directions.

We further assume that the horizontal components of the velocity, v_x and v_y, do not depend on the vertical coordinate (z). The vertical component of the velocity (v_z) is generated by a change, in time, of the shallow water height *in the reference frame of a fluid element,*

$$v_z = \frac{Dh}{Dt} = \frac{\partial h}{\partial t} + v_x \frac{\partial h}{\partial x} + v_y \frac{\partial h}{\partial y}. \tag{10.3}$$

The preceding expression also assumes that horizontal advection dominates over its vertical counterpart and/or that the shallow water height does not depend on z. It follows that

$$\frac{\partial h}{\partial t} + v_x \frac{\partial h}{\partial x} + v_y \frac{\partial h}{\partial y} + h \left(\frac{\partial v_x}{\partial x} + \frac{\partial v_y}{\partial y} \right) = 0. \tag{10.4}$$

You will find that a similar expression obtains when using spherical coordinates.

Reverting to vector notation, we obtain

$$\frac{\partial h}{\partial t} + \nabla_{\mathrm{h}}.\,(h\vec{v}) = 0. \tag{10.5}$$

In the preceding expression, it is important to note that the gradient operator is defined only in the horizontal directions—hence, the "h" subscript on ∇_{h}.

Since the shallow water system is incompressible *by construction*, it means that sound waves are excluded or "filtered out" and no perturbations of the mass density may exist. Formally, the sound speed is infinite and supersonic flow may never occur.

10.2.2 The implications of hydrostatic equilibrium: Substituting the pressure gradient with the height gradient

We again use Cartesian coordinates. Hydrostatic equilibrium states that

$$\frac{\partial P}{\partial z} = -\rho g. \tag{10.6}$$

For a given column of atmosphere with a height h, we denote the pressure corresponding to the water surface as P_0. Integrating over the preceding expression yields

$$P = P_0 + \rho g \left(h - z \right). \tag{10.7}$$

We have assumed that the mass density (ρ) is constant across height. Unlike in the usual situation, where the mass density is not constant and obeys an ideal gas law, hydrostatic equilibrium yields a linear relationship between the pressure and shallow water height.

With the expression for the pressure in hand, we may apply the horizontal gradient to it, yielding

$$\frac{\nabla_{\mathrm{h}} P}{\rho} = g \nabla_{\mathrm{h}} h. \tag{10.8}$$

Notice how the pressure, mass density and temperature are "absorbed" into the shallow water height. In one stroke, we have replaced all three quantities by h.

We then substitute this expression into the Navier-Stokes equation *defined only in the horizontal directions*,

$$\frac{\partial \vec{v}}{\partial t} + \vec{v}.\nabla_{\mathrm{h}}\vec{v} = -g\nabla_{\mathrm{h}}h - 2\vec{\Omega} \times \vec{v} + \nu\nabla_{\mathrm{h}}^2\vec{v} + \vec{F}_{\mathrm{others}}. \tag{10.9}$$

We set the term associated with viscous compression to zero,

$$\frac{\nu}{3}\nabla\left(\nabla.\vec{v}\right) = 0, \tag{10.10}$$

because we have previously assumed incompressibility.

10.2.3 Adding other physical effects

Collectively, equations (10.5) and (10.9) describe the shallow water system, which is intrinsically two-dimensional. We now add other physical effects to it. In the momentum equation, these extra effects are contained in the term,

$$\vec{F}_{\mathrm{others}} = -\frac{\vec{v}}{t_{\mathrm{drag}}} + \frac{\vec{B}.\nabla\vec{B}}{4\pi\rho}. \tag{10.11}$$

In the atmospheric science community, the first term after the equality describes *Rayleigh drag*, which mimics macroscopic friction. We will see later that Rayleigh drag is *scale-free*, meaning that it damps small- and large-small motions equally. The second term, as described towards the end of Chapter

9, is associated with *magnetic tension*, which is the intrinsic resistance offered by ambient magnetic field lines to being bent or stretched. We will employ a sleight of hand: although the shallow water momentum equation is not defined in the vertical directions, we consider the idealized case of a purely vertical, background magnetic field, which requires its vertical gradient to be defined.

In the equation for the shallow water height, we may also add a heating/cooling term (Q),

$$\frac{\partial h}{\partial t} + \nabla_{\mathrm{h}} \cdot (h\vec{v}) = Q = \frac{h_{\mathrm{eq}} - h}{t_{\mathrm{rad}}}. \tag{10.12}$$

Note that Q does not have the same physical units as, for example, its counterpart in the thermodynamic equation.

Physically, this term mimics the effect of radiative heating and cooling. An atmosphere in *radiative equilibrium* has an equilibrium shallow water height ($h = h_{\mathrm{eq}}$). When it is perturbed away from radiative equilibrium ($h \neq h_{\mathrm{eq}}$), it attempts to return to it by heating up or cooling down on a radiative time scale (t_{rad}). The shallow water height behaves almost like a temperature. When $h > h_{\mathrm{eq}}$, we have $Q < 0$ and cooling occurs. When $h < h_{\mathrm{eq}}$, we have $Q > 0$ and heating occurs. In the shallow water approximation, an atmosphere that departs from radiative equilibrium also becomes transiently compressible.

In the sections that follow, we will explore and study various physical effects using the shallow water system.

10.3 GRAVITY AS THE RESTORING FORCE: THE GENERATION OF GRAVITY WAVES

We consider the simplest situation possible: a *free system* (which has a complete absence of forcing or friction) in one dimension. No magnetic field is present. Since planetary rotation is an intrinsically two-dimensional phenomenon, it is also absent. The governing equations become

$$\begin{aligned} \frac{\partial h}{\partial t} + \frac{\partial}{\partial x}(hv) &= 0, \\ \frac{\partial v}{\partial t} + v\frac{\partial v}{\partial x} &= -g\frac{\partial h}{\partial x}. \end{aligned} \tag{10.13}$$

In this idealized situation, atmospheric motions may only be seeded by external perturbations and the restoring force is gravity.

We will now perform a *linear perturbation analysis*. Consider a background state,[2] at rest, described by a constant shallow water height (H). This state of rest is perturbed,

$$v = U + v', \quad h = H + h'. \tag{10.14}$$

---

[2]We denote the shallow water height, associated with the background state, also by H, because it should be on the order of the pressure scale height.

We set $U = 0$, since the background state is at rest.[3] We assume that the perturbations are small ($h' \ll H$). The perturbed governing equations become

$$\frac{\partial h'}{\partial t} + H\frac{\partial v'}{\partial x} = 0,$$
$$\frac{\partial v'}{\partial t} = -g\frac{\partial h'}{\partial x}. \tag{10.15}$$

We have retained only the first-order terms, which have at most one small or perturbed quantity. The advection term ($v'\frac{\partial v'}{\partial x}$) is neglected because it is second order in magnitude—in other words, it is not linear.

The pair of perturbed equations may be combined to yield a single equation,

$$\frac{\partial^2 h'}{\partial t^2} = gH\frac{\partial^2 h'}{\partial x^2}. \tag{10.16}$$

It is a wave equation describing the propagation of *gravity waves* with a speed of

$$c_0 = (gH)^{1/2}. \tag{10.17}$$

Shallow water systems are incompressible by definition and do not admit sound waves. Instead, gravity waves play the role of sound waves in shallow water systems—they are responsible for communicating perturbations in the shallow water height throughout the fluid.

A more general way of analyzing the shallow water system is to seek its *normal modes* by considering the following wave solutions,

$$v' = v_0 e^{i(k_x x - \omega t)}, \ h' = h_0 e^{i(k_x x - \omega t)}, \tag{10.18}$$

with $i \equiv \sqrt{-1}$ and k_x being the wavenumber in the x-direction. The wavelength is $2\pi/k_x$. Physically, the normal modes describe the resonant frequencies or wavelengths of a system. Normal modes are orthogonal to one another in the sense that exciting one will not cause the excitation of another. Atmospheric motions, as mediated through waves, are generally superpositions of normal modes. We will use the terms "wave" and "normal mode" interchangeably, although it should be understood that a wave may generally be composed of a series of normal modes.

Generally, ω is a complex number with real and imaginary components. The real component is the wave frequency, while the imaginary component is the growth or decay frequency. Since we are currently dealing with a free system, we expect ω to be real.

By plugging equation (10.18) into (10.15), we obtain the *dispersion relation*, which relates the wave frequency, wavenumber and quantities associated with any physical effect present in the system (in this case, gravity),

$$\omega^2 = gHk_x^2. \tag{10.19}$$

---

[3]It is also possible to explore other, more complicated background states, but we will not pursue this train of thought.

The waves in the system move with a group velocity[4] given by

$$c_g \equiv \frac{\partial \omega}{\partial k_x} = (gH)^{1/2} = c_0. \tag{10.20}$$

Again, we conclude that the only waves present in the system are gravity waves—waves for which the restoring force is solely gravity. Gravity waves in shallow water systems have the special property that they are *non-dispersive*, meaning that they move at the same speed regardless of their size, i.e., c_0 is independent of k_x. This property vanishes when the atmosphere becomes deep. Generally, waves are dispersive.

When atmospheric flows are faster than the speed of sound, discontinuities called *shocks* may form (see Chapter 11). Essentially, the flow before and after the shock becomes causally disconnected, because information about the fluid cannot be transmitted faster than the sound speed. In shallow water systems, *hydraulic jumps* may form if the atmospheric flow speed exceeds the gravity wave speed; an analogous causal disconnection occurs between parts of the flow separated by the hydraulic jump.

10.4   FRICTION IN AN ATMOSPHERE: MOLECULAR VISCOSITY AND RAYLEIGH DRAG

Real atmospheres have some form of friction present within them, e.g., caused by turbulence or fluid instabilities. We use the shallow water model to study the effects of molecular viscosity (ν) and Rayleigh drag, the latter of which we associate with a constant drag time scale (t_{drag}). Such a treatment is certainly simplistic and idealized, but it allows us to cleanly understand what it does to the atmospheric flow. The one-dimensional governing equations inherit additional terms,

$$\begin{aligned} &\frac{\partial h}{\partial t} + \frac{\partial}{\partial x}(hv) = 0, \\ &\frac{\partial v}{\partial t} + v\frac{\partial v}{\partial x} = -g\frac{\partial h}{\partial x} + \nu\frac{\partial^2 v}{\partial x^2} - \frac{v}{t_{\text{drag}}}. \end{aligned} \tag{10.21}$$

We will focus on each type of friction in turn.

If we consider a system only with molecular viscosity ($\nu \neq 0$, $t_{\text{drag}}^{-1} = 0$) and perform the linear perturbation analysis previously described for pure gravity waves, then we obtain the dispersion relation,

$$i\omega^2 - \nu\omega k_x^2 = igHk_x^2. \tag{10.22}$$

The wave frequency (ω) is now complex: it contains real (ω_{R}) and imaginary (ω_{I}) parts,

$$\omega = \omega_{\text{R}} + i\omega_{\text{I}}. \tag{10.23}$$

---

[4]Note that the subscript "g" in c_g refers to the group velocity.

For the normal modes, the wave solutions become

$$v' = v_0 e^{\omega_I t} e^{i(k_x x - \omega_R t)}, \quad h' = h_0 e^{\omega_I t} e^{i(k_x x - \omega_R t)}. \tag{10.24}$$

It is apparent that ω_R is associated with the oscillatory behavior of the waves. By contrast, ω_I accounts for either the growth ($\omega_I > 0$) or decay ($\omega_I < 0$) of the waves. Physically, we expect friction to be associated with decay; we will confirm this expectation in a moment.

The dispersion relation may be separated into its real and imaginary components,

$$\omega_R^2 = gHk_x^2 - \left(\frac{\nu k_x^2}{2}\right)^2, \quad \omega_I = -\frac{\nu k_x^2}{2}. \tag{10.25}$$

As we expected, molecular viscosity damps the amplitude of the gravity waves and reduces their frequency (i.e., they oscillate more slowly). Molecular viscosity also has a scale dependence: it acts more strongly on smaller scales (larger k_x).

More physical intuition may be gleaned from the dispersion relations. There is a length scale on which molecular viscosity acts so strongly that it completely damps out the gravity waves. This occurs when $\omega_R = 0$, which yields an expression for the *viscous length scale*,

$$l_\nu = \frac{\pi \nu}{c_0} \sim 0.1 \ \mu\text{m} \left(\frac{\nu}{0.1 \ \text{cm}^2 \ \text{s}^{-1}}\right) \left(\frac{g}{10 \ \text{m s}^{-1}} \frac{H}{10 \ \text{km}}\right)^{-1/2}. \tag{10.26}$$

We have plugged in numbers that are typical for the Earth's atmosphere, but the value of l_ν remains tiny ($l_\nu \ll H \ll R$) even when other Solar System bodies and exoplanets are considered. This estimate informs us that molecular viscosity is important only on very small scales and generally plays no role in the large-scale dynamics of atmospheres. We may also estimate the Reynolds number associated with the viscous length scale,

$$\mathcal{R}_e = \frac{c_0 l_\nu}{\nu} = \pi \sim 1. \tag{10.27}$$

In line with our intuition, the Reynolds number is about unity when molecular viscosity acts strongly.

What happens when molecular viscosity is replaced by Rayleigh drag? The dispersion relations look very similar,

$$\omega_R^2 = gHk_x^2 - \left(\frac{1}{2t_{\text{drag}}}\right)^2, \quad \omega_I = -\frac{1}{2t_{\text{drag}}}. \tag{10.28}$$

Rayleigh drag also damps and slows down the gravity waves, but the damping acts *equally on all scales* if t_{drag} is taken to be constant. In other words, ω_I has no dependence on k_x. However, for each value of the drag time scale, there is a length scale on which Rayleigh drag acts most strongly to slow down the gravity waves,

$$l_{\text{drag}} = 4\pi c_0 t_{\text{drag}} \sim 10^4 \ \text{km} \left(\frac{g}{10 \ \text{m s}^{-1}} \frac{H}{100 \ \text{m}}\right)^{1/2} \left(\frac{t_{\text{drag}}}{1 \ \text{day}}\right). \tag{10.29}$$

In plugging in numbers for Earth, we have focused on the planetary boundary layer (the frictional layer between the atmosphere and the surface), interpreted H to be its thickness and assumed Rayleigh drag to act typically on a time scale of a day. Our estimates tell us that the planetary boundary layer significantly affects the atmospheric dynamics of Earth on large/planetary scales.

Next, we subsume molecular viscosity and Rayleigh drag into a single drag frequency,

$$\omega_\nu = \nu k_x^2 + \frac{1}{t_{\text{drag}}}. \tag{10.30}$$

The wave solutions previously stated in equation (10.24) are difficult to visualize and apply, as they are complex.[5] To obtain expressions we can actually work with, we use Euler's formula[6] and take the real parts of these expressions,

$$v' = v_0 e^{-\omega_\nu t/2} \cos{(k_x x - \omega_{\text{R}} t)},$$
$$h' = \frac{v_0}{g k_x} e^{-\omega_\nu t/2} \left[\omega_{\text{R}} \cos{(k_x x - \omega_{\text{R}} t)} - \frac{\omega_\nu}{2} \sin{(k_x x - \omega_{\text{R}} t)} \right]. \tag{10.31}$$

One may also take the imaginary parts of the expressions in (10.24) and these correspond to an identical solution branch that is simply translated backwards or forwards in time. Physically, the solutions in equation (10.31) describe *damped gravity waves*.

We may learn several lessons from examining equation (10.31). First, the wave amplitudes for the velocity and shallow water height perturbations are specified to some arbitrary normalization (v_0). Second, if friction was absent ($\omega_\nu = 0$), the amplitudes would remain unchanged in time. Both the velocity and shallow water height would be in phase (since they are both described by cosine waves). Third, when friction is "switched on," an out-of-phase component (the part in the expression for h' associated with a sine wave) is present in the shallow water height, which causes it to oscillate out of sync with the velocity. As time progresses, the waves gradually die out. This is exactly what we expect: without an external source of energy to maintain it, friction eventually kills off any atmospheric motion after several viscous/drag time scales.

10.5   FORCING THE ATMOSPHERE: STELLAR IRRADIATION

We now neglect all sources of friction and consider an external energy input: heating by the parent or host star of the exoplanet. First, we define a source term,

$$S = \frac{h_{\text{eq}} - H}{t_{\text{rad}}}. \tag{10.32}$$

---

[5]Meaning that they contain real and imaginary components, rather than that they are complicated.

[6]For some x, Euler's formula states that $e^{ix} = \cos x + i \sin x$.

In general, it may have an arbitrary functional form, but for simplicity we assume it to have the same functional form as the shallow water height perturbation,

$$S = S_0 h', \tag{10.33}$$

with S_0 being a normalization constant. The heating/cooling term becomes

$$Q = S - \frac{h'}{t_{\text{rad}}} = F_0 h', \tag{10.34}$$

where we have defined the *forcing* to be $F_0 \equiv S_0 - 1/t_{\text{rad}}$. The forcing vanishes when radiative heating and cooling cancel each other out, i.e., $S_0 = 1/t_{\text{rad}}$. The dispersion relations are

$$\omega_{\text{R}}^2 = gHk_x^2 - \left(\frac{F_0}{2}\right)^2, \quad \omega_{\text{I}} = \frac{F_0}{2}. \tag{10.35}$$

While forcing also decreases the frequency of gravity waves, it enhances their amplitude. On a time scale $\sim 1/F_0$, the wave solutions,

$$v' = v_0 e^{F_0 t/2} \cos\left(k_x x - \omega_{\text{R}} t\right),$$
$$h' = \frac{v_0}{gk_x} e^{F_0 t/2} \left[\omega_{\text{R}} \cos\left(k_x x - \omega_{\text{R}} t\right) - \frac{F_0}{2} \sin\left(k_x x - \omega_{\text{R}} t\right)\right], \tag{10.36}$$

experience exponential growth in their amplitudes and become unstable. Similar to the case with friction, forcing introduces an out-of-phase component to the shallow water height perturbation.

When both stellar heating and friction are included, we obtain

$$\omega_{\text{I}} = \frac{F_0 - \omega_\nu}{2}. \tag{10.37}$$

As you may have anticipated, if forcing and friction are present with the same strength, they balance each other out. No growing or decaying waves are present. The atmosphere attains a state of equipoise, where the energy input from the star is first converted into mechanical energy (atmospheric motion) and eventually dissipated as heat. One may visualize an atmosphere as being a heat engine, as several researchers have previously done [73, 182, 186]. It is a pleasant surprise that our simple, one-dimensional shallow water model exhibits this property.

10.6 LIKE PLUCKING A STRING: ALFVÉN WAVES

So far, we have used a simple, one-dimensional shallow water model to develop intuition for atmospheres without magnetic fields. What happens when magnetic tension is added to the analysis? We consider the idealized situation of a purely vertical magnetic field, which has a field strength of B_0. It is perfectly coupled to the fluid and provides a restoring force when atmospheric motion is present. This is Lenz's law at play: Nature abhors changes in the background, vertical magnetic field, so it creates a perturbed, horizontal magnetic field that generates an opposing force to the original motion. In other words, it is really like a taut string is being plucked: the motion of the fluid stretches the magnetic field line and it responds by exerting tension. In the limit that the magnetic field is arbitrarily strong, all atmospheric motion is ceased.

The horizontal magnetic field has no background component and only exists as a perturbation. We denote the perturbed, horizontal magnetic field as b'. Ignoring stellar heating and friction, the perturbed governing equations are

$$
\begin{aligned}
\frac{\partial h'}{\partial t} + H \frac{\partial v'}{\partial x} &= 0, \\
\frac{\partial v'}{\partial t} &= -g \frac{\partial h'}{\partial x} - \frac{B_0 b'}{4\pi \rho H}, \\
\frac{\partial b'}{\partial t} &= \frac{B_0 v'}{H}.
\end{aligned}
\tag{10.38}
$$

We have approximated the vertical gradient of the magnetic field as $-b'/H$; the minus sign ensures that a restoring force is present. A plus sign would result in atmospheric motion always reinforcing itself—an unphysical phenomenon.

Gravity and magnetic tension couple to produce *magnetogravity waves*, which are described by the following dispersion relation,

$$
\omega_{\mathrm{R}}^2 = gH k_x^2 + \left(\frac{v_{\mathrm{A}}}{H} \right)^2.
\tag{10.39}
$$

The *Alfvén speed*,

$$
v_{\mathrm{A}} = \frac{B_0}{2 (\pi \rho)^{1/2}} \sim 10^3 \ \mathrm{cm \ s^{-1}} \left(\frac{B_0}{1 \ \mathrm{G}} \right) \left(\frac{T}{1000 \ \mathrm{K}} \right)^{1/2} \left(\frac{m}{2m_{\mathrm{H}}} \frac{P}{1 \ \mathrm{bar}} \right)^{-1/2},
\tag{10.40}
$$

plays the role of the sound speed for the magnetic fields. When a magnetic field line is being stretched, the perturbation is communicated at the Alfvén wave speed. When gravity is absent, the only waves present are *Alfvén waves*.

The velocity and shallow water height perturbations are in phase, since no sources of forcing or friction are present. However, the magnetic field perturbation is 90 degrees out of phase with the velocity and shallow water height, as is evident from the wave solutions,

$$
\begin{aligned}
v' &= v_0 \cos \left(k_x x - \omega_{\mathrm{R}} t \right), \\
h' &= \frac{v_0 k_x H}{\omega_{\mathrm{R}}} \cos \left(k_x x - \omega_{\mathrm{R}} t \right), \\
b' &= -\frac{v_0 B_0}{\omega_{\mathrm{R}} H} \sin \left(k_x x - \omega_{\mathrm{R}} t \right),
\end{aligned}
\tag{10.41}
$$

That the magnetic field is out of phase with the other quantities is the mathematical expression of Lenz's law, which we discussed earlier.

10.7   ROTATION: THE GENERATION OF POINCARÉ AND ROSSBY WAVES

In the rotating frame, the Coriolis force deflects the motion of a parcel of atmosphere and causes it to change its trajectory. Mathematically, it is a two-dimensional effect, because it is associated with the term $-2\vec{\Omega} \times \vec{v}$ and vector cross products are not meaningful in one dimension. On a sphere, the angular frequency $(\vec{\Omega})$ is not constant locally and varies across latitude, which in turn leads to a gradient in the Coriolis force. It is precisely this gradient that gives rise to *Rossby waves*, as we will explore in this section. The algebra associated with solving the shallow water equations in spherical coordinates is tedious and messy (see Problem 10.11.6) and is not a recommended first approach for developing physical intuition. Instead, we use a trick long established in the atmospheric/geophysical community known as the *β-plane approximation* [105, 134, 244]. We work in Cartesian coordinates, but rewrite the Coriolis force (per unit mass) as

$$-\vec{f} \times \vec{v} = f v_y \hat{x} - f v_x \hat{y}. \tag{10.42}$$

The unit vectors in the x-, y- and z-directions are $\hat{x}$, $\hat{y}$ and $\hat{z}$, respectively. The velocity is expressed as $\vec{v} = (v_x, v_y)$. For now, we note that $\vec{f} = f\hat{z}$. The quantity f is often termed the *Coriolis parameter*.

The essence of the β-plane approximation is to allow f to have a functional dependence in y by subjecting it to a series expansion truncated at the linear term [105, 134, 244],

$$f = f_0 + \beta y, \tag{10.43}$$

where we have

$$f_0 \equiv 2\Omega \cos\theta, \quad \beta = \frac{2\Omega \sin\theta}{R}. \tag{10.44}$$

The polar angle (θ), in standard spherical coordinates, is also known as the *co-latitude*.[7] At the poles, we have $\beta = 0$ and $f = f_0$, which is known as the *f-plane limit*. In this limit, the Coriolis force is constant. Near the equator, we have $\beta \neq 0$ and the Coriolis force varies across y, which is a proxy for the latitude. The Coriolis force vanishes exactly at the equator.

We again perform a linear analysis by perturbing the shallow water system about a state of rest. To not cloud our intuition and isolate the effects of rotation, we ignore forcing, friction and magnetic fields. The perturbed governing equations become

$$\frac{\partial h'}{\partial t} + H\left(\frac{\partial v_x'}{\partial x} + \frac{\partial v_y'}{\partial y}\right) = 0,$$

$$\frac{\partial v_x'}{\partial t} = -g\frac{\partial h'}{\partial x} + f v_y', \tag{10.45}$$

$$\frac{\partial v_y'}{\partial t} = -g\frac{\partial h'}{\partial y} - f v_x'.$$

---

[7] If we denote the latitude by θ', then we have $\theta' = 90° - \theta$.

As usual, we seek wave solutions but in two dimensions,

$$v'_x = v_{x_0} e^{i(k_x x + k_y y - \omega t)}, \quad v'_y = v_{y_0} e^{i(k_x x + k_y y - \omega t)}, \quad h' = h_0 e^{i(k_x x + k_y y - \omega t)}. \quad (10.46)$$

The wavenumber has both x- and y-components, denoted by k_x and k_y, respectively. The total wavenumber (k) obtains from $k^2 = k_x^2 + k_y^2$. Since we are again dealing with a free system, we expect that $\omega = \omega_R$.

The dispersion relation reveals the existence of two types of waves,

$$\omega_R^3 - \omega_R \left(gHk^2 + f^2 \right) - gHk_x \beta = 0. \quad (10.47)$$

When rotation is absent ($\Omega = 0$ and thus $f = \beta = 0$), we recover the familiar expression for gravity waves. When the Coriolis force is constant ($\beta = 0$), we see that the wave frequency is modified by the presence of rotation,

$$\omega_R^2 = gHk^2 + f^2. \quad (10.48)$$

Constant rotation has the effect of speeding up the frequency at which gravity waves oscillate—this occurs equally in all directions. These rotationally-modified gravity waves have a special name: *Poincaré waves*, named after the French polymath, Henri Poincaré. Since ω_R has both positive and negative roots and the group velocity of waves, in the x-direction, is given by

$$c_{g_x} \equiv \frac{\partial \omega_R}{\partial k_x} = \pm \frac{gHk_x}{\omega_R}, \quad (10.49)$$

Poincaré waves may move in opposing directions across longitude. Alternatively, the dispersion relation for Poincaré waves obtains approximately from demanding that we are only interested in fast waves (large ω_R) and considering only the first two terms in the dispersion relation.

Conversely, if we are only interested in slow (small ω_R) and large (small k) waves, we instead obtain

$$\omega_R \approx -\frac{gHk_x \beta}{f^2}. \quad (10.50)$$

These slow waves are distinct from Poincaré waves. First, they only exist when the Coriolis force has a gradient across latitude ($\beta \neq 0$). Second, they move only in one direction (across longitude)—against the direction of rotation (as is evident from the minus sign), since

$$c_{g_x} \approx -\frac{gH\beta}{f^2}. \quad (10.51)$$

These slow waves are called *Rossby waves*, after the Swedish-American meteorologist Carl-Gustaf Rossby. Rossby waves are a consequence of the conservation of potential vorticity, as we will explore in Problem 10.11.3.

10.8   GENERAL COUPLING OF PHYSICAL EFFECTS

So far, we have performed the analogue of controlled experiments in the laboratory by considering each physical effect in turn and studying how they couple to
gravity. We have seen that Alfvén, magnetogravity, Poincaré and Rossby waves
result from each "experiment." Generally, gravity, friction, forcing, rotation and
magnetic fields may couple to produce multiple wave modes, several of which
have not yet been named [93].

If we consider a two-dimensional, Cartesian, shallow water system, then the
perturbed governing equations become

$$
\begin{aligned}
\frac{\partial h'}{\partial t} + H \left(\frac{\partial v'_x}{\partial x} + \frac{\partial v'_y}{\partial y} \right) &= F_0 h', \\[2mm]
\frac{\partial v'_x}{\partial t} &= -g \frac{\partial h'}{\partial x} + f v'_y + \nu \left(\frac{\partial^2 v'_x}{\partial x^2} + \frac{\partial^2 v'_x}{\partial y^2} \right) - \frac{v'_x}{t_{\text{drag}}} - \frac{B_0 b'_x}{4\pi\rho H}, \\[2mm]
\frac{\partial v'_y}{\partial t} &= -g \frac{\partial h'}{\partial y} - f v'_x + \nu \left(\frac{\partial^2 v'_y}{\partial x^2} + \frac{\partial^2 v'_y}{\partial y^2} \right) - \frac{v'_y}{t_{\text{drag}}} - \frac{B_0 b'_y}{4\pi\rho H}, \\[2mm]
\frac{\partial b'_x}{\partial t} &= \frac{B_0 v'_x}{H} + \eta \left(\frac{\partial^2 b'_x}{\partial x^2} + \frac{\partial^2 b'_x}{\partial y^2} \right), \\[2mm]
\frac{\partial b'_y}{\partial t} &= \frac{B_0 v'_y}{H} + \eta \left(\frac{\partial^2 b'_y}{\partial x^2} + \frac{\partial^2 b'_y}{\partial y^2} \right).
\end{aligned}
\tag{10.52}
$$

To be able to study the effects of a gradient in the Coriolis force, we need to recast
one of the governing equations in a form where β explicitly appears by itself.
This is because when β appears as part of f, it is typically subdominant ($f \approx f_0$)
near the equator as both β and y are nearly zero. The trick is to differentiate
one of the governing equations for the perturbed velocity with respect to y [85],
since $\frac{\partial f}{\partial y} = \beta$, which yields

$$
\begin{aligned}
\frac{\partial^2 v'_x}{\partial t \partial y} = &-g \frac{\partial^2 h'}{\partial x \partial y} + \beta v'_y + f \frac{\partial v'_y}{\partial y} + \nu \left(\frac{\partial^3 v'_x}{\partial x^2 \partial y} + \frac{\partial^3 v'_x}{\partial y^3} \right) \\[2mm]
&- \frac{1}{t_{\text{drag}}} \frac{\partial v'_x}{\partial y} - \frac{B_0}{4\pi\rho H} \frac{\partial b'_x}{\partial y}.
\end{aligned}
\tag{10.53}
$$

Another useful trick is to eliminate b'_x and b'_y from the perturbed governing
equations and end up with a trio of equations for h', v'_x and v'_y. We then apply
the procedure, which should be well-established to you by now, of seeking normal
modes. Here, I will introduce an alternative way of deriving the dispersion
relation. With three governing equations, one should obtain three equations
involving v_{x_0}, v_{y_0} and h_0. This trio of equations may be re-arranged into the
following form,

$$
\hat{M} \begin{pmatrix} v_{x_0} \\ v_{y_0} \\ h_0 \end{pmatrix} = 0.
$$

A trivial solution of this equation is $v_{x_0} = v_{y_0} = h_0 = 0$. Non-trivial solutions require the determinant of the matrix $\hat{M}$ to be zero, i.e., $\det \hat{M} = 0$. The matrix has the form

$$\hat{M} = \begin{pmatrix} k_y \omega_{B_0} & -(\beta + if k_y) & -g k_x k_y \\ f & -i\omega_{B_0} & igk_y \\ k_x H & k_y H & -\omega_F \end{pmatrix}.$$

We have made the algebraic more compact by defining a series of complex frequencies, some of which are embedded in one another [93],

$$\omega_F \equiv \omega - iF_0,$$

$$\omega_\nu \equiv \nu k^2 + \frac{1}{t_{\text{drag}}},$$

$$\omega_\eta \equiv \omega + i\eta k^2, \tag{10.54}$$

$$\omega_B \equiv \omega_\nu + \frac{i}{\omega_\eta} \left(\frac{v_A}{H} \right)^2,$$

$$\omega_{B_0} \equiv \omega + i\omega_B.$$

Physically, the forcing is expressed through ω_F, whereas all of the terms associated with friction and magnetic tension are compacted into a single generalized frequency (ω_{B_0}), which we term the *generalized friction*. What is particularly elegant about such an approach is that when magnetic fields are absent, we simply recover the hydrodynamic friction. The mathematical form of $\hat{M}$ remains unchanged.

Evaluating $\det \hat{M} = 0$ yields a complex expression,

$$ik_y \omega_{B_0}^2 \omega_F - igHk_y \left(k^2 \omega_{B_0} + k_x \beta \right) - \beta \omega_F f - i f^2 k_y \omega_F = 0. \tag{10.55}$$

This expression alone does not allow us to clearly see the oscillatory and growing/decaying behavior of the waves in the system. To proceed, we need to write $\omega = \omega_R + i\omega_I$ and separate out the real and imaginary parts of the preceding expression. We will not pursue the algebra here as it is tedious and does not contribute much to our physical intuition. Instead, I encourage you to work through Problem 10.11.4. In the end, one obtains two coupled equations describing ω_R and ω_I as functions of k_x, k_y, g, F_0, ω_ν, β, v_A and η. These equations describe the general coupling between gravity, forcing, friction, rotation, magnetic tension and magnetic diffusivity. There are no analytical solutions for ω_R and ω_I; instead, the pair of equations needs to be solved numerically using a computer. These solutions describe a general set of waves resulting from the different couplings between the physical effects in the system.

10.9   SHALLOW ATMOSPHERES AS QUANTUM HARMONIC OSCILLATORS

So far, we have sought oscillatory solutions in both horizontal directions. While this is good enough for studying the response of a system to perturbations and developing an initial intuition for the waves present, atmospheres tend to be a little more complicated. Specifically, one should instead seek wave modes with an arbitrary functional dependence across latitude. Physically, we expect waves to oscillate continuously across longitude, but are bounded between the poles of an exoplanet across latitude.

To render the governing equations more elegant, we cast them in dimensionless units in terms of characteristic length and time scales [165],

$$l_0 = \left(\frac{c_0}{\beta} \right)^{1/2}, \ t_{\mathrm{dyn}} = \frac{1}{(c_0 \beta)^{1/2}}. \tag{10.56}$$

As we have implied by the subscript "dyn," t_{dyn} may be interpreted as being a dynamical time scale. The characteristic speed is c_0, the gravity wave speed. It follows that the perturbed governing equations, in the β-plane approximation, are

$$\frac{\partial h'}{\partial t} = -\frac{\partial v'_x}{\partial x} - \frac{\partial v'_y}{\partial y} + F_0 h',$$

$$\frac{\partial v'_x}{\partial t} = -\frac{\partial h'}{\partial x} + y v'_y + \frac{1}{\mathcal{R}_e} \left(\frac{\partial^2 v'_x}{\partial x^2} + \frac{\partial^2 v'_x}{\partial y^2} \right) - \frac{v'_x}{t_{\mathrm{drag}}} - \Gamma b'_x,$$

$$\frac{\partial v'_y}{\partial t} = -\frac{\partial h'}{\partial y} - y v'_x + \frac{1}{\mathcal{R}_e} \left(\frac{\partial^2 v'_y}{\partial x^2} + \frac{\partial^2 v'_y}{\partial y^2} \right) - \frac{v'_y}{t_{\mathrm{drag}}} - \Gamma b'_y, \tag{10.57}$$

$$\frac{\partial b'_x}{\partial t} = \epsilon v'_x + \frac{1}{\mathcal{R}_B} \left(\frac{\partial^2 b'_x}{\partial x^2} + \frac{\partial^2 b'_x}{\partial y^2} \right),$$

$$\frac{\partial b'_y}{\partial t} = \epsilon v'_y + \frac{1}{\mathcal{R}_B} \left(\frac{\partial^2 b'_y}{\partial x^2} + \frac{\partial^2 b'_y}{\partial y^2} \right).$$

All of the physical effects discussed so far have been included: forcing, friction (including molecular viscosity and Rayleigh drag), rotation and magnetic fields (including magnetic tension and magnetic diffusivity). This is the most sophisticated system we have considered so far, and it allows us to study the coupling between these different physical effects. In non-dimensionalizing the term associated with magnetic diffusivity, we have defined the *magnetic Reynolds number*,

$$\mathcal{R}_B = \frac{c_0 l_0}{\eta}. \tag{10.58}$$

Notice how it is exactly the magnetic analogue of the Reynolds number with the molecular viscosity being replaced by the magnetic diffusivity. Unlike for its hydrodynamic counterpart, the magnetic Reynolds number may be of the order of unity or smaller in astrophysical situations. Furthermore, we have defined

$$\epsilon \equiv \frac{l_0}{H}, \ \Gamma \equiv \epsilon \left(\frac{v_A}{c_0} \right)^2, \tag{10.59}$$

such that their product is the square of the ratio of the dynamical to the Alfvén time scales, $\epsilon\Gamma = (t_{\mathrm{dyn}}/t_{\mathrm{A}})^2$.

Note that we have not used distinct notation for the non-dimensional quantities in order to avoid cluttering. It should be understood that h', v'_x, v'_y, b'_x, b'_y, x, y, F_0 and t_{drag} in equation (10.57), and for the rest of this section, are cast in quantities that have been normalized by the characteristic length, speed and time scales, as well as by the vertical background magnetic field (B_0).

We apply the usual recipe. We seek normal modes of the following form,

$$\vec{X}' = \vec{X}_0 e^{i(k_x x - \omega t)}, \tag{10.60}$$

where $\vec{X}' = (h', v'_x, v'_y, b'_x, b'_y)$ and $\vec{X}_0 = (h_0, v_{x_0}, v_{y_0}, b_{x_0}, b_{y_0})$. Note that $\vec{X}'$, $\vec{X}_0$, k_x, x, ω and t are all dimensionless quantities. The normalizations $\vec{X}_0$ have an unknown functional dependence in y (which is also dimensionless), which we will solve for. It follows that one may reduce the set of five perturbed governing equations into three equations describing the wave amplitudes [93],

$$
\begin{aligned}
v_{x_0} &= \frac{1}{\omega_{B_0}} \left(k_x h_0 + i y v_{y_0} + \frac{i}{\mathcal{R}_e} \frac{\partial^2 v_{x_0}}{\partial y^2} + \frac{\Gamma}{\omega_\eta \mathcal{R}_B} \frac{\partial^2 b_{x_0}}{\partial y^2} \right), \\
v_{x_0} &= \frac{1}{y} \left(i \omega_{B_0} v_{y_0} - \frac{\partial h_0}{\partial y} + \frac{1}{\mathcal{R}_e} \frac{\partial^2 v_{y_0}}{\partial y^2} - \frac{i\Gamma}{\omega_\eta \mathcal{R}_B} \frac{\partial^2 b_{y_0}}{\partial y^2} \right), \\
v_{x_0} &= \frac{1}{k_x} \left(h_0 \omega_{\mathrm{F}} + i \frac{\partial v_{y_0}}{\partial y} \right).
\end{aligned}
\tag{10.61}
$$

As before, there is a series of complex frequencies present in these equations. We again have $\omega_{\mathrm{F}} \equiv \omega - i F_0$. The quantity ω_{B_0} contains a series of embedded complex frequencies,

$$\omega_\nu \equiv \frac{k_x^2}{\mathcal{R}_e} + \frac{1}{t_{\mathrm{drag}}}, \quad \omega_\eta \equiv \omega + \frac{i k_x^2}{\mathcal{R}_B}, \quad \omega_B \equiv \omega_\nu + \frac{i\epsilon\Gamma}{\omega_\eta}, \quad \omega_{B_0} \equiv \omega + i\omega_B, \tag{10.62}$$

which have all been non-dimensionalized.

What follows is a series of tricks of algebra, which I will not relegate to an exercise as it is not obvious how to develop them without staring at the equations for a long time. First, one needs to equate the first and third equations in (10.61) to eliminate v_{x_0} and obtain an expression for h_0,

$$h_0 = i \left(\frac{k_x}{\omega_{B_0}} - \frac{\omega_{\mathrm{F}}}{k_x} \right)^{-1} \left(\frac{1}{k_x} \frac{\partial v_{y_0}}{\partial y} - \frac{y v_{y_0}}{\omega_{B_0}} - \frac{1}{\omega_{B_0} \mathcal{R}_e} \frac{\partial^2 v_{x_0}}{\partial y^2} + \frac{i\Gamma}{\omega_{B_0} \omega_\eta \mathcal{R}_B} \frac{\partial^2 b_{x_0}}{\partial y^2} \right). \tag{10.63}$$

Second, we combine the first and second equations in (10.61), together with the preceding expression for h_0, to obtain a somewhat tedious expression for the derivative of h_0,

$$
\begin{aligned}
\left(\frac{k_x}{\omega_{B_0}} - \frac{\omega_F}{k_x}\right)\frac{\partial h_0}{\partial y} = {}& iv_{y_0}\left(\omega_{B_0} - \frac{y^2}{\omega_{B_0}}\right)\left(\frac{k_x}{\omega_{B_0}} - \frac{\omega_F}{k_x}\right) - \frac{iy}{\omega_{B_0}}\frac{\partial v_{y_0}}{\partial y} \\
& + \frac{iy^2 k_x v_{y_0}}{\omega_{B_0}^2} + \frac{1}{\mathcal{R}_e}\frac{\partial^2 v_{y_0}}{\partial y^2}\left(\frac{k_x}{\omega_{B_0}} - \frac{\omega_F}{k_x}\right) \\
& + \frac{iy\omega_F}{\omega_{B_0}\mathcal{R}_e k_x}\frac{\partial^2 v_{x_0}}{\partial y^2} + \frac{\Gamma k_x y}{\omega_{B_0}^2 \omega_\eta \mathcal{R}_B}\frac{\partial^2 b_{x_0}}{\partial y^2} \\
& - \frac{\Gamma}{\omega_\eta \mathcal{R}_B}\left(\frac{k_x}{\omega_{B_0}} - \frac{\omega_F}{k_x}\right)\left(\frac{y}{\omega_{B_0}}\frac{\partial^2 b_{x_0}}{\partial y^2} + i\frac{\partial^2 b_{y_0}}{\partial y^2}\right).
\end{aligned}
\tag{10.64}
$$

Finally, we recognize that differentiating the third equation in (10.61), with respect to y, forms the basis of the final equation we want. This equation requires expressions for $\frac{\partial h_0}{\partial y}$ and $\frac{\partial v_{x_0}}{\partial y}$, the former of which we have previously obtained and the latter of which is obtained by differentiating the first equation in (10.61) with respect to y. Putting this all together, we obtain a general governing equation for v_{y_0},

$$
\begin{aligned}
& \frac{\partial^2 v_{y_0}}{\partial y^2}\left[1 + \frac{i}{\mathcal{R}_e}\left(\frac{k_x^2}{\omega_{B_0}} - \omega_F\right)\right] + \left(\omega_{B_0}\omega_F - k_x^2 - \frac{k_x}{\omega_{B_0}} - \frac{y^2\omega_F}{\omega_{B_0}}\right)v_{y_0} \\
& - \frac{1}{\omega_{B_0}\mathcal{R}_e}\left(y\omega_F\frac{\partial^2 v_{x_0}}{\partial y^2} + k_x\frac{\partial^3 v_{x_0}}{\partial y^3}\right) + \frac{i\Gamma k_x}{\omega_{B_0}\omega_\eta \mathcal{R}_B}\left(\frac{k_x y}{\omega_{B_0}}\frac{\partial^2 b_{x_0}}{\partial y^2} + \frac{\partial^3 b_{x_0}}{\partial y^3}\right) \\
& - \frac{\Gamma}{\omega_\eta \mathcal{R}_B}\left(\frac{k_x^2}{\omega_{B_0}} - \omega_F\right)\left(\frac{iy}{\omega_{B_0}}\frac{\partial^2 b_{x_0}}{\partial y^2} - \frac{\partial^2 b_{y_0}}{\partial y^2}\right) = 0.
\end{aligned}
\tag{10.65}
$$

This intimidating equation is difficult to solve, and we have neither the desire nor the need to do so. Instead, we examine the inviscid ($\mathcal{R}_e \to \infty$) and ideal magnetohydrodynamic ($\mathcal{R}_B \to \infty$) limit,

$$
\frac{\partial^2 v_{y_0}}{\partial y^2} + \left(\omega_{B_0}\omega_F - k_x^2 - \frac{k_x}{\omega_{B_0}} - \frac{y^2\omega_F}{\omega_{B_0}}\right)v_{y_0} = 0.
\tag{10.66}
$$

If we further demand that the system is unforced, possesses no sources of friction and is non-magnetized, we obtain the equation for the quantum harmonic oscillator [165],

$$
\frac{\partial^2 v_{y_0}}{\partial y^2} + \left(\omega^2 - k_x^2 - \frac{k_x}{\omega} - y^2\right)v_{y_0} = 0,
\tag{10.67}
$$

if the following combination of quantities is "quantized,"

$$
\omega^2 - k_x^2 - \frac{k_x}{\omega} = 2n + 1,
\tag{10.68}
$$

where the quantity n is an integer.

In quantum mechanics, quantization is an intrinsic property of nature—between energy and time, position and momentum, etc. In our shallow water model, quantization refers to the "alphabet" by which all waves in the system may be constructed. The integer n is a proxy for how large a normal mode is across latitude, analogous to the number of octaves on a piano. In the β-plane approximation, the zonal wavenumber k_x is a continuous quantity. We will see, in Problem 10.11.6, that it becomes an integer when the shallow water model is cast on a sphere. Generally, any wave in the atmosphere may be constructed from some linear combination of wave modes, each with its own value of n and k_x.

Is the governing equation for the meridional velocity *generally* that of the quantum harmonic oscillator? It turns out that it is if the suitable coordinate transformation is made [93]. Specifically, the latitude y is transformed to

$$\tilde{y} \equiv \alpha y \text{ where } \alpha \equiv \left(\frac{\omega_F}{\omega_{B_0}} \right)^{1/4}. \tag{10.69}$$

This transformation results in a different form of the governing equation,

$$\frac{\partial^2 v_{y_0}}{\partial \tilde{y}^2} + \left[\left(\omega_F \omega_{B_0} - k_x^2 - \frac{k_x}{\omega_{B_0}} \right) \left(\frac{\omega_{B_0}}{\omega_F} \right)^{1/2} - \tilde{y}^2 \right] v_{y_0} = 0. \tag{10.70}$$

This is the governing equation for the *damped quantum harmonic oscillator*, since the coefficients involved are complex. It requires that the following combination of quantities is quantized,

$$\left(\omega_F \omega_{B_0} - k_x^2 - \frac{k_x}{\omega_{B_0}} \right) \left(\frac{\omega_{B_0}}{\omega_F} \right)^{1/2} = 2n + 1. \tag{10.71}$$

Notice that it involves the square root of the ratio of complex frequencies—complex quantities raised to fractional powers are generally multi-valued. In our case, the preceding expression is double-valued. To get out of this dilemma, we have to demand that it reduces to the correct expression in the free, hydrodynamic limit. We will relegate this task to Problem 10.11.5, as the algebra involved is tedious.

We return to equation (10.70). Can it be solved analytically? It turns out that quantum physicists (and mathematicians before them) have long realized that the solution belongs to a class of special functions known as *parabolic cylinder functions*,

$$v_{y_0} = v_0 e^{-\tilde{y}^2/2} \tilde{\mathcal{H}}_n, \tag{10.72}$$

where v_0 is an arbitrary normalization constant. The quantity $\mathcal{H}_n$ is yet another class of special functions known as *Hermite polynomials* (of the nth degree or order),[8] whose properties have been well-studied for decades, if not centuries. We have written $\tilde{\mathcal{H}}_n \equiv \mathcal{H}_n(\tilde{y})$ as a shorthand.

---

[8] We have used the "physicists' Hermite polynomials," where $\mathcal{H}_1(x) = 2x$, rather than x.

To obtain the complete solution for the perturbed meridional velocity (and not just its ampltitude), we need to multiply v_{y_0} by $e^{i(k_x x - \omega t)}$ and take the real part of the expression. The algebra is generally intractable for the time-dependent situation. However, in the limit of a steady state, we can set $t = 0$ and perform this step. It follows that [93]

$$v_y' = v_0 e^{-\tilde{y}^2/2} \tilde{\mathcal{H}}_n \cos(k_x x),$$

$$h' = -\frac{v_0 k_x \zeta_0}{k_x^2 + \zeta_0 |F_0|} e^{-\tilde{y}^2/2} \left[\frac{\alpha \left(2n\tilde{\mathcal{H}}_{n-1} - \tilde{y}\tilde{\mathcal{H}}_n\right)}{k_x} \cos(k_x x) - \frac{y\tilde{\mathcal{H}}_n}{\zeta_0} \sin(k_x x) \right],$$

$$v_x' = -\frac{v_0}{k_x^2 + \zeta_0 |F_0|} e^{-\tilde{y}^2/2} \left[-y|F_0| \tilde{\mathcal{H}}_n \cos(k_x x) \right.$$
$$\left. + \alpha k_x \left(2n\tilde{\mathcal{H}}_{n-1} - \tilde{y}\tilde{\mathcal{H}}_n\right) \sin(k_x x) \right],$$

$$(10.73)$$

where we have defined

$$\zeta_0 \equiv \frac{1}{t_{\text{drag}}} + \frac{k_x^2}{\mathcal{R}_e} + \frac{\epsilon \Gamma \mathcal{R}_B}{k_x^2}. \qquad (10.74)$$

Why do the Reynolds number and its magnetic counterpart appear in the expression for ζ_0? In deriving the governing equation and its solution, we have employed a trick: we have set $\mathcal{R}_e$, $\mathcal{R}_B \to \infty$ in equation (10.70), but not within ω_ν and ω_η. Physically, we are assuming that molecular viscosity and magnetic diffusivity do not exert a strong influence on the global structure of an atmosphere (and thus allowing the quantum harmonic oscillator equation to persist), but do contribute additional drag forces. This simplification allows us to study their effects without sacrificing algebraic tractability. With this approximation, one can see that ζ_0 represents a generalized friction, in the steady-state limit, that includes Rayleigh drag, molecular viscosity, magnetic tension and magnetic diffusivity.

In the steady-state limit, we have $\alpha \equiv (|F_0|/\zeta_0)^{1/4}$. We choose $F_0 = -|F_0|$, because this allows for a positive height perturbation to be associated with radiative cooling. Negative height perturbations are then associated with radiative heating. If $F_0 > 0$, then runaway cooling or heating would prevail, which would be unphysical.

We should not let the tedious algebra distract us from the physical intuition we may glean from this exercise. The solutions for the perturbed water height and velocity components are similar in spirit to what we derived in the one-dimensional limit. In the frictionless limit ($\zeta_0 \to \infty$), the height perturbation and zonal velocity are in phase. Rotation renders the meridional velocity out of phase with the height perturbation and zonal velocity. When generalized friction exists, additional out-of-phase components exist. The expression for ζ_0 informs us that molecular viscosity acts on small scales, while magnetic diffusivity acts on large scales.

10.10   SHALLOW WATER SYSTEMS AND EXOPLANETARY ATMOSPHERES

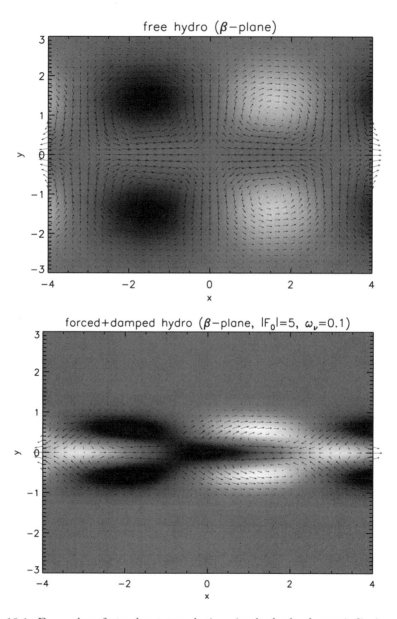

Figure 10.1: Examples of steady-state solutions in the hydrodynamic limit on a β-plane. Taken from Heng & Workman [93]. The free, hydrodynamic solution was originally found by Matsuno [165].

In this chapter, we have introduced the shallow water system and the mathematical machinery of linear perturbation analysis, which allows us to derive the global structure of an atmosphere and the waves present within it. We have learned that gravity, Alfvén, Poincaré and Rossby waves may exist within an atmosphere, each the outcome of different physical effects coupling together. Our simple one-dimensional models have shown that molecular viscosity acts predominantly on small scales and when the Reynolds number is on the order of unity, whereas Rayleigh drag (with a constant drag time scale) acts equally on all scales. When we allow for the functional form of the fluid solutions, across latitude, to be arbitrary, we find that the meridional velocity obeys the same governing equation as a damped quantum harmonic oscillator. Figure 10.1 shows two examples of the global structure of an atmosphere with and without forcing and friction/damping.

The techniques associated with the linear perturbation analysis may be used for other purposes. For example, they may be used to understand the momentum transport properties of waves [219]. More generally, solving for the imaginary component of the wave frequency (ω_I) allows one to understand if instabilities may exist within the atmospheric flow, a technique we will put to use in Chapter 12. Rather than cover all of these applications exhaustively, I have chosen to highlight the salient properties of the shallow water model and the techniques involved. These techniques may also be used for other limiting forms of the equations of fluid dynamics or even for the Navier-Stokes equation itself.

Shallow water systems may also be explored on a sphere, cast in spherical coordinates. We have relegated this generalization to Problem 10.11.6 for the free, hydrodynamic limit. For the forced, magnetized case with friction, I refer the reader to Heng & Workman [93] as the algebra involved becomes formidable.

10.11   PROBLEM SETS

10.11.1   The isothermal Euler equation

Show that the isothermal Euler equation, in two dimensions and without gravity, has the same mathematical structure as the shallow water equation for momentum conservation. What is the analogous expression for the gravity wave speed?

10.11.2   Damped gravity waves

(a) Starting from equation (10.21), derive the perturbed governing equations for a one-dimensional shallow water system with both molecular viscosity and Rayleigh drag present.
(b) Consider a background state at rest. By performing a linear perturbation analysis, obtain the dispersion relations when both molecular viscosity and Rayleigh drag are present. In other words, obtain the expressions for ω_I and ω_R accompanying the wave solutions in equation (10.31).

(c) What is the kinetic energy of the system? How about the potential energy? Is the total energy (the sum of the kinetic and potential energy) changing with time?

10.11.3 Rossby waves and absolute vorticity conservation

Shallow water systems are barotropic by construction, which simplifies the statement of the conservation of potential vorticity. Specifically, it is the *absolute vorticity* [134] that is conserved,

$$\frac{D}{Dt}(\omega_z + f) = 0, \tag{10.75}$$

where ω_z is the z-component of the vorticity and f is the Coriolis parameter. We wish to show that the linearized version of this equation leads to the dispersion relation for Rossby waves.

(a) By considering a background state of rest, show that

$$\frac{\partial}{\partial t}\left(\frac{\partial v_y'}{\partial x} - \frac{\partial v_x'}{\partial y}\right) = -\beta v_y', \tag{10.76}$$

where primed quantities are (small) perturbations of the velocity.

(b) Cast the components of the velocity perturbations in terms of a streamfunction (Ψ),

$$v_x' = \frac{\partial \Psi}{\partial y}, \quad v_y' = -\frac{\partial \Psi}{\partial x}. \tag{10.77}$$

What is the divergence of the velocity? Is the system incompressible?

(c) Consider normal modes of the streamfunction,

$$\Psi = \Psi_0 e^{i(k_x x + k_y y - \omega t)}, \tag{10.78}$$

where Ψ_0 is an arbitrary normalization constant, k_x is the wavenumber in the x-direction, k_y is the wavenumber in the y-direction and ω is the wave frequency. Show that

$$\omega = -\frac{\beta k_x}{k^2}, \tag{10.79}$$

where $k^2 = k_x^2 + k_y^2$. Convince yourself that ω is real (and has no imaginary component).

10.11.4 Obtaining the dispersion relation using separation functions

Consider the forced, dragged, magnetized shallow water system in two dimensions, where sinusodial functional dependences were sought in both directions. We previously showed that the dispersion relation takes the form

$$ik_y \omega_{B_0}^2 \omega_F - igHk_y \left(k^2 \omega_{B_0} + k_x \beta\right) - \beta \omega_F f - if^2 k_y \omega_F = 0. \tag{10.80}$$

Our task is to separate the preceding expression into real and imaginary components and then demand that each component vanish independently. The difficulty of the task comes from separating the complex frequency $\omega_{B_0} = \omega + i\omega_B$ into real and imaginary components, which first involves doing the same for ω_B.
(a) First, write $\omega_B = \zeta_R + i\zeta_I$ without worrying about what the actual functional forms of the *separation functions* ζ_R and ζ_I are. By using this expression for ω_B, show that two non-complex/real expressions for the dispersion relation are obtained [93],

$$\omega_I^3 - 3\omega_R^2\omega_I - (2\zeta_R - F_0)\left(\omega_R^2 - \omega_I^2\right) + 4\zeta_I\omega_R\omega_I - \omega_I\left[\zeta_R\left(2F_0 - \zeta_R\right) + \zeta_I^2\right]$$

$$- 2\zeta_I\left(F_0 - \zeta_R\right)\omega_R + gHk^2\left(\omega_I + \zeta_R\right) - \frac{\beta f\omega_R}{k_y} + f^2\left(\omega_I - F_0\right)$$

$$- \left(\zeta_R^2 - \zeta_I^2\right)F_0 = 0,$$

$$\omega_R^3 - 3\omega_R\omega_I^2 - 2\left(2\zeta_R - F_0\right)\omega_R\omega_I - 2\zeta_I\left(\omega_R^2 - \omega_I^2\right) + \omega_R\left[\zeta_R\left(2F_0 - \zeta_R\right) + \zeta_I^2\right]$$

$$- 2\zeta_I\left(F_0 - \zeta_R\right)\omega_I + gHk^2\left(\zeta_I - \omega_R\right) - f^2\omega_R + \frac{\beta f}{k_y}\left(F_0 - \omega_I\right) - 2\zeta_R\zeta_I F_0$$

$$- gHk_x\beta = 0.$$

$$(10.81)$$

where we again have $k^2 = k_x^2 + k_y^2$.
(b) Next, we wish to derive the functional forms of ζ_R and ζ_I. By recalling the definition of ω_B, show that

$$\zeta_R = \omega_\nu + \left(\frac{v_A}{H}\right)^2 \frac{\eta k^2 + \omega_I}{\omega_R^2 + \left(\eta k^2 + \omega_I\right)^2},$$

$$\zeta_I = \left(\frac{v_A}{H}\right)^2 \frac{\omega_R}{\omega_R^2 + \left(\eta k^2 + \omega_I\right)^2}.$$

$$(10.82)$$

(c) In the free, non-rotating, hydrodynamic limit, show that the dispersion relation for gravity waves is recovered.

10.11.5 Obtaining the dispersion relation using De Moivre's formula

When the functional dependence across latitude is non-sinusoidal, a different, complex dispersion relation ensues,

$$\left(\omega_F\omega_{B_0} - k_x^2 - \frac{k_x}{\omega_{B_0}}\right)\left(\frac{\omega_{B_0}}{\omega_F}\right)^{1/2} = 2n + 1,$$

$$(10.83)$$

as we previously described. When it is multiplied by ω_F throughout, the term $\left(\omega_F\omega_{B_0}\right)^{1/2}$ is present and has to be separated into its real and imaginary components.
(a) Using the separation functions ζ_R and ζ_I, we first write $\omega_F\omega_{B_0} = \zeta_R + i\zeta_I$. To

evaluate $(\omega_F \omega_{B_0})^{1/2}$, we need to use De Moivre's formula [93]. First, we need to realize that for any complex number $\zeta_R + i\zeta_I$, the amplitude is $\zeta = \sqrt{\zeta_R^2 + \zeta_I^2}$. If we denote the phase by ψ, then we may write $\zeta_R = \zeta \cos\psi$ and $\zeta_I = \zeta \sin\psi$. For a real number N, De Moivre's formula states that [7]

$$(\cos\psi + i\sin\psi)^N = \cos(N\psi) + i\sin(N\psi) \tag{10.84}$$

Use De Moivre's formula with $N = 1/2$ to show that

$$(\zeta_R + i\zeta_I)^{1/2} = \zeta^{1/2}\left[\cos\left(\frac{\psi}{2}\right) + i\sin\left(\frac{\psi}{2}\right)\right] \tag{10.85}$$

and

$$\cos\left(\frac{\psi}{2}\right) = \left(\frac{1 + \zeta_R/\zeta}{2}\right)^{1/2}, \quad \sin\left(\frac{\psi}{2}\right) = \left(\frac{1 - \zeta_R/\zeta}{2}\right)^{1/2}. \tag{10.86}$$

(d) Next, we need to relate these separation functions to the other parameters of the system. Show that

$$\omega_{B_0} = \zeta_- \omega_R + i\left(\zeta_0 + \zeta_+ \omega_I\right), \tag{10.87}$$

where the various separation functions are

$$\begin{aligned}
\zeta_\pm &= 1 \pm \frac{\epsilon\Gamma}{\omega_R^2 + \left(k_x^2/\mathcal{R}_B + \omega_I\right)^2}, \\
\zeta_0 &= \omega_\nu + \frac{\epsilon\Gamma k_x^2}{\mathcal{R}_B\left[\omega_R^2 + \left(k_x^2/\mathcal{R}_B + \omega_I\right)^2\right]}, \\
\zeta_R &= \zeta_- \omega_R^2 - \left(\omega_I - F_0\right)\left(\zeta_0 + \zeta_+ \omega_I\right), \\
\zeta_I &= \omega_R\left[\zeta_0 + \zeta_+ \omega_I + \zeta_-\left(\omega_I - F_0\right)\right].
\end{aligned} \tag{10.88}$$

(e) Finally, with the expressions for $(\omega_F \omega_{B_0})^{1/2}$ and ω_{B_0} in hand, show that

$$\begin{aligned}
&\zeta_-^2 \omega_R^3 - \omega_R\left(\zeta_0 + \zeta_+ \omega_I\right)^2 - 2\zeta_- \omega_R\left(\omega_I - F_0\right)\left(\zeta_0 + \zeta_+ \omega_I\right) - k_x^2 \zeta_- \omega_R - k_x \\
&- (2n+1)\left(\frac{\zeta + \zeta_R}{2}\right)^{1/2} = 0, \\
&2\zeta_-\left(\zeta_0 + \zeta_+ \omega_I\right)\omega_R^2 + \zeta_-^2 \omega_R^2\left(\omega_I - F_0\right) - \left(\omega_I - F_0\right)\left(\zeta_0 + \zeta_+ \omega_I\right)^2 - k_x^2\left(\zeta_0 + \zeta_+ \omega_I\right) \\
&- (2n+1)\left(\frac{\zeta - \zeta_R}{2}\right)^{1/2} = 0.
\end{aligned} \tag{10.89}$$

(f) What is the dispersion relation in the free, hydrodynamic limit? Which types of waves are present in the system?

10.11.6 Free, hydrodynamic shallow water waves on a sphere

The shallow water system yields analytical solutions even when cast on a sphere. In this problem, we revisit the classic solutions of the oceanographer Michael Longuet-Higgins, who derived them in a seminal paper in 1968 [149]. In standard spherical coordinates, the shallow water equations are

$$
\begin{aligned}
\frac{\partial v'_\theta}{\partial t} &= -\frac{g}{R}\frac{\partial h'}{\partial \theta} + 2\Omega v'_\phi \cos\theta, \\
\frac{\partial v'_\phi}{\partial t} &= -\frac{g}{R\sin\theta}\frac{\partial h'}{\partial \phi} - 2\Omega v'_\theta \cos\theta, \\
\frac{\partial h'}{\partial t} &+ \frac{H}{R\sin\theta}\left[\frac{\partial}{\partial \theta}\left(v'_\theta \sin\theta\right) + \frac{\partial v'_\phi}{\partial \phi}\right] = 0,
\end{aligned}
\tag{10.90}
$$

where θ is the polar angle, ϕ is the azimuthal angle, v'_θ and v'_ϕ are the polar and azimuthal components of the perturbed velocity, respectively, and R is the radius of the exoplanet. To render the governing equations amenable to analytical solution, one has to perform a set of less-than-obvious mathematical transformations,

$$
\begin{aligned}
v''_{\theta,\phi} &\equiv v'_{\theta,\phi}\sin\theta, \\
h'_{\rm v} &\equiv \frac{gh'}{2\Omega R}, \\
\mu &\equiv \cos\theta, \\
\hat{D} &\equiv -\sin\theta\frac{\partial}{\partial \theta} = \left(1-\mu^2\right)\frac{\partial}{\partial \mu}, \\
t_0 &\equiv 2\Omega t.
\end{aligned}
\tag{10.91}
$$

Furthermore, one has to define the Rossby number as $\mathcal{R}_0 \equiv c_0/2\Omega R$ and write *Lamb's parameter* as $\xi \equiv 1/\mathcal{R}_0^2$.

(a) Using the mathematical transformations described above, show that the governing equations become

$$
\begin{aligned}
-i\frac{\partial}{\partial t_0}\left(iv''_\theta\right) - \mu v''_\phi - \hat{D}h'_{\rm v} &= 0, \\
i\mu\left(iv''_\theta\right) - \frac{\partial v''_\phi}{\partial t_0} - \frac{\partial h'_{\rm v}}{\partial \phi} &= 0, \\
\left(1-\mu^2\right)\frac{\partial h'_{\rm v}}{\partial t_0} + \mathcal{R}_0^2\left[i\hat{D}\left(iv''_\theta\right) + \frac{\partial v''_\phi}{\partial \phi}\right] &= 0.
\end{aligned}
\tag{10.92}
$$

(b) Next, seek wave solutions of the following form,

$$
iv''_\theta = v_{\theta_0}e^{i(m\phi - \omega t_0)}, \quad v''_\phi = v_{\phi_0}e^{i(m\phi - \omega t_0)}, \quad h'_{\rm v} = h_{v_0}e^{i(m\phi - \omega t_0)},
\tag{10.93}
$$

where m is the *zonal wavenumber*,[9] and prove that the wave amplitudes obey the following governing equations,

$$\omega v_{\theta_0} + \mu v_{\phi_0} + \hat{D} h_{v_0} = 0,$$
$$\mu v_{\theta_0} + \omega v_{\phi_0} - m h_{v_0} = 0, \qquad (10.94)$$
$$\hat{D} v_{\theta_0} + m v_{\phi_0} - \omega \xi \left(1 - \mu^2\right) h_{v_0} = 0.$$

Analogous to the β-plane case, the amplitudes have a functional dependence on θ that we will solve for.

(c) Prove that this trio of wave-amplitude equations may be reduced to

$$\left(\omega \hat{D} + \mu m\right) \left[\frac{\left(\omega \hat{D} - \mu m\right) v_{\theta_0}}{\omega^2 \xi \left(1 - \mu^2\right) - m^2} \right] + \left(\omega^2 - \mu^2\right) v_{\theta_0} = 0. \qquad (10.95)$$

Subsequently, prove that

$$\frac{\partial}{\partial \mu} \left[\left(1 - \mu^2\right) \frac{\partial}{\partial \mu} \right] v_{\theta_0} - \frac{m v_{\theta_0}}{\omega} - \frac{m^2 v_{\theta_0}}{1 - \mu^2} + \xi \left(\omega^2 - \mu^2\right) v_{\theta_0}$$
$$+ \frac{2 \omega \xi \mu}{\omega^2 \xi \left(1 - \mu^2\right) - m^2} \left(\omega \hat{D} - \mu m\right) v_{\theta_0} = 0. \qquad (10.96)$$

This is the general governing equation for the polar component of the amplitude of the perturbed velocity. It has analytical solutions only in certain limits.

(d) In the limit of slow rotation ($\xi \to 0$), the governing equation for v_{θ_0} reduces to the associated Legendre equation [2, 7],

$$\left(1 - \mu^2\right) \frac{\partial^2 v_{\theta_0}}{\partial \mu^2} - 2\mu \frac{\partial v_{\theta_0}}{\partial \mu} - m \left[\frac{1}{\omega} + \frac{m}{\left(1 - \mu^2\right)} \right] v_{\theta_0} = 0. \qquad (10.97)$$

By recognizing that the following combination of quantities is discretized in terms of an integer l (following the convention used in quantum mechanics),

$$-\frac{m}{\omega} = l \left(l + 1\right), \qquad (10.98)$$

derive the dispersion relations, i.e., expressions for ω_{R} and ω_{I}.

(e) Show that, in the slowly-rotating limit, the solutions for the wave amplitudes are

$$v_{\theta_0} = v_0 \mathcal{P}_l^m,$$
$$h_{v_0} = \frac{v_0}{m^2} \left\{ \omega_{\mathrm{R}} \left(l - m + 1\right) \mathcal{P}_{l+1}^m - \mu \left[\omega_{\mathrm{R}} \left(l + 1\right) - m \right] \mathcal{P}_l^m \right\}, \qquad (10.99)$$
$$v_{\phi_0} = \frac{v_0}{m} \left[\left(l - m + 1\right) \mathcal{P}_{l+1}^m - \mu \left(l + 1\right) \mathcal{P}_l^m \right],$$

---

[9]This notation follows the tradition for describing spherical harmonics in quantum mechanics and is only used for this problem. In other circumstances, we use m to denote the mean molecular mass.

where $\mathcal{P}_l^m$ is the associated Legendre function [2, 7]. It is worth noting that, short of a normalization factor, $\mathcal{P}_l^m e^{im\phi}$ are spherical harmonics. Hence, prove that the time-dependent wave solutions are

$$v_\theta' = \frac{v_0 \mathcal{P}_l^m}{\sin\theta} \sin(m\phi - \omega_R t_0),$$

$$h_v' = \frac{v_0}{m^2} \left\{ \omega_R \left(l - m + 1 \right) \mathcal{P}_{l+1}^m - \mu \left[\omega_R \left(l + 1 \right) - m \right] \mathcal{P}_l^m \right\} \cos(m\phi - \omega_R t_0),$$

$$v_\phi' = \frac{v_0}{m \sin\theta} \left[(l - m + 1) \mathcal{P}_{l+1}^m - \mu \left(l + 1 \right) \mathcal{P}_l^m \right] \cos(m\phi - \omega_R t_0).$$

$$(10.100)$$

Do the solutions blow up when we have $\theta = 0°$?

(f) In the limit of fast rotation ($\xi \to \infty$), show that, by making the transformation,

$$\tilde{\mu} = \alpha\mu, \quad \alpha \equiv \xi^{1/4}, \qquad (10.101)$$

we obtain the following governing equation for v_{θ_0} *near the equator*,

$$\frac{\partial^2 v_{\theta_0}}{\partial \tilde{\mu}^2} + \left[\left(\xi\omega^2 - m^2 - \frac{m}{\omega} \right) \xi^{-1/2} - \tilde{\mu}^2 \right] v_{\theta_0} = 0. \qquad (10.102)$$

By demanding that the following combination of quantities is quantized,

$$\left(\xi\omega^2 - m^2 - \frac{m}{\omega} \right) \xi^{-1/2} = 2l + 1, \qquad (10.103)$$

derive the dispersion relations. Is $\omega_I = 0$ formally a solution of the dispersion relations? When do the dispersion relations reduce to their counterparts on the β-plane?

(g) Prove that the time-dependent wave solutions, in the rapidly-rotating limit, are

$$v_\theta' = \frac{v_0}{\sin\theta} e^{-\tilde{\mu}^2/2} \tilde{\mathcal{H}}_l \sin(m\phi - \omega_R t_0),$$

$$h_v' = \frac{v_0}{\omega_R^2 \xi - m^2} e^{-\tilde{\mu}^2/2} \cos(m\phi - \omega_R t_0) \left[2l\omega_R \alpha \tilde{\mathcal{H}}_{l-1} - \tilde{\mathcal{H}}_l \left(\omega_R \alpha\tilde{\mu} + \mu m \right) \right],$$

$$v_\phi' = \frac{v_0}{(\omega_R^2 \xi - m^2) \sin\theta} e^{-\tilde{\mu}^2/2} \cos(m\phi - \omega_R t_0) \left[2ml\alpha \tilde{\mathcal{H}}_{l-1} - \tilde{\mathcal{H}}_l \left(\tilde{\mu}m\alpha + \mu\omega_R \xi \right) \right],$$

$$(10.104)$$

where we have defined $\tilde{\mathcal{H}}_l \equiv \mathcal{H}_l(\tilde{\mu})$ and $\mathcal{H}_l(\tilde{\mu})$ is the Hermite polynomial of the lth order and with an argument $\tilde{\mu}$. Are the wave solutions finite when $\theta = 0°$? Why or why not?

Chapter Eleven

The de Laval Nozzle and Shocks

11.1 WHAT IS THE DE LAVAL NOZZLE?

In the 19th century, the Swedish industrialist Gustav de Laval became interested in designing steam turbines, which led to a construction now known as the *de Laval nozzle*, as shown in Figure 11.1. It was later widely adapted for use in rocket engines. In its most basic and idealized form, it is a tunnel that starts out with some width and continuously narrows to some minimum value of its cross-sectional area, known as the *throat*, before widening again in a geometrically symmetric fashion. Despite its apparent disconnect to exoplanetary atmospheres, we study the de Laval nozzle for several reasons: it provides a deep insight into the transition between subsonic and supersonic flow, it teaches us about the formation of shocks, and it forms the basis for understanding atmospheric escape in the hydrodynamic limit (Chapter 13).

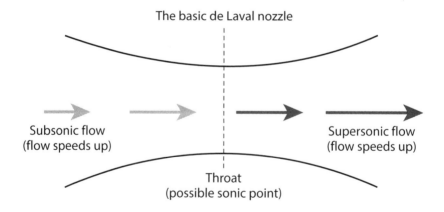

The basic de Laval nozzle

Subsonic flow
(flow speeds up)

Supersonic flow
(flow speeds up)

Throat
(possible sonic point)

Figure 11.1: The de Laval nozzle in its idealized form. Is the flow a mirror reflection of itself about the throat? (Surprisingly, it turns out that it is not.)

Despite its innocuous construction, the de Laval nozzle has a number of surprising—and pedagogically useful—features. Intuitively, one expects from everyday experience—some would call it "common sense"[1]—that the flow should

---

[1] To highlight its impediment to progress in physics, Einstein has been quoted as saying,

speed up as the nozzle narrows. If the mass flux of fluid through the nozzle is $\dot{M} = \rho v A$, where ρ is the mass density of the fluid, v is its flow velocity and A is the cross-sectional area of the nozzle, then it is easy to convince oneself that when $\dot{M}$ and ρ are held fixed, a decreasing A leads to an increasing v.

It turns out that this explanation is partially wrong, or at least incomplete. The mistake lies in assuming that the mass density is constant. More generally, differentiating the expression for $\dot{M}$ yields

$$\frac{d\dot{M}}{\dot{M}} = \frac{dA}{A} + \frac{dv}{v} + \frac{d\rho}{\rho} = 0. \tag{11.1}$$

The change in pressure[2] is $dP = -\rho v \, dv$. Using the definition for the sound speed $(c_s^2 = \frac{\partial P}{\partial \rho})$, we obtain

$$\frac{d\rho}{\rho} = -\mathcal{M}^2 \frac{dv}{v}, \tag{11.2}$$

where $\mathcal{M} \equiv v/c_s$ is the Mach number of the fluid flow. This expression already teaches us something: when the flow is very subsonic ($\mathcal{M} \ll 1$), which is typical of the liquids we encounter in everyday life, the change in density is negligible ($d\rho/\rho \approx 0$). In other words, incompressibility is directly related to the speed of the flow, where "fast" or "slow" is measured relative to the sound speed. Since changes in density are communicated via sound waves, in a system where sound waves are very fast ($\mathcal{M} \ll 1$) they occur almost instantaneously. Substituting the expression for $d\rho/\rho$ back into the expression for $d\dot{M}/\dot{M}$ gives

$$-\frac{dA}{A} = \frac{dv}{v}\left(1 - \mathcal{M}^2\right). \tag{11.3}$$

For very subsonic flows, the expression we just derived informs us that $dA/A = -dv/v$: a constriction of the nozzle ($dA < 0$) leads to an increase in the flow speed ($dv > 0$), which fits our "common sense" intuition. However, when the flow is supersonic ($\mathcal{M} > 1$), the opposite happens: the flow *slows down* when the nozzle narrows.

Equation (11.3) also informs us that, at the throat ($dA = 0$), the flow *always* crosses the sonic point ($\mathcal{M} = 1$). This is only true if the pressure and mass density obey the appropriate conditions, as we will see later when we develop a more sophisticated model of the de Laval nozzle. The key word is "always"—our simple model does not show us when this statement breaks down. For now, we will assume that the sonic point does indeed reside at the throat of the de Laval nozzle.

Already, our simple toy model of a de Laval nozzle teaches us something important: *a geometrically symmetric tunnel may give rise to a physically asymmetric flow.* Before the throat, subsonic flow means that the constriction of the

---

"Common sense is the collection of prejudices acquired before age eighteen."

[2]The minus sign is needed, because when the fluid slows down ($dv < 0$) it should lead to compression ($dP > 0$).

nozzle leads to an increasing flow speed. After the throat, the flow becomes supersonic, which has the consequence that, as the nozzle re-widens, it speeds up
as well. Conceptually, the de Laval nozzle highlights the use of compressibility
to accelerate a flow with the sound speed playing a key role. We will defer a
discussion of its relevance to understanding shocks to a later section.

11.2 WHAT ARE SHOCKS?

The speed of light is a well-known speed limit of Nature—it is the limit at
which electromagnetic signals may be communicated across space. Its lesser
known cousin is the speed of sound, which is the limit at which changes in
density are communicated. When two parcels of fluid have a relative speed that
is greater than the sound speed, they become causally disconnected.[3] A *shock
front* forms, separating the pre- and post-shock flow.

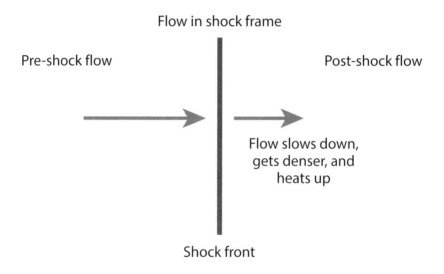

Figure 11.2: In certain situations, supersonic flow leads to the formation of a
shock, which causally separates the pre- and post-shock flow in an atmosphere.
The shock slows down the flow, renders it denser and heats it up.

What is the effect of the shock on the flow? To answer this question, we need
to derive the *Rankine-Hugoniot jump conditions* [261], which describe how the
velocity, density, pressure and temperature change across the shock front. They
are usually expressed in the frame of the shock, i.e., the shock is stationary in

---

[3]Two points in space separated by a great distance may only communicate at the speed
of light via electromagnetic signals. Superluminal communication is not allowed.

this reference frame. The jump conditions are nothing more than a restatement of the conservation of mass, momentum and energy.

Let the pre- and post-shock quantities be subscripted by "1" and "2," respectively. We adopt the usual notation: ρ for the mass density, v for the velocity, P for the pressure and T for the temperature. Consider a one-dimensional system with the distance denoted by x. The mass continuity equation in steady state,

$$\frac{\partial}{\partial x}(\rho v) = 0, \tag{11.4}$$

naturally leads to the jump condition,

$$\rho_1 v_1 = \rho_2 v_2. \tag{11.5}$$

This is by far the easiest jump condition to derive. Physically, it is nothing more than the conservation of the flux of mass per unit volume.

For the momentum equation in steady state and ignoring gravity, we have

$$\frac{\partial P}{\partial x} + \rho v \frac{\partial v}{\partial x} = 0. \tag{11.6}$$

However, we recognize that

$$\frac{\partial}{\partial x}(\rho v^2) = \rho v \frac{\partial v}{\partial x} + v \frac{\partial}{\partial x}(\rho v), \tag{11.7}$$

where the second term after the first equality vanishes due to mass conservation. The second jump condition follows,

$$P_1 + \rho_1 v_1^2 = P_2 + \rho_2 v_2^2. \tag{11.8}$$

This combination of quantities $(P + \rho v^2)$ is sometimes known as *Bernoulli's constant*. One may use it to understand why, for example, fast winds engulfing a house causes it to *explode*, rather than implode, because the pressure external to the house becomes *lower* than its value inside it.

For the final jump condition, we may just write down the conservation of energy (per unit volume),

$$\frac{\partial}{\partial x}\left(\frac{\rho v^2}{2} + \rho g z + E_{\text{int}}\right) = 0, \tag{11.9}$$

as the sum of the kinetic, potential and internal energies, where $E_{\text{int}} = P/(\gamma_{\text{ad}} - 1)$ is the internal energy, γ_{ad} is the adiabatic gas index and z is the height at which the shock is located. In hydrostatic equilibrium, we have $P = \rho g z$, which implies that

$$E_{\text{int}} + P = \frac{\gamma_{\text{ad}} P}{\gamma_{\text{ad}} - 1}. \tag{11.10}$$

The final jump condition is

$$\frac{v_1^2}{2} + \frac{\gamma_{\rm ad}}{\gamma_{\rm ad} - 1}\frac{P_1}{\rho_1} = \frac{v_2^2}{2} + \frac{\gamma_{\rm ad}}{\gamma_{\rm ad} - 1}\frac{P_2}{\rho_2}. \tag{11.11}$$

Collectively, these three jump conditions are known as the Rankine-Hugoniot conditions. We have derived them for a plane-parallel shock, but they may also be derived for an oblique shock (Problem 11.5.2). Furthermore, they may be generalized to magnetohydrodynamic shocks (Problem 11.5.3).

In the form in which we have derived them, the jump conditions are not very useful. It is more useful to cast them solely in terms of the Mach number of the pre-shock flow. For an adiabatic gas ($P_1 \propto \rho_1^{\gamma_{\rm ad}}$), we have the Mach number of the pre-shock flow being described by

$$\mathcal{M}^2 = \frac{\rho_1 v_1^2}{\gamma_{\rm ad} P_1}. \tag{11.12}$$

It follows that the changes in mass density, pressure and Mach number, across the shock front, are

$$\frac{\rho_2}{\rho_1} = \frac{v_1}{v_2} = \frac{(\gamma_{\rm ad} + 1)\,\mathcal{M}^2}{2 + (\gamma_{\rm ad} - 1)\,\mathcal{M}^2},$$

$$\frac{P_2}{P_1} = \frac{2\gamma_{\rm ad}\mathcal{M}^2 - \gamma_{\rm ad} + 1}{\gamma_{\rm ad} + 1}, \tag{11.13}$$

$$\mathcal{M}'^2 = \frac{2 + (\gamma_{\rm ad} - 1)\,\mathcal{M}^2}{2\gamma_{\rm ad}\mathcal{M}^2 - \gamma_{\rm ad} + 1}.$$

It is easy to verify that no shock exists when $\mathcal{M} = 1$. Physically, the pre- and post-shock flow may communicate when the flow is not supersonic and no causal discontinuity exists. For a strong shock ($\mathcal{M} \gg 1$), the density contrast becomes $\rho_2/\rho_1 = (\gamma_{\rm ad}+1)/(\gamma_{\rm ad}-1)$, which is 4 for a monoatomic gas and 6 for a diatomic one. The velocity contrast is the reciprocal of the density contrast.

The Mach number of the *post-shock flow* is denoted by $\mathcal{M}'$; the jump condition on $\mathcal{M}'$ follows naturally from the ones on the mass density and pressure. It is easy to verify that $\mathcal{M}' = 1$ when $\mathcal{M} = 1$. More importantly, we always have $\mathcal{M}' < 1$ when $\mathcal{M} > 1$ (Figure 11.3). In other words, the Mach number always decreases to less than unity across a shock. Our derivation and study of the Rankine-Hugoniot jump conditions have taught us two important lessons: supersonic flow is a necessary condition for shock formation and, once a shock forms or develops, the Mach number of the flow decreases across it.

In the limit of a strong shock, the jump condition involving the pressure may be rewritten yet again into one involving the post-shock temperature,

$$T_2 = \frac{2\,(\gamma_{\rm ad} - 1)}{(\gamma_{\rm ad} + 1)^2}\frac{m v_1^2}{k_{\rm B}}, \tag{11.14}$$

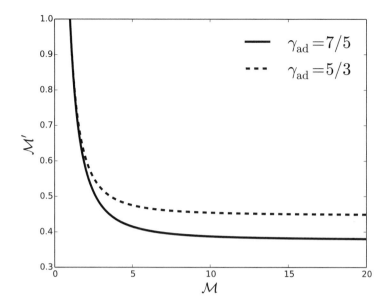

Figure 11.3: Post-shock Mach number as a function of the pre-shock Mach number. When $\mathcal{M} = 1$, we get $\mathcal{M}' = 1$ and no shock develops. Notice how $\mathcal{M}'$ is always less than unity and asymptotes to a constant value as the pre-shock Mach number becomes large.

where m is the mean molecular mass of the flow and k_B is the Boltzmann constant. Since the kinetic energy of the pre-shock flow is given by $mv_1^2/2$, we may deduce that $3/8 = 37.5\%$ of it is converted to heat by the shock, if the gas is monoatomic. For a diatomic gas, this percentage goes down to $5/18 \approx 28\%$.

So far, we neglected to mention an important caveat: it is assumed that the pre-shock flow eventually encounters a zero-velocity boundary condition—in other words, it hits a wall. In such a situation, $\mathcal{M} > 1$ is a necessary and sufficient condition for a shock to form. Generally, we will see that $\mathcal{M} > 1$ is a necessary but insufficient condition for shock formation. A more in-depth study of the de Laval nozzle will allow us to understand this issue.

11.3   WHAT DOES THE DE LAVAL NOZZLE TEACH US ABOUT SHOCKS?

The simple model of a de Laval nozzle, which we constructed earlier in the chapter, teaches us that it allows flows to transition from being subsonic to supersonic. Our derivation of the Rankine-Hugoniot conditions teaches us that supersonic flow is necessary for a shock to form. But is it sufficient? And does the transition to supersonic flow, within a de Laval nozzle, always happen? To answer these questions, we need a more sophisticated model of the de Laval nozzle.

11.3.1   The change in flow conditions across the de Laval nozzle

It turns out that this problem has been extensively studied by engineers working on aerodynamics and designing wind tunnels, because it is desirable to *avoid* shock formation in these situations [142]. Let the flow at the entry of the de Laval nozzle be described by a pressure (P_0), mass density (ρ_0), temperature (T_0) and velocity (v_0). It is safe to assume that the entry flow is reasonably slow ($v_0 \approx 0$). The sum of the potential and internal energies, per unit mass, may be written as $c_P T_0$, where c_P is the specific heat capacity at constant pressure. The conservation of specific[4] energy then allows us to write

$$\frac{v^2}{2} + c_P T = c_P T_0, \tag{11.15}$$

where v and T are the velocity and temperature, at some arbitrary point within the de Laval nozzle, respectively. An alternative expression involving the sound speed, which only depends on $c_P T$, is

$$c_s^2 = (\gamma_{\text{ad}} - 1) c_P T. \tag{11.16}$$

Denoting the sound speed of the flow at the entry point by c_{s_0}, energy conservation becomes

$$\frac{v^2}{2} + \frac{c_s^2}{\gamma_{\text{ad}} - 1} = \frac{c_{s_0}^2}{\gamma_{\text{ad}} - 1}. \tag{11.17}$$

A more useful form of the preceding equation is

$$\frac{T_0}{T} = 1 + \frac{(\gamma_{\text{ad}} - 1) \mathcal{M}^2}{2}, \tag{11.18}$$

where the Mach number, $\mathcal{M} \equiv v/c_s$, is defined for an arbitrary point within the nozzle.

Expressing energy conservation as the ratio of temperatures is useful, because we can then use it to derive expressions for how the pressure and mass density change from the entry point to any point within the de Laval nozzle. For an adiabatic and ideal gas, we have

$$\frac{P_0}{P} = \left(\frac{T_0}{T}\right)^{\gamma_{\text{ad}}/(\gamma_{\text{ad}}-1)} = \left[1 + \frac{(\gamma_{\text{ad}} - 1) \mathcal{M}^2}{2}\right]^{\gamma_{\text{ad}}/(\gamma_{\text{ad}}-1)},$$
$$\frac{\rho_0}{\rho} = \left(\frac{P_0}{P}\right)^{1/\gamma_{\text{ad}}} = \left[1 + \frac{(\gamma_{\text{ad}} - 1) \mathcal{M}^2}{2}\right]^{1/(\gamma_{\text{ad}}-1)}. \tag{11.19}$$

---

[4]Short for "per unit mass."

We are now able to relate the conditions of the flow upon its entry into the nozzle to any point, farther downstream, within it, placing us in a position to evaluate the conditions at the throat.

Suppose we now *demand* that the throat *is* the sonic point, as a means to derive its values of pressure and mass density. Setting $\mathcal{M} = 1$, we obtain

$$
\begin{aligned}
\frac{P_{\text{throat}}}{P_0} &= \left(\frac{2}{\gamma_{\text{ad}} + 1} \right)^{\gamma_{\text{ad}}/(\gamma_{\text{ad}} - 1)}, \\
\frac{\rho_{\text{throat}}}{\rho_0} &= \left(\frac{2}{\gamma_{\text{ad}} + 1} \right)^{1/(\gamma_{\text{ad}} - 1)}.
\end{aligned}
\tag{11.20}
$$

If we plug in $\gamma_{\text{ad}} = 7/5$ for a diatomic gas, it means that the flow transitions from subsonic to supersonic exactly at the throat if the pressure decreases to about 53% of its entry value. Correspondingly, the mass density drops to about 63% of its entry value.

By demanding that the flux of mass is constant (i.e., $\dot{M} = \rho v A$ is invariant) and denoting the cross-sectional area at the sonic point by A_s, one may show that

$$
\frac{A}{A_s} = \frac{1}{\mathcal{M}} \left\{ \frac{2}{\gamma_{\text{ad}} + 1} \left[1 + \frac{(\gamma_{\text{ad}} - 1)\,\mathcal{M}^2}{2} \right] \right\}^{(\gamma_{\text{ad}} + 1)/2(\gamma_{\text{ad}} - 1)}.
\tag{11.21}
$$

One may use the expression we derived earlier, for P_0/P, to eliminate the Mach number and thus obtain

$$
\begin{aligned}
\frac{A}{A_s} &= \left(\frac{\gamma_{\text{ad}} - 1}{2} \right)^{1/2} \left(\frac{2}{\gamma_{\text{ad}} + 1} \right)^{(\gamma_{\text{ad}} + 1)/2(\gamma_{\text{ad}} - 1)} \left(\frac{P}{P_0} \right)^{-1/\gamma_{\text{ad}}} \\
&\quad \times \left[1 - \left(\frac{P}{P_0} \right)^{1 - 1/\gamma_{\text{ad}}} \right]^{-1/2}.
\end{aligned}
\tag{11.22}
$$

It is desirable to recast equations (11.21) and (11.22) as expressions for the Mach number and ratio of pressures in terms of A/A_s. Algebraically, this would be messy. It is somewhat easier to simply plot curves of $\mathcal{M}$ and P/P_0 as functions of A/A_s, as shown in Figure 11.4. Generally, we see that there are two solution branches: subsonic and supersonic. The sonic point is depicted by a filled circle. If the pressure does not decrease enough, relative to its entry-point value, then the sonic point is never established. The de Laval nozzle becomes both geometrically and physically symmetric—the flow before and after the throat are identical in structure, and it remains subsonic throughout.

If the critical value of the pressure contrast is reached at the throat (i.e., $P/P_0 \approx 0.53$ for a diatomic gas), then the throat becomes the sonic point and something curious happens. The flow before and after the throat becomes physically asymmetric, because it becomes subsonic and supersonic, respectively. Generally, the entry and exit points of the de Laval nozzle need not be constructed in a geometrically symmetric fashion. For example, one may decide to

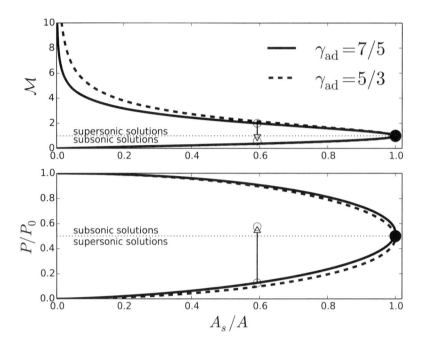

Figure 11.4: Mach number and ratio of pressures as functions of the normalized cross-sectional area of the de Laval nozzle. The filled circles represent the sonic point, while the empty circles depict the example of a Mach 2 shock for a diatomic gas. Adapted from Liepmann & Roshko [142].

build a nozzle which starts with $A_s/A = 0.1$, but ends with $A_s/A \approx 0.6$. In this case, the exit pressure may simply be read off the curve in Figure 11.4.

Since we constructed the curves for P/P_0 and $\mathcal{M}$ from the relations for an adiabatic gas, they are also *isentropic*, meaning that entropy is conserved for any solution along these curves. From our study of the Rankine-Hugoniot jump conditions, we have seen that, for a shock to develop, two things need to happen: the flow needs to be supersonic and the Mach number needs to decrease. Along these isentropic curves, the Mach number is able to smoothly transition from being subsonic to being supersonic, passing en route through the sonic point.

This already informs us that no shock forms, because one of the conditions is not fulfilled—since entropy depends on the temperature and mass density and these quantities change across a shock, we expect the entropy to change as well. It also tells us that *supersonic flow is a necessary but insufficient condition for shock formation.* This statement is not so surprising—after all, one may always transform to a frame of reference in which a supersonic fluid becomes stationary.

What ultimately matters is the *relative* speed between two parcels of fluid. In a reference frame in which a parcel of fluid is moving at supersonic speeds, what counts is that the Mach number of its neighbor is less than unity. Our sophisticated model of the de Laval nozzle teaches us something very important: it is possible to *avoid* shock formation, even when the flow is supersonic.

In Figure 11.4, we provide an example of a "Mach 2" ($\mathcal{M} = 2$) shock,[5] which causes the Mach number to decrease to about 0.6. In doing so, it picks out a solution *off the curve*, which is—by definition—non-isentropic. It informs us that shocks cause a change in the entropy of the flow. Correspondingly, the solution picks out a *higher* pressure than if it was residing on the isentropic curve. Thus, another way to visualize the situation is that shocks arise from the recompression of a supersonic flow.

An alternative way of elucidating the conditions under which a shock will form involves the *method of characteristics* [142, 261]. The mathematical machinery for answering this question is mature: essentially, first-order[6] partial differential equations may be recast as ordinary differential equations residing on special curves known as *characteristics*. The technique is formally used to solve the Riemann problem. When these characteristics intersect, the solution becomes multi-valued and shocks form. For the one-dimensional Euler equation, our characteristics are straight lines in the distance-time plane (Problem 11.5.5). Rather than restate the method of characteristics here, we have opted for a more physically intuitive approach to understanding shock formation.

11.4   APPLICATIONS TO, AND CONSEQUENCES FOR, EXOPLANETARY ATMOSPHERES

The physical intuition we have developed in this section is useful for understanding exoplanetary atmospheres, especially those that are tidally locked and experience strong dayside-only heating from their stars. This heating profile creates a temperature gradient between the dayside and nightside hemispheres, which produces winds. For the sake of argument, assume that the winds are constant in velocity and are described by an adiabatic gas (i.e., they lose no heat through radiative cooling). The Mach number of the winds is

$$\mathcal{M} \propto T^{-1/2} \propto P^{-(1-\gamma_{\mathrm{ad}})/2\gamma_{\mathrm{ad}}}. \tag{11.23}$$

For a diatomic gas (e.g., an atmosphere dominated by hydrogen or nitrogen), we obtain $\mathcal{M} \propto P^{-1/7}$. Even in the simplistic case where the wind velocity is constant across hemispheres, the Mach number decreases due to recompression of the atmospheric flow as it travels from the dayside to the nightside.

---

[5] For the example of the Mach 2 shock, the qualitative trends depicted in Figure 1 of [90] are correct, but the post-shock Mach number and pressure are discrepant with the calculations here. This arises from a coding error that mixed up γ_{ad} and $\mathcal{M}$ as inputs for computing A/A_s. Since these errors do not alter the conclusions or results of [90], I have decided to describe them here (and not as an erratum to the paper).

[6] In time t.

Typical astrophysical situation   **Exoplanetary atmosphere**

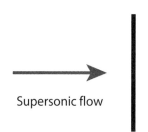

Supersonic flow slams into
a **zero-velocity boundary
condition** "wall"
(example: interstellar medium)

Continuous flow, across
hemispheres, with no "wall"
(but with temperature contrast
between hemispheres)

Figure 11.5: Schematic comparing the classical astrophysical situation (e.g., a supernova explosion) and an exoplanetary atmosphere with strong dayside-only heating.

Such a situation may be contrasted with the scenario typically encountered in astrophysics: a supersonic flow slams into a boundary condition that enforces a zero velocity (Figure 11.5). This is commonly encountered in the Universe at large [86]: the death of massive stars (supernovae), neutron stars streaking across the sky at ~ 1000 km s^{-1} (pulsar wind nebulae) and accretion disks around black holes inducing high-speed jets. In all of these cases, an energetic source drives a supersonic flow into the surrounding medium, which eventually brings the flow to a halt.

In exoplanetary atmospheres, there is no zero-velocity "wall." Instead, it is the contrast in temperature—and therefore pressure—between hemispheres that causes the Mach number of the flow to decrease. In other words, highly-irradiated atmospheres create their own "wall." Such a configuration is unfamiliar in the study of the atmospheres of Earth and the other Solar System bodies, because the flows encountered are never supersonic and induced by strong dayside-only heating.

11.5 PROBLEM SETS

11.5.1 Basic Rankine-Hugoniot jump conditions

(a) Derive the basic jump conditions in equation (11.13).
(b) By assuming an ideal gas, derive the jump condition for the temperature, as stated in equation (11.14).

(c) Recall the third equation in (11.13), which relates the pre- and post-shock Mach numbers, denoted by $\mathcal{M}$ and $\mathcal{M}'$, respectively. By assuming that $\mathcal{M}' < \mathcal{M}$, show that one always obtains $\mathcal{M} > 1$.

(d) Derive the jump condition for the entropy of the flow. Does the entropy increase or decrease across the shock front?

11.5.2   Oblique and bow shocks

Consider a plane-parallel shock, but with the flow approaching the shock front at an angle of β, where $\beta \leq 90°$. Let the spatial coordinate perpendicular and parallel to the shock front be denoted by x and y, respectively. Let the perpendicular and parallel[7] components of the velocity be denoted by $v_\perp$ and $v_\parallel$. We wish to generalize the Rankine-Hugoniot jump conditions to include the two different components of the velocity.

(a) By considering the mass continuity equation, show that $\rho v_\perp$ is invariant across the shock front. (Hint: consider a "pill box" encapsulating the shock front and let it become infinitesimally thin such that $\frac{\partial}{\partial y} = 0$.)

(b) By considering the parallel component of the momentum equation, show that $\rho v_\perp v_\parallel$ is conserved. Hence, show that $v_\parallel$ is invariant.

(c) Show that the perpendicular component of the momentum equation allows us to recover Bernoulli's constant, $P + \rho v_\perp^2$.

(d) Consider a *bow shock*, where the shock front is curved. Does the flow bend towards or away from the shock front?

(e) Show that the jump conditions for an oblique shock generalize to [142]

$$\frac{\rho_2}{\rho_1} = \frac{(\gamma_{\mathrm{ad}} + 1)\,\mathcal{M}^2 \sin^2 \beta}{2 + (\gamma_{\mathrm{ad}} - 1)\,\mathcal{M}^2 \sin^2 \beta},$$

$$\frac{P_2}{P_1} = \frac{2\gamma_{\mathrm{ad}}\mathcal{M}^2 \sin^2 \beta - \gamma_{\mathrm{ad}} + 1}{\gamma_{\mathrm{ad}} + 1}, \qquad (11.24)$$

$$\mathcal{M}'^2 = \frac{2 + (\gamma_{\mathrm{ad}} - 1)\,\mathcal{M}^2 \sin^2 \beta}{2\gamma_{\mathrm{ad}}\mathcal{M}^2 \sin^2 \beta - \gamma_{\mathrm{ad}} + 1}.$$

(Hint: the Mach number is defined using the *total* magnitude of the velocity.)

11.5.3   Magnetohydrodynamic (MHD) shocks

The jump conditions for a magnetized shock may be derived from the conservation of mass and momentum, as well as from the induction equation and by requiring that magnetic monopoles do not exist [54].

(a) By enforcing $\nabla \cdot \vec{B} = 0$, show that

$$\frac{\partial B_\perp}{\partial x} = 0. \qquad (11.25)$$

---

[7]Be aware that some authors prefer to define "parallel" as being parallel to the *normal* of the shock front, i.e., this is what we call "perpendicular."

Magnetic field lines passing through the shock front are unaltered by it.
(b) Show that $\rho v_\perp$ remains invariant even when a magnetic field is present.
(c) By considering the induction equation in the ideal MHD limit, show that

$$\frac{\partial}{\partial x}\left(v_\perp B_\| - B_\perp v_\|\right) = 0. \tag{11.26}$$

(d) By requiring the conservation of momentum, show that

$$\frac{\partial}{\partial x}\left(P + \rho v_\perp^2 + \frac{B_\|^2}{8\pi}\right) = 0,$$
$$\frac{\partial}{\partial x}\left(\rho v_\perp v_\| - \frac{B_\perp B_\|}{4\pi}\right) = 0. \tag{11.27}$$

Physically, Bernoulli's constant is augmented by the presence of both magnetic pressure and tension, while the parallel component of the velocity is now modified by the presence of magnetic tension.
(e) Consider the idealized situation where the magnetic field threads the fluid only in a direction parallel to the shock front. If the fluid approaches the shock front at an angle, does it get deflected towards or away from the front in the post-shock flow? Does the magnetic field strength increase or decrease across the shock?
(f) Next, consider a magnetic field that is purely perpendicular to the shock front. Again, consider fluid approaching the front at an angle. What effect does the magnetic field have on the fluid?

11.5.4   The de Laval nozzle

(a) For a diatomic gas, is the sound speed of the flow at the throat higher or lower than its value at the entry point? By how much? How about for a monoatomic gas?
(b) By assuming $\rho v A$ to be constant and using it to relate the sonic point with an arbitrary point in the flow, write A/A_s as consisting of a set of ratios of mass densities and velocities. Hence, derive equation (11.21).
(c) Show that the Mach number may be written as

$$\mathcal{M}^2 = \frac{2}{\gamma_{\mathrm{ad}} - 1}\left[\left(\frac{P_0}{P}\right)^{(\gamma_{\mathrm{ad}}-1)/\gamma_{\mathrm{ad}}} - 1\right]. \tag{11.28}$$

(d) Hence, derive equation (11.22).
(e) Imagine a de Laval nozzle in which the sonic point is reached *before* the throat. What do you think will happen to the flow?

11.5.5   The 1D Euler equation: The method of characteristics

It is challenging to derive analytical solutions via the method of characteristics, even for the one-dimensional Euler equation without gravity. We will "cheat" a little and consider the following pair of equations,

$$\frac{\partial \rho}{\partial t} + \frac{\partial}{\partial x}(\rho v) = 0,$$
$$\frac{\partial v}{\partial t} + v\frac{\partial v}{\partial x} = -\frac{1}{\rho_0}\frac{\partial P}{\partial x}. \tag{11.29}$$

We have taken an approximation on the pressure gradient term: we assert that changes to it are small enough that the mass density is always close to a background value denoted by ρ_0.

(a) By considering a characteristic speed v_0, as described by

$$v_0^2 = \frac{\rho \mathcal{R} T}{\rho_0}, \tag{11.30}$$

show that the pair of equations becomes

$$2\frac{\partial v_0}{\partial t} + 2v\frac{\partial v_0}{\partial x} + v_0\frac{\partial v}{\partial x} = 0,$$
$$\frac{\partial v}{\partial t} + v\frac{\partial v}{\partial x} + 2v_0\frac{\partial v_0}{\partial x} = 0. \tag{11.31}$$

(b) By adding and subtracting the preceding pair of equations in turn, show that it reduces to the *transport equation*,

$$\frac{\partial f}{\partial t} + (v \pm v_0)\frac{\partial f}{\partial x} = 0. \tag{11.32}$$

What is the expression for the quantity, f, being transported?

(c) Mathematically and independent of the physics, the total derivative of $f = f(x, t)$ is

$$\frac{df}{dt} = \frac{\partial f}{\partial t} + \frac{\partial f}{\partial x}\frac{dx}{dt}. \tag{11.33}$$

Using this expression, show that $\frac{df}{dt} = 0$ if

$$x = (v \pm v_0)t + x_0, \tag{11.34}$$

where x_0 is some constant. Hence, the quantity f is conserved along straight lines in the x-t plane. The pair of partial differential equations is reduced to a single ordinary differential equation with a trivial solution.

(d) Show that the same machinery outlined here applies exactly to the shallow water wave system without the need for the approximation involving the pressure gradient term.

Chapter Twelve

Convection, Turbulence and Fluid Instabilities

12.1 FLUID MOTION INDUCED BY PHYSICALLY UNSTABLE CONFIGURATIONS

Atmospheric motion may generally be seeded by unstable configurations of temperature, density or entropy. Convection is probably the most common instability encountered in an atmosphere. It is also a very generic instability, because it does not rely on the atmosphere being approximated as a fluid and instead derives directly from thermodynamics. Turbulence is a ubiquitous feature of fluids on large scales and yet it remains poorly understood.

In this chapter, we will study the conditions under which convection is triggered and how it is (approximately) modeled. Specifically, we examine mixing length theory and the convective adjustment scheme, which are often used in one- and three-dimensional models of an atmosphere, respectively. Mixing length theory is sometimes used to mimic the effects of turbulence. Finally, we will study the Rayleigh-Taylor, Kelvin-Helmholtz and baroclinic instabilities, which are generic instabilities that are expected to be present in atmospheres.

12.2 HOT AIR RISES AND COLD AIR SINKS: SCHWARZSCHILD'S CRITERION FOR CONVECTIVE STABILITY

Convection refers specifically to the motion of fluid induced by an unstable entropy configuration. In everyday life, we experience it as the mechanism causing hot air to rise and cold air to sink. In atmospheres, we are interested in how it acts on larger scales, where a background gradient of temperature and pressure exists.

To begin, we invoke the first law of thermodynamics, which states that the heat gained or lost by a system (dQ) is the sum of the change in its internal energy (dE_{int}) and the work done on it (dW),

$$dQ = dE_{\text{int}} + dW. \tag{12.1}$$

We have cast all of the quantities in terms of energy per unit mass (and not per unit volume). The work done is associated with the change in volume of the system, PdV, with P being the pressure; the volume per unit mass is $V = 1/\rho$ with ρ being the mass density. The change in internal energy is

$dE_{\mathrm{int}} = c_V \, dT$, where c_V is the specific heat capacity at constant volume and T is the temperature. The heat gained or lost is directly related to the change in the specific entropy of the system,

$$dQ = T dS. \tag{12.2}$$

By using the ideal gas law, we obtain an expression for the change in specific entropy,

$$\frac{dS}{c_P} = d\left(\ln T\right) - \kappa_{\mathrm{ad}} \, d\left(\ln P\right), \tag{12.3}$$

where $\kappa_{\mathrm{ad}} \equiv \mathcal{R}/c_P$ is the adiabatic coefficient, $\mathcal{R}$ is the specific gas constant and c_P is the specific gas capacity at constant pressure. If κ_{ad} is constant, it follows that the entropy (per unit mass) is

$$S = \int c_P \, d\left(\ln \Theta\right). \tag{12.4}$$

The quantity Θ is the potential temperature,

$$\Theta \equiv T \left(\frac{P}{P_0}\right)^{-\kappa_{\mathrm{ad}}}, \tag{12.5}$$

where we have interpreted the constant of integration to be $\ln P_0^{\kappa_{\mathrm{ad}}}$ and P_0 is some reference pressure. If c_P is constant, then short of a constant factor the specific entropy and potential temperature are equivalent quantities. As mentioned in an earlier chapter, the potential temperature is the generalization of the temperature and accounts for adiabatic changes in it due to pressure variations. In the limit of a constant pressure, the potential temperature and temperature may be visualized interchangeably. This is the typical situation in everyday life: fluid heated from below becomes buoyant and rises; over the short length scale on which this occurs, the change in pressure is negligible. Over much larger length scales, the change in pressure has to be accounted for and the potential temperature is the appropriate quantity to consider.

An equivalently useful quantity is the *potential density*, which is the pressure-adjusted density of a parcel of atmosphere,[1]

$$\rho_\Theta = \rho \left(\frac{P}{P_0}\right)^{\kappa_{\mathrm{ad}}}. \tag{12.6}$$

It may be obtained by applying the ideal gas law and replacing T by Θ. It is useful to visualize the potential temperature and density in tandem, since a higher potential temperature corresponds to a lower potential density (and vice versa). If one compares two parcels of air at different altitudes or pressures, the

---

[1] Our definition departs from that of Vallis [244], who defined the potential density as the density of a fluid parcel if it was compressed to the reference pressure P_0.

comparison is only fair if it is the potential density, and not the density, that is being scrutinized.

Generally, we expect a gradient of entropy to exist within a fluid. When is this gradient stable to convection? And what gives the fluid a sense of direction for which to determine if the gradient is positive or negative? The answer to the second question is gravity. In an atmosphere, gravity works against pressure support to establish hydrostatic equilibrium: the pressure is generally higher towards the surface of the exoplanet and decreases with increasing altitude. Do the following thought experiment: displace a parcel of atmosphere and ask what will happen to its entropy? If the parcel gets displaced downwards, the increase in pressure *decreases* its entropy; displacing it upwards does the opposite. If the background entropy of the atmosphere *increases* with altitude, this is a stable situation.

Conversely, we expect the configuration of decreasing entropy (increasing potential density) with increasing altitude to be unstable. A parcel displaced upwards will decrease its potential density. Since it is more tenuous than its environment, it continues to rise. If the parcel is displaced downwards, its enhanced potential density, compared to its surroundings, causes it to keep sinking. Convection requires a pressure gradient to act and it is gravity that establishes it. Generally, one can imagine that other mechanisms may act to establish the pressure gradient, but for historical reasons we do not usually term the resulting motion as being convective.

This thought experiment immediately tells us that the atmosphere is stable against convection if

$$\frac{\partial S}{\partial z} \geq 0 \implies \frac{\partial \Theta}{\partial z} \geq 0. \tag{12.7}$$

In other words, the pressure-adjusted temperature needs to increase with altitude. It resolves a paradox that defies "common sense": over large scales, the Earth's atmosphere is hotter near the surface, becomes colder with altitude and yet is convectively stable (in most regions), defying our expectation that hot air rises and cold air sinks. Over atmospheric scales, "hot" and "cold" are described by the potential temperature and not the temperature.

We may use this property to derive a stability condition on the temperature gradient of a hydrostatic atmosphere. By differentiating the expression for Θ, we obtain

$$\frac{T}{\Theta} \frac{\partial \Theta}{\partial z} = \frac{\partial T}{\partial z} + \frac{g}{c_P}. \tag{12.8}$$

By demanding that $\frac{\partial \Theta}{\partial z} \geq 0$, we obtain (Karl) *Schwarzschild's criterion*,

$$\Gamma \leq \Gamma_{\text{ad}} \tag{12.9}$$

The quantity $\Gamma \equiv -\frac{\partial T}{\partial z}$ is typically termed the *lapse rate*, while the *adiabatic lapse rate* is given by

$$\Gamma_{\text{ad}} \equiv \frac{g}{c_P}. \tag{12.10}$$

In the absence of convection, the adiabatic lapse rate gives the drop in temperature with altitude. For example, we estimate that $\Gamma_{ad} \approx 10$ K km^{-1} on Earth.
In reality, moisture and fluid motion account for a gentler drop in temperature.

In the absence of moisture, Schwarzschild's criterion sets a limit on the steep-
ness of the temperature gradient in an atmosphere. If the temperature decreases
too rapidly, then the atmosphere becomes unstable to convection. In such a situation, we say that the lapse rate is *super-adiabatic*.

12.3   A SIMPLIFIED "THEORY" OF CONVECTION: MIXING LENGTH THEORY

So far, we have described *why* convection exists and *how* it is triggered. Now, we
will quantify its influence on the atmospheric flow. The German engineer Ludwig
Prandtl first conceived of an approximate way to describe convection known
as *mixing length theory*. He visualized a convectively unstable parcel of fluid
traveling some characteristic distance before dissolving and merging with its
surroundings. Consider a parcel of atmosphere that has a density contrast $(\delta\rho)$
with its surroundings, such that the force (per unit volume), due to buoyancy,
on it is

$$F = -\delta\rho\, g. \tag{12.11}$$

The minus sign is present because a positive force (upward motion) requires a
negative density contrast, i.e., the parcel needs to be less dense than its surroundings.

Using the ideal gas law, we obtain

$$\frac{\delta\rho}{\rho} = \frac{\delta P}{P} - \frac{\delta T}{T}. \tag{12.12}$$

In traveling a mixing length (l_{mix}), we assume that the mean molecular mass of
the parcel is unchanged. We also assume that it comes into pressure equilibrium
with its surroundings $(\delta P = 0)$ instantaneously.[2]

The key assumption involved in mixing length theory is that the temperature
contrast between the parcel and its environment is caused by a super-adiabatic
temperature gradient,

$$\delta T = (\Gamma - \Gamma_{ad})\, l_{mix}, \tag{12.13}$$

such that $\delta T = 0$ when $\Gamma \le \Gamma_{ad}$. This implies that the force due to buoyancy is

$$F = \frac{\rho g l_{mix}}{T} (\Gamma - \Gamma_{ad}) \tag{12.14}$$

To relate the buoyancy force with a characteristic speed of convection (v_{conv}),
we invoke the *virial theorem*, which states that half of the potential energy of

---

[2]This assumption ensures that $v_{conv} = 0$ when $\Gamma = \Gamma_{ad}$.

the system is transformed into kinetic energy,

$$\frac{Fl_{\mathrm{mix}}}{2} = \frac{\rho v_{\mathrm{conv}}^2}{2}. \tag{12.15}$$

We interpret the potential energy to be the work done on the parcel to displace it by a mixing length. It follows that the convective speed is

$$v_{\mathrm{conv}} = l_{\mathrm{mix}} \left[\frac{g}{T} \left(\Gamma - \Gamma_{\mathrm{ad}} \right) \right]^{1/2}, \tag{12.16}$$

and the convective time scale is $t_{\mathrm{conv}} = l_{\mathrm{mix}}/v_{\mathrm{conv}}$.

Mixing length "theory" is not really a theory in a strict sense, because l_{mix} cannot be calculated from first principles. In practice, three-dimensional numerical simulations of convection are performed and the value of l_{mix} is calibrated to match the results.

If we visualize the atmosphere as consisting of a collection of layers, each with a thickness $\sim H$, then convection may be approximated as diffusion if $l_{\mathrm{mix}} \ll H$. In this limit, we may approximately derive a diffusion coefficient, often termed the *eddy diffusion coefficient*,

$$K_{\mathrm{zz}} \sim v_{\mathrm{conv}} l_{\mathrm{mix}} = l_{\mathrm{mix}}^2 \left[\frac{g}{T} \left(\Gamma - \Gamma_{\mathrm{ad}} \right) \right]^{1/2}. \tag{12.17}$$

Generally, convection is hardly a diffusive process, and we may have $l_{\mathrm{mix}} \sim H$ in some situations, but the K_{zz} approach is widely used because it is an easy way to describe the strength of convection once one guesses the value of l_{mix}.

Mixing length theory is a convenient approach to including convection, but it remains a local description at best—meaning it is unable to capture large-scale structures formed by convection. Generally, short of performing numerical simulations, this is a very hard calculation to make and few analytical solutions exist.[3]

12.4   IMPLEMENTING CONVECTION IN NUMERICAL CALCULATIONS: CONVECTIVE ADJUSTMENT SCHEMES

Generally, convective motions occur on small scales. Numerical simulations of the governing equations of a fluid formally allow for convection to occur, but the fine resolutions needed may bottleneck the computational time step and render the simulation time infeasibly long. One way out of this dilemma is to implement a *convective adjustment scheme*, which assumes that convection occurs instantaneously to return the atmosphere to its stable, adiabatic lapse rate while conserving energy [157].

---

[3] *Rayleigh-Bénard convection* provides a classic example of self-organizing structures forming in the flow, but we will not pursue it here.

What do we exactly mean by "energy"? If we recall the first law of thermo-dynamics, the heat gained or lost, in units of energy per unit mass, is

$$dQ = c_V \, dT + P \, d\left(\frac{1}{\rho}\right). \tag{12.18}$$

By invoking the ideal gas law and hydrostatic equilibrium $(dP = -\rho g dz)$, we obtain

$$dQ = c_P \, dT + g \, dz. \tag{12.19}$$

Since the coefficients are now constants, we may simply integrate the preceding expression to obtain

$$Q = c_P T + gz, \tag{12.20}$$

which is known as the *specific enthalpy*. In the atmospheric and climate sciences, we sometimes refer to Q as the *dry static energy* [251]. Physically, our expectation is that convection occurs so quickly that any adjustments caused by it are adiabatic. Consequently, Q is invariant.

To implement convective adjustment, the relevant quantity is not the specific enthalpy, but the enthalpy per unit area in a column of atmosphere. It is the product of the enthalpy and the column mass,

$$Q\tilde{m} = \frac{QP}{g}. \tag{12.21}$$

Consider a pair of layers in an atmosphere labeled "1" and "2." Let Layer 1 reside at a lower pressure than Layer 2. Each layer is associated with its own height (z_1 and z_2), temperature (T_1 and T_2) and thickness (expressed as the incremental column mass associated with each layer, $\Delta\tilde{m}_1$ and $\Delta\tilde{m}_2$). The distance between the midpoints of each layer is $\Delta z = z_1 - z_2$.

If the temperature gradient between the layers is super-adiabatic, then the temperatures are replaced by T_1' and T_2', such that

$$\frac{T_2' - T_1'}{\Delta z} = \Gamma_{\text{ad}}. \tag{12.22}$$

For the total enthalpy of the pair of layers to be conserved, we must have

$$\Delta P_1 T_1 + \Delta P_2 T_2 = \Delta P_1 T_1' + \Delta P_2 T_2'. \tag{12.23}$$

Since convective adjustment leaves z_1 and z_2 unchanged, the potential energy per unit area $(gz\Delta\tilde{m})$ does not feature in the preceding expression. The pair of constraints may be solved to obtain the convectively-adjusted temperatures (T_1' and T_2') in terms of the old temperatures,

$$T_1' = \frac{\Delta P_1 T_1 + \Delta P_2 \left(T_2 - \Gamma_{\text{ad}}\Delta z\right)}{\Delta P_1 + \Delta P_2}, \tag{12.24}$$
$$T_2' = T_1' + \Gamma_{\text{ad}}\Delta z.$$

All pairs of atmospheric layers are checked for convective instability and their temperatures are adjusted, according to the procedure outlined, if necessary. The process is repeated until the entire atmosphere is convectively stable. This is known as a *dry convective adjustment scheme*. When moisture and condensation are important, moist convective adjustment schemes are needed [19, 20].

12.5   A SIMPLE "THEORY" OF TURBULENCE: THE SCALING LAWS OF KOLMOGOROV

Both laboratory experiments and numerical simulations bear witness to the fact that at some critical, large value of the Reynolds number, the orderly and laminar flow of a fluid gives way to disordered and *turbulent* flow. In atmospheres, the onset of turbulence serves as a source of friction and dissipation and also influences the dynamical structure. Entire tomes have been written on the subject of turbulence and it is not my intention to explore it exhaustively. Rather, I wish to touch upon some salient principles associated with turbulence, so as to give the reader a basis upon which to pursue the topic further, if desired. The subject of turbulence has puzzled great minds such as Heisenberg[4] and remains an unsolved problem of theoretical physics.

The basis upon which to think about turbulence comes largely from the Soviet mathematician Andrey Kolmogorov, who made seminal contributions to several branches of mathematics and physics. Drawing upon deep physical insights, he derived scaling laws to describe turbulence that essentially amount to nothing more than dimensional analysis [126, 127]. While the mathematics involved was straightforward or even trivial, the physical consequences of these scaling laws are profound.

In isolation, a fluid cannot remain turbulent indefinitely, as turbulence is inherently a dissipative process. The turbulent velocity field may be visualized as being composed of eddies or vortices of different sizes. Since vorticity is generally not a conserved quantity in three dimensions, energy is transferred from the largest eddies to the smallest ones, a process known as an *energy cascade*, an idea usually attributed to the English physicist Lewis Richardson.[5] At the smallest scale (l_{small}), the Reynolds number is of the order of unity and molecular viscosity becomes an important effect—energy is dissipated by these small eddies. A fluid may remain persistently turbulent only if the energy injected into, and dissipated by, it remain in equipoise. Energy injection is assumed to occur at the largest scale (l_{large}).

At intermediate scales, the eddies are described by a speed v and a size l. Dimensional analysis implies that the energy per unit mass and time cascading

---

[4]Legend has it that the German Nobel laureate and physicist, Werner Heisenberg, on his deathbed, said, "When I meet God, I am going to ask him two questions: Why relativity? And why turbulence? I really believe he will have an answer for the first." However, this quote has also been attributed to the English fluid dynamicist Horace Lamb.

[5]Richardson composed the verse, "Big whirls have little whirls that feed on their velocity, and little whirls have lesser whirls and so on to viscosity."

through these intermediate eddies is

$$\epsilon \propto \frac{v^3}{l}.$$ (12.25)

Since we expect ϵ to be constant, we must have

$$\frac{v_{\text{large}}^3}{l_{\text{large}}} = \frac{v_{\text{small}}^3}{l_{\text{small}}}.$$ (12.26)

But at the same time, because the Reynolds number at the smallest scale is of the order of unity, we expect that the molecular viscosity is

$$\nu \sim v_{\text{small}} l_{\text{small}}.$$ (12.27)

These considerations imply that there is a relationship between the largest and smallest scales in the turbulent cascade,

$$l_{\text{large}} \sim \mathcal{R}_e^{3/4} l_{\text{small}}.$$ (12.28)

In other words, the Reynolds number associated with the largest eddies $(\mathcal{R}_e)$ sets the scale of the smallest ones. The higher the Reynolds number, the wider the range of length scales spanned by the turbulent cascade.

The collection of length scales in between the smallest and largest $(l_{\text{small}} \leq l \leq l_{\text{large}})$ is known as the *inertia range*. A natural question to ask is: how is energy distributed across the inertia range? Instead of a length scale, physicists usually like to visualize the distribution of energy across wavenumber, $k \sim 1/l$. If the *power spectrum* (energy per unit mass and wavenumber) only depends on the rate of energy cascading through wavenumber, then dimensional analysis yields

$$E_k \propto \epsilon^{2/3} k^{-5/3}.$$ (12.29)

Remarkably, experiments have corroborated this basic behavior [34].

Since the seminal insights of Kolmogorov, a body of literature has emerged that has built on his work, including the exploration of turbulence in two dimensions. Key results include the finding that, while energy cascades to smaller scales in three dimensions, an *inverse cascade* of energy exists in two-dimensional turbulence [129, 130]. A large body of work also exists on attempting to understand if the zonal jets of Jupiter are the outcome of self-organizing, turbulent structures [67].

Analogous to convection, mixing length theory has also been used to describe turbulence, usually in the form of an eddy diffusion coefficient [134]. Again, the assumption being made is that the typical mixing length[6] associated with turbulence is much smaller than the pressure scale height of the atmosphere, such that turbulent transport resembles diffusion. In addition to having to derive a

---

[6]Also termed a *correlation length* [244].

mixing length from first principles, one has to also calculate the characteristic speed of the turbulent flow. When convection and turbulence are both present, the difficulty of the task is exacerbated.

The idea of approximately describing turbulent transport as a diffusive process is not restricted to the atmospheric sciences. In astrophysics, an influential description of accretion disks (around black holes, stars, etc.) is the "α-disk model" [213], which approximates the turbulent diffusion coefficient as

$$\nu_{\text{turb}} = \alpha c_s H, \tag{12.30}$$

where c_s is the sound speed and H is the pressure scale height of the disk. By comparing the α-disk model with submillimeter observations of $\sim 10^6$-year-old protoplanetary disks, it is estimated that $\alpha \sim 0.01$ [5], although there is no reason to expect that α will remain constant throughout the lifetime of a disk.

Nevertheless, the theory of turbulence remains incomplete. Outstanding questions include: how do we calculate the transitional or critical value of the Reynolds number from first principles? Is the Reynolds number even the correct quantity for predicting the transition from laminar to turbulent flow? How do all of the different scaling laws derive from the Navier-Stokes equation? How do we calculate and predict all of the coherent and incoherent structures forming at different scales? And what role does intermittency play? What is certain is that these questions will continue to puzzle theoretical physicists for decades to come.

12.6   WATER OVER OIL: THE RAYLEIGH-TAYLOR INSTABILITY

When a denser layer of fluid resides over a less dense one, we intuitively expect the situation to be unstable. Unlike in the case of convection, gravity plays a direct role in giving the system a sense of direction. In this section, we aim to describe the *Rayleigh-Taylor instability* using the simplest possible model and derive the time and length scales on which it occurs. To begin, we consider two layers of fluid with constant densities of ρ_1 and ρ_2; the layer labeled "1" sits on top of the one labeled "2." Initially, the interface separating the two layers resides at $z = 0$, where z is the vertical spatial direction. We assume the fluids to be incompressible ($\nabla.\vec{v} = 0$) and irrotational ($\nabla \times \vec{v} = 0$). While such assumptions may seem simplistic or severe, they do illustrate a point: this is a very generic instability that manifests itself independent of the effects of compressibility or the presence of vorticity.

The joint constraints of incompressibility and irrotationality lend themselves to a special simplification that we may employ: it allows for a streamfunction (Ψ) to be defined, such that

$$\vec{v} = \nabla \Psi. \tag{12.31}$$

The preceding expression automatically satisfies the condition of irrotationality. When applied to incompressibility, it yields Laplace's equation,

$$\nabla^2 \Psi = 0. \tag{12.32}$$

At this point, we appeal to our physical intuition. We demand that the vertical component of the velocity, given by $\frac{\partial \Psi}{\partial z}$, must vanish far away from the interface between the two fluid layers. This allows us to *guess* the form of the streamfunction,

$$\Psi = \begin{cases} U_1 x + \Psi_{0,1}\, e^{-kz}\, e^{i(kx - \omega t)}, & z > 0, \\ U_2 x + \Psi_{0,2}\, e^{kz}\, e^{i(kx - \omega t)}, & z < 0, \end{cases} \tag{12.33}$$

where, for simplicity, we have assumed that the wavenumbers in both the horizontal (x) and vertical (z) directions are equal and may be represented by k. Here, ω denotes the wave frequency. The quantities $\Psi_{0,1}$ and $\Psi_{0,2}$ are normalization factors that we will eventually eliminate. It is easy to check that this form of the streamfunction satisfies Laplace's equation. We put in the terms U_1 and U_2 by hand in order to elucidate the Kelvin-Helmholtz instability later—they allow for the two fluid layers to have non-zero and distinct horizontal speeds. In other words, U_1 and U_2 are the background speeds of each fluid layer.

We represent the vertical position of the interface by χ and allow it to oscillate or grow/decay in the horizontal direction,

$$\chi = \chi_0\, e^{i(kx - \omega t)}, \tag{12.34}$$

where χ_0 is a normalization constant. As a boundary condition, we assert that the vertical component of the velocity is given by the rate of change of the position of the interface,

$$\frac{\partial \Psi}{\partial z} = \frac{D\chi}{Dt}. \tag{12.35}$$

At $z = 0$, this boundary condition yields a pair of constraints relating $\Psi_{0,1}$, $\Psi_{0,2}$ and χ_0,

$$\begin{aligned} \Psi_{0,1} k &= i\chi_0\, (\omega - kU_1), \\ \Psi_{0,2} k &= i\chi_0\, (kU_2 - \omega). \end{aligned} \tag{12.36}$$

What is the governing equation into which we insert these relations? Recasting the Euler equation, excluding planetary rotation, in terms of the streamfunction leads to an elegant expression,

$$\nabla \left(\frac{\partial \Psi}{\partial t} + \frac{v^2}{2} + gz + \frac{P}{\rho} \right) = 0. \tag{12.37}$$

It tells us that the quantity enclosed by the parentheses is conserved across the interface. If we evaluate $v^2 = v_x^2 + v_z^2$ using the streamfunction and examine only the zeroth order terms, we arrive at the conclusion that the kinetic energy, per unit volume, is continuous across the interface,

$$\frac{\rho_1 U_1^2}{2} = \frac{\rho_2 U_2^2}{2}, \tag{12.38}$$

if the pressure is constant.

Retaining the first-order terms and applying the previously-derived relations in equation (12.36) allow us to obtain the dispersion relation,

$$\rho_1 \left(\omega - kU_1\right)^2 + \rho_2 \left(\omega - kU_2\right)^2 + \left(\rho_1 - \rho_2\right) gk = 0. \tag{12.39}$$

We are finally ready to examine the dispersion relation for the Rayleigh-Taylor instability, which occurs when the layers are at rest ($U_1 = U_2 = 0$),

$$\omega^2 = -gk\delta, \tag{12.40}$$

where we have written the density contrast as $\delta \equiv (\rho_1 - \rho_2)/(\rho_1 + \rho_2)$. When $\rho_1 > \rho_2$ or $\delta > 0$, the frequency (ω) becomes imaginary and the instability is triggered. When gravity is absent, the instability vanishes. If we examine the fluid layers on a length scale $l \sim 1/k$, then the time scale for the Rayleigh-Taylor instability to be triggered is

$$t_{\mathrm{RT}} \sim \left(\frac{l}{g\delta}\right)^{1/2}. \tag{12.41}$$

It is apparent that the instability acts more quickly on smaller scales.

12.7 SHEARING FLUIDS: THE KELVIN-HELMHOLTZ INSTABILITY

The *Kelvin-Helmholtz instabilty* is a variation on a theme of the Rayleigh-Taylor instability: two fluid layers, again with different densities, are moving with different speeds past each other. Having already done the hard work of deriving the dispersion relation in equation (12.36), we can simply re-arrange the expression to obtain a quadratic equation,

$$\omega^2 \left(\rho_1 + \rho_2\right) - 2k\omega \left(\rho_1 U_1 + \rho_2 U_2\right) + k^2 \left(\rho_1 U_1^2 + \rho_2 U_2^2\right) + \left(\rho_1 - \rho_2\right) gk = 0. \tag{12.42}$$

By writing down the solution for ω using the standard formula for solving a quadratic equation and demanding that ω is imaginary, we obtain a constraint on the wavenumber,

$$k > \frac{\left(\rho_2^2 - \rho_1^2\right) g}{\rho_1 \rho_2 \left(U_1 - U_2\right)^2}. \tag{12.43}$$

This constraint already allows us to make a powerful statement: in the absence of gravity, the Kelvin-Helmholtz instability acts on all length scales, albeit not with the same time scale. When gravity is present and the layers have a non-zero relative velocity between them, a finite density contrast between the layers implies the existence of a maximum length scale above which the Kelvin-Helmholtz instability does not act. Generally, the Kelvin-Helmholtz instability is bounded from above by gravity and from below by surface tension (see Problem 12.9.4). In other words, it is stabilized or neutralized by these physical effects.

The Kelvin-Helmholtz instability is an example of a barotropic instability, in which surfaces of constant density and pressure coincide.

12.8   WEATHER AT MID-LATITUDES: THE BAROCLINIC INSTABILITY

Earlier in the chapter, we convinced ourselves that the definition of the potential temperature does *not* depend on whether gravity is present—it arises solely from thermodynamics. As long as a mechanism acts to establish a gradient in pressure, gradients in the potential temperature and potential density may persist. Therefore, we generally expect gradients in the potential temperature to be unstable, regardless of whether they are in the horizontal or vertical directions. Horizontal gradients give rise to fluid motion that is essentially convective, except that gravity does not act to dictate if the horizontal gradient of the potential temperature should be positive or negative. The mechanism behind such "horizontal or slanted convection" is more formally known as the *baroclinic instability* [190]. On Earth, it is responsible for the weather at mid-latitudes and is caused by the gradient in insolation across latitude.

We wish to elucidate the conditions under which the baroclinic instability is triggered. Recall, from Chapter 9, the definition of the potential vorticity on isentropic surfaces,

$$\Phi = \frac{\vec{\omega} + 2\vec{\Omega}}{\rho}.\nabla\Theta. \tag{12.44}$$

Note that we now switch to using $\vec{\omega}$ to denote the vorticity (and not the wave frequency). This expression already tells us that the potential vorticity is seeded by gradients in the entropy or potential temperature. In the special case of an incompressible fluid, the potential vorticity may be defined on isopycnic or isodensity surfaces and be seeded by density gradients.

Without loss of generality, we will assume that the atmosphere has a neutral lapse rate ($\frac{\partial\Theta}{\partial z} = 0$), i.e., convection in the traditional sense plays no role in triggering the baroclinic instability. Also, we set $\vec{\Omega} = 0$ and $\frac{\partial\Theta}{\partial x} = 0$ to demonstrate that rotation and a zonal gradient of entropy are not essential ingredients. If the vertical component of the velocity is subdominant, then the potential vorticity becomes

$$\Phi = \frac{1}{\rho}\frac{\partial v_x}{\partial z}\frac{\partial\Theta}{\partial y}. \tag{12.45}$$

Let the mass density (ρ) be a function of z only. Differentiating the potential vorticity with respect to latitude (y) gives

$$\rho \frac{\partial \Phi}{\partial y} = \frac{\partial^2 v_x}{\partial y \partial z} \frac{\partial \Theta}{\partial y} + \frac{\partial v_x}{\partial z} \frac{\partial^2 \Theta}{\partial y^2}. \tag{12.46}$$

We now integrate the system over all values of z,

$$\int_0^\infty \rho \frac{\partial \Phi}{\partial y} \, dz = \left[\frac{\partial v_x}{\partial y} \frac{\partial \Theta}{\partial y} \right]_0^\infty - \int_0^\infty \frac{\partial v_x}{\partial y} \frac{\partial^2 \Theta}{\partial y \partial z} \, dz + \int_0^\infty \frac{\partial v_x}{\partial z} \frac{\partial^2 \Theta}{\partial y^2} \, dz. \tag{12.47}$$

The first term after the equality vanishes if we demand that the top and bottom of the atmosphere are stable to the baroclinic instability. The second term vanishes because we have earlier assumed a neutral lapse rate.

If we further integrate the system over all latitudes, then we obtain

$$\int_{-\infty}^\infty \int_0^\infty \rho \frac{\partial \Phi}{\partial y} - \frac{\partial v_x}{\partial z} \frac{\partial^2 \Theta}{\partial y^2} \, dz \, dy = 0. \tag{12.48}$$

Consider the simplest situation in which an unstable horizontal gradient of the potential temperature is present: $\frac{\partial \Theta}{\partial y} = C$, where C is a constant. In this case, we have

$$\int_{-\infty}^\infty \int_0^\infty \rho \frac{\partial \Phi}{\partial y} \, dz \, dy = 0. \tag{12.49}$$

Since the mass density is always positive, it means that $\frac{\partial \Phi}{\partial y}$ must change sign somewhere across latitude and height. In other words, within the spatial domain the potential temperature must have an inflection point. This is the most important of the *Charney-Stern-Pedlosky* conditions for the baroclinic instability to occur.

Typically, these conditions are derived under the approximation of a *quasi-geostrophic* atmosphere and/or using a linear analysis [105, 244], whereas we have chosen to perform a more heuristic derivation. Furthermore, the quasi-geostrophic approximation requires that the Rossby number is much less than unity, such that the force balance is primarily between the Coriolis force and the pressure gradient. On exoplanets, we do not generally expect the quasi-geostrophic approximation to hold, since the contribution of advection to the force balance cannot be ignored.

It is also easy to see why the Charney-Stern-Pedlosky conditions are necessary but insufficient. For example, if $\frac{\partial^2 \Theta}{\partial y^2} \neq 0$ or if meridional components of the Coriolis force are present, then a flip in the sign of the meridional gradient of the potential vorticity may not imply the presence of the baroclinic instability.

What happens if we wish to define the potential vorticity on isobaric surfaces? In this case, the equation governing the potential vorticity becomes (see Chapter 9)

$$\frac{D\Phi}{Dt} = \frac{\vec{\omega} + 2\vec{\Omega}}{\rho} . \nabla \left(\frac{DP}{Dt} \right), \tag{12.50}$$

and Φ is not a conserved quantity since there is generally no reason to assume that $\frac{DP}{Dt} = 0$. It is possible to show that the preceding expression may be recast into a form in which a quantity known as the "quasi-geostrophic potential vorticity" is conserved [105, 244].

It is worth noting that the term "baroclinic instability" is also used in the study of protoplanetary disks to describe a somewhat different and strictly non-linear mechanism [140].

12.9 PROBLEM SETS

12.9.1 Schwarzschild's criterion

Show that Schwarzschild's criterion for convective stability is equivalently written as

$$\frac{\partial (\ln T)}{\partial (\ln P)} \leq \kappa_{\text{ad}}. \tag{12.51}$$

In stellar astrophysics, κ_{ad} is often termed the *adiabatic gradient* [108], exactly for the reason that it is a dimensionless gradient when expressed in the preceding manner. Is the criterion more or less restrictive for a monoatomic versus a diatomic gas?

12.9.2 Convective adjustment for N atmospheric layers

Consider a region in an atmosphere that is convectively unstable and divided into N layers. Forcing each pair of unstable layers to the convective adiabat yields $N - 1$ algebraic equations, while enforcing the conservation of enthalpy (per unit area) yields the Nth algebraic equation.

(a) By forcing each pair of layers to have a lapse rate that is equal to the convective adiabat (Γ_{ad}), show that

$$T'_{i+1} - T'_i = C_0, \qquad i = 1, 2, 3, ..., N, \tag{12.52}$$

where T'_i is the convectively-adjusted temperature of the ith layer, $C_0 \equiv \Gamma_{\text{ad}}\Delta z$ and Δz is the distance between the midpoints of each layer. We assume that Δz is constant throughout the atmosphere.

(b) Prove that the conservation of enthalpy or dry static energy yields

$$\sum_{i=1}^{N} \Delta P_i\, T_i = \sum_{i=1}^{N} \Delta P_i\, T'_i, \tag{12.53}$$

where ΔP_i is the pressure difference between the top and bottom boundaries of the ith layer.

(c) The constraints derived in (a) and (b) may be compactly rewritten in the form of an equation involving a matrix $\hat{A}$,

$$\hat{A}\vec{T}'_i = \vec{C}, \tag{12.54}$$

where $\vec{T}_i'$ is the column matrix of convectively-adjusted temperatures. The column matrix $\vec{C}$ has C_0 as its first $N - 1$ elements. Derive the exact form of the matrix $\hat{A}$. What is the Nth element of $\vec{C}$?

(d) Consider the simple case of $N = 2$. What is the exact form of the matrix $\hat{A}$? What is the form of its inverse $(\hat{A}^{-1})$?

12.9.3   Rayleigh-Taylor instability

(a) Show that, if only the first-order terms are considered, then

$$v^2 = \begin{cases} 2ikU_1 \Psi_{0,1} \; e^{-kz} \; e^{i(kx - \omega t)}, & z > 0, \\ 2ikU_2 \Psi_{0,2} \; e^{kz} \; e^{i(kx - \omega t)}, & z < 0. \end{cases} \tag{12.55}$$

Hence, derive the dispersion relation in equation (12.39).

(b) A natural generalization of our model of the Rayleigh-Taylor instability involves allowing the pressures on either side of the interface to be unequal. Specifically, an additional boundary condition is imposed,

$$P_1 - P_2 = b_0 \frac{\partial^2 \chi}{\partial x^2}, \tag{12.56}$$

which describes the *surface tension* at the interface. The quantity b_0 is a constant of proportionality. What is the new dispersion relation? Does surface tension act more strongly on small or large scales? Does it reinforce or impede the Rayleigh-Taylor instability?

12.9.4   Kelvin-Helmholtz instability

Does the Kelvin-Helmholtz instability act more quickly on small or large scales? What is the influence of surface tension, at the interface, on the instability?

12.9.5   The baroclinic instability in the quasi-geostrophic limit

Reproduce the derivations of the Charney-Stern-Pedlosky conditions under the approximation that the flow is nearly geostrophic, as described in the textbooks of Holton [105] and Vallis [244]. List all of the assumptions invoked in the derivations.

Chapter Thirteen

Atmospheric Escape

13.1   THE KNUDSEN NUMBER AND JEANS PARAMETER

An exoplanet needs to be sufficiently massive or cold to retain an atmosphere. However, even in the presence of gravity this retention is imperfect and the atmosphere may slowly leak into space over geological or cosmic time scales. This process is known as *atmospheric escape* and has been studied extensively for the Earth and Solar System bodies, including Mars, Venus and Titan. It involves the investigation of their upper atmospheres, which is known as *aeronomy* [36].

Generally, the governing equation of atmospheric escape is the Boltzmann equation with various source and sink terms that account for the transport of particles, collisions between them and heating and cooling processes [215]. The Boltzmann equation is generally difficult to solve, but simplified governing equations, derived from it in various limits, may be solved instead. The *Knudsen number*,

$$\mathcal{K}_n \equiv \frac{l_{\mathrm{mfp}}}{H}, \tag{13.1}$$

where l_{mfp} is the mean free path for collisions between the constituent particles in the atmosphere and H is the pressure scale height, demarcates these limits. It describes how collisional a localized patch of atmosphere is. When $\mathcal{K}_n \ll 1$, it is collisional and may be described using fluid dynamics. When $\mathcal{K}_n \gg 1$, the governing equation is instead that of collisionless kinetics.

The *exopause* (or exobase) is the location in the atmosphere corresponding to $\mathcal{K}_n = 1$. In practice, it is not a sharp boundary, but rather a gradual transition. The *exosphere* sits above the exopause and is generally collisionless. Within the exosphere, particles may exist on a variety of orbits, but non-thermal processes may introduce collisional corrections to this picture [215]. The part of the upper atmosphere below the exopause is sometimes known as the *thermosphere*, and it is the region where photo-ionization plays a significant role in producing ions from atoms and molecules [68, 119, 120, 258].

Unlike the Knudsen number, the *Jeans parameter* describes the two physical regimes of atmospheric escape: *Jeans escape* and *hydrodynamic escape* [177, 248]. It is the ratio of the gravitational potential to the local thermal energy of the atmosphere,

$$\lambda_{\mathrm{J}} \equiv \frac{GM}{c_P T r}, \tag{13.2}$$

where G is Newton's gravitational constant, M is the mass of the exoplanet, c_P is the specific heat at constant pressure, T is the temperature and r is the radial coordinate (with its origin located at the center of the exoplanet). Jeans escape occurs when $\lambda_{\mathrm{J}} \gtrsim 1$ and the particles have relatively small thermal energies—and thus small thermal speeds—compared to their potential energies. In this situation, most of the particles below the exobase are slow. If these particles exist in a velocity distribution, then a minority of them in the high-velocity "tail" of this distribution may escape to space if they have speeds exceeding the *escape speed*,

$$v_{\mathrm{esc}} = \left(\frac{2GM}{r}\right)^{1/2} \approx 8 \text{ km s}^{-1} \left(\frac{M}{M_\oplus}\right)^{1/2} \left(\frac{r}{R_\oplus}\right)^{-1/2}. \tag{13.3}$$

Hydrodynamic escape, on the other hand, occurs when $\lambda_{\mathrm{J}} \lesssim 1$ and a radial outflow of particles becomes possible. For Earth, we have $\lambda_{\mathrm{J}} \sim 10\text{–}100$ if we take $T = 1000$ K, which clearly puts its atmosphere in the Jeans-escape regime. For super Earths with hydrogen-dominated atmospheres, we have $\lambda_{\mathrm{J}} \sim 0.1\text{–}1$, implying that hydrodynamic escape is a more appropriate description.

Jeans and hydrodynamic escape may also be understood from the perspective of characteristic speeds. Individual particles may escape the gravitational well of the exoplanet if they have thermal speeds,

$$v_{\mathrm{th}} \sim \left(\frac{k_{\mathrm{B}}T}{m}\right)^{1/2} \approx 0.3 \text{ km s}^{-1} \left(\frac{T}{300 \text{ K}}\right)^{1/2} \left(\frac{m}{29m_{\mathrm{H}}}\right)^{-1/2}, \tag{13.4}$$

with k_{B} being the Boltzmann constant and m the mean molecular mass, that are larger than the escape speed. (The exact order-of-unity coefficient in the preceding expression depends on if one is referring to the mean, root-mean-square or most probable speed.) Since the sound speed,

$$c_s = \left(\frac{\gamma_{\mathrm{ad}}k_{\mathrm{B}}T}{m}\right)^{1/2}, \tag{13.5}$$

where γ_{ad} is the adiabatic gas index, is comparable to the thermal speed, it follows that a particle may escape if its speed is supersonic provided that the sound speed exceeds the escape speed[1], which occurs when sufficient external heat is deposited in the atmosphere. In this manner, the sound speed provides a natural benchmark in the problem of atmospheric escape.

In this picture, hydrodynamic escape occurs when the sonic point of the flow is situated below the exobase—by the time the particles reach the exosphere and become collisionless, they already have sufficient thermal speeds to escape to space. In other words, a hydrodynamic flow remains collisional up to and beyond the sonic point. By contrast, Jeans escape occurs when the sonic point sits above the exobase, meaning that most of the particles do not attain thermal speeds that exceed the escape speed before they become collisionless.

---

[1] This is the case when $\lambda_{\mathrm{J}} \lesssim 1$.

In this chapter, we will build basic intuition by examining toy models of Jeans and hydrodynamic escape. Towards the end of the chapter, we will discuss non-thermal processes associated with atmospheric escape.

13.2   JEANS ESCAPE

The simplest model of Jeans escape [112], a mechanism named after the Englishman James H. Jeans, assumes that the constituent particles of the atmosphere may be described by a Maxwell-Boltzmann distribution,

$$f_{\mathrm{M}} = \left(\frac{a}{\pi}\right)^{3/2} n e^{-av^2}, \tag{13.6}$$

where $a \equiv m/2k_{\mathrm{B}}T$, n is the number density of particles and v is the magnitude of the velocity. The first moment of this distribution yields the escaping flux of the particles (Problem 13.6.1), known as the *Jeans escape flux* [215],

$$F_{\mathrm{J}} = n \left(\frac{2k_{\mathrm{B}}T}{\pi m}\right)^{1/2} (1 + \lambda_{\mathrm{esc}}) e^{-\lambda_{\mathrm{esc}}}, \tag{13.7}$$

where the *escape parameter* is defined as

$$\lambda_{\mathrm{esc}} \equiv a v_{\mathrm{esc}}^2 = \frac{m v_{\mathrm{esc}}^2}{2k_{\mathrm{B}}T}. \tag{13.8}$$

If we normalize the Jeans escape flux by $n v_{\mathrm{esc}}$, then $F_{\mathrm{J}}/n v_{\mathrm{esc}}$ becomes a one-parameter function that only depends on the escape parameter. The quantity $n v_{\mathrm{esc}}$ may be regarded as the idealized flux of particles per unit area and time, which implies that $F_{\mathrm{J}}/n v_{\mathrm{esc}}$ measures the efficiency of Jeans escape. Figure 13.1 shows the graphical solution of $F_{\mathrm{J}}/n v_{\mathrm{esc}}$ and places Mars, Earth and Venus on the same plot. Jeans escape works reasonably well for Mars and Earth and is believed to be partially responsible for the loss of water on the latter [215]. By contrast, classical (thermal) Jeans escape is inefficient on Venus.

13.3   THE CLASSICAL PARKER WIND SOLUTION

The traditional starting point of hydrodynamic escape is the *Parker wind solution*, which was originally formulated by the American solar physicist Eugene Parker to understand the solar wind [184, 185]. In its simplest form, it assumes an isothermal atmosphere, an assumption we will make in this section.

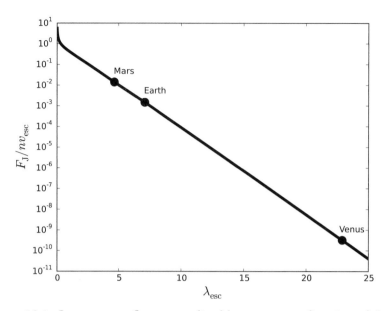

Figure 13.1: Jeans escape flux, normalized by nv_{esc}, as a function of the escape parameter. We have placed Mars, Earth and Venus on this plot using $\lambda_{esc} = 4.65, 7.06$ and 22.89, respectively, following the values calculated by Shizgal & Arkos [215].

To understand why the Parker wind solution is a simplification of the full problem, it helps to state the set of governing equations in a steady state,

$$\frac{\partial}{\partial r}\left(\rho v r^2\right) = 0,$$
$$v\frac{\partial v}{\partial r} = -\frac{1}{\rho}\frac{\partial P}{\partial r} - \frac{GM}{r^2},$$
$$\frac{\partial}{\partial r}\left(\frac{v^2}{2} + c_P T - \frac{GM}{r} - Q\right) = 0,$$

(13.9)

which is nothing more than the conservation of mass, momentum and energy along the radial coordinate and assuming spherical symmetry. The term Q accounts for all sources of heating and cooling and generally depends on r and time. It is challenging to solve this set of equations in a general sense, even though they are in a steady state.

The Parker wind solution "cheats" to obtain an answer, in the sense that the energy equation is ignored and one works only with mass and momentum conservation. Mass conservation leads to an expression for the gradient of the mass density (ρ),

$$\frac{1}{\rho}\frac{\partial \rho}{\partial r} = -\frac{1}{v}\frac{\partial v}{\partial r} - \frac{2}{r}. \tag{13.10}$$

In the momentum equation, the pressure may be eliminated for the sound speed by using

$$c_s^2 = \frac{\partial P}{\partial \rho}, \tag{13.11}$$

which leads to the momentum equation being

$$v\frac{\partial v}{\partial r} = -\frac{c_s^2}{\rho}\frac{\partial \rho}{\partial r} - \frac{GM}{r^2}. \tag{13.12}$$

Combining equations (13.10) and (13.12) yields a single governing equation,

$$\frac{\partial \left(v^2\right)}{\partial r}\left(1 - \frac{c_s^2}{v^2}\right) = \frac{4c_s^2}{r} - \frac{2GM}{r^2}. \tag{13.13}$$

Written in this form, it is easy to see that a general solution for v is difficult to obtain, because the sound speed generally depends on r. Since $c_s \propto T^{1/2}$ and the atmosphere is assumed to be isothermal, we end up with a situation where each side of the preceding expression may be integrated independently of the other. If the integration is performed from the sonic point to some arbitrary point, then we end up with

$$\mathcal{M}^2 - \ln\left(\mathcal{M}^2\right) = 4\ln x + 4\left(\frac{1}{x} - 1\right), \tag{13.14}$$

where we have defined the Mach number as $\mathcal{M} \equiv v/c_s$ and the normalized spatial coordinate as $x \equiv r/r_s$. The radius corresponding to the sonic point is given by

$$r_s = \frac{GM}{2c_s^2}. \tag{13.15}$$

Instead of a differential equation for the velocity, equation (13.14) is a transcendental, algebraic equation for the Mach number. It has five solution branches [21, 184], but three of them are unphysical: two of these branches are double-valued and the third is always supersonic (and never passes through the sonic point). A fourth branch is always subsonic and yields the *breeze solution*. The final branch consists of a pair of monotonic curves that pass through the sonic point: one of them is the Parker wind solution [184, 185], while the other describes *Bondi accretion* [21]. Outflows and accretion are opposite sides of the same coin.

It is illuminating to examine the asymptotic solutions of equation (13.14). If we consider a situation where the sonic point is located far below the exobase, then the solution corresponding to hydrodynamic escape has $x \gg 1$ and $\mathcal{M} \gg 1$,

$$v \approx 2c_s \left(\ln x\right)^{1/2}. \tag{13.16}$$

If we denote the mass density at the sonic point by ρ_s, then integrating over equation (13.10) from r_s to r yields

$$\rho = \rho_s \left(\frac{v}{c_s} \right)^{-1} x^{-2}. \tag{13.17}$$

Thus, the mass loss rate due to hydrodynamic escape is

$$\dot{M} = 4\pi r^2 \rho v = \frac{\pi \rho_s G^2 M^2}{c_s^3}. \tag{13.18}$$

Unfortunately, this simple model is not predictive, as we do not know how to calculate ρ_s from first principles and instead have to assume a value for it.

For completeness, we list the other asymptotic solutions from equation (13.14) as well. The accretion and breeze solutions are approximately described, at large distances from the exoplanet ($x \gg 1$), by

$$v \approx c_s x^{-2}. \tag{13.19}$$

At small distances ($x \ll 1$), the breeze solution is

$$v \approx c_s e^{-2/x}, \tag{13.20}$$

while the accretion solution is

$$v \approx 2 c_s x^{-1/2}. \tag{13.21}$$

13.4   NON-ISOTHERMAL PARKER WINDS: USING THE NOZZLE SOLUTIONS

Another way of simplifying the set of governing equations in (13.9) is to deal only with mass and energy conservation. This is mathematically identical to the de Laval nozzle solutions we previously discussed in Chapter 11, and we have seen that these solutions pass smoothly through the sonic point of the flow without producing shocks. The shortcoming of the isothermal Parker wind solution is that it does not allow us to inject energy into the system. On the other hand, the nozzle solutions do not allow us to load momentum into the wind, because there is no connection to the momentum equation. To gain further insight into the problem, we need to somehow combine both solutions such that the conservation of mass, momentum and energy are simultaneously considered.

Consider the conservation of energy between the base of the wind and an arbitrary point in the flow,

$$\frac{v_0^2}{2} + c_P T_0 - \frac{GM}{r_0} = \frac{v^2}{2} + c_P T - \frac{GM}{r}, \tag{13.22}$$

where quantities subscripted by zero refer to the base of the wind. We have assumed that the specific heat capacity at constant pressure remains invariant throughout the atmosphere. For simplicity, we absorb all sources of heating into $c_P T_0$ and assume that all heating occurs at the base of the wind. To render the algebra tractable, we further assume the gravitational potential energy of the wind to be roughly constant throughout, which yields

$$\frac{GM}{r} \approx \frac{GM}{r_0}. \tag{13.23}$$

Gravity still appears in the momentum equation to counteract the pressure gradient.

Reshuffling equation (13.22) yields an expression for T in terms of v,

$$T = T_0 \left[1 + \frac{1}{2c_P T_0} \left(v_0^2 - v^2 \right) \right]. \tag{13.24}$$

Using $c_s^2 = (\gamma_{\rm ad} - 1) c_P T$, we then insert $T(v)$ into equation (13.13) to obtain

$$\int_{v_0^2}^{v^2} \frac{\gamma_{\rm ad} + 1}{2} - \frac{\gamma_{\rm ad} - 1}{v^2} \left(c_P T_0 + \frac{v_0^2}{2} \right) d \left(v^2 \right)$$
$$= 2GM \left(\frac{1}{r} - \frac{1}{r_0} \right) + 4 \left(\gamma_{\rm ad} - 1 \right) c_P \int_{r_0}^{r} \frac{T}{r} \, dr. \tag{13.25}$$

To proceed, one needs to know the function $T(r)$. Strictly speaking, it is an outcome of solving the entire set of equations self-consistently. However, we will "cheat" by assuming an ad hoc functional form for the temperature,

$$T = T_0 \left(\frac{r}{r_0} \right)^{\alpha}, \tag{13.26}$$

where the index α is a free parameter of the model. By adopting different values for α, we may study the effects of different heating profiles on the wind at the expense of a self-consistent and fully predictive model for the atmosphere. By following through on equation (13.25), we obtain

$$\frac{\gamma_{\rm ad} + 1}{2} \left[\left(\frac{v}{v_0} \right)^2 - 1 \right] - \frac{\gamma_{\rm ad} - 1}{2} \left(1 + \frac{1}{\lambda_{\rm esc}} \right) \ln \left(\frac{v}{v_0} \right)^2$$
$$= \frac{\lambda_{\rm J}}{\lambda_{\rm esc}} \left(\frac{1}{x} - 1 \right) + \frac{2 \left(\gamma_{\rm ad} - 1 \right)}{\alpha \lambda_{\rm esc}} \left(x^{\alpha} - 1 \right), \tag{13.27}$$

where we have defined $x \equiv r/r_0$. To cast the problem in terms of dimensionless input parameters, we have defined the Jeans and escape parameters (at the base of the wind) to be, respectively,

$$\lambda_{\rm J} \equiv \frac{GM}{c_P T_0 r_0}, \quad \lambda_{\rm esc} \equiv \frac{v_0^2}{2c_P T_0}. \tag{13.28}$$

Just as in the case of the isothermal Parker wind solution, we may seek the asymptotic solution when $v \gg v_0$ and $x \gg 1$,

$$v \approx v_0 \left\{ \frac{2}{(\gamma_{\mathrm{ad}} + 1) \lambda_{\mathrm{esc}}} \left[\frac{2(\gamma_{\mathrm{ad}} - 1)}{\alpha} \left(\frac{r}{r_0} \right)^\alpha - \lambda_{\mathrm{J}} \right] \right\}^{1/2}. \tag{13.29}$$

The corresponding solution for the mass density may be obtained using equation (13.17). As expected, a steeper increase of the temperature with altitude (i.e., larger α) leads to higher wind speeds. Decreasing either the Jeans or escape parameter amounts to enhanced heating at the base of the wind, which increases the wind speed.

To confirm these trends, we compute exact solutions of equation (13.27), based on a numerical trick that is described in Problem 13.6.3, which allows you to obtain v for any x. Figure 13.2 shows examples of the non-isothermal Parker wind solutions for the wind speed and mass density. To check our intuition, I have varied the temperature index, Jeans parameter and escape parameter.

13.5 DETAILED PROCESSES: PHOTO-IONIZATION, RADIATIVE COOLING AND NON-THERMAL MECHANISMS

The toy models I have presented elucidate the basic features of Jeans and hydro-dynamic escape, but do not capture important details such as radiative heating and cooling. It is challenging—if not impossible—to include these effects in simple analytical models. There are also *non-thermal mechanisms* that have to do with atomic and molecular physics. In this section, I will summarize, in a heuristic way, the basic mechanisms associated with radiation and non-thermal processes that play a significant role in sculpting the structure of an escaping flow.

13.5.1 Extreme ultraviolet and X-ray radiation

In our toy models of hydrodynamic escape, heat is deposited at the base of the wind only. In reality, it may also be deposited throughout the wind—from beyond the sonic point to the base. The details of this deposition are the subject of radiative transfer, but generally speaking the incident stellar heating may be grouped into extreme ultraviolet (EUV) and X-ray radiation. The wind may be EUV- or X-ray-driven, depending on the relative positions of the *ionization front* created versus the sonic point (which is really an altitude or layer within the wind). Generally, the atmospheric escape of strongly-irradiated exoplanets is X-ray driven at early times (stellar age $\lesssim 100$ Myr), when the star is younger and more X-ray luminous, and EUV-driven at late times [181]. Later-type stars also tend to be more X-ray luminous than earlier-type ones.

An ionization front is the layer where the optical depth to EUV radiation and X-rays is on the order of unity. EUV radiation has sufficient photon energies to

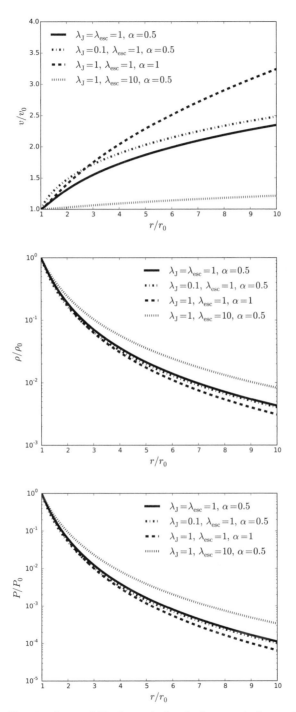

Figure 13.2: Non-isothermal Parker wind solutions: wind speed, mass density and pressure as functions of the radial coordinate. All quantities have been normalized by their values at the base of the wind. An atmosphere containing atoms has been assumed ($\gamma_{ad} = 5/3$).

ionize atomic hydrogen, which requires 13.6 eV. X-rays are mostly absorbed by, and photoionize, metals (e.g., atomic carbon and oxygen). Radiative cooling is mediated by Lyα radiation associated with hydrogen atoms [177] and the spectral lines of metals [181]. If the ionization front sits above the sonic point in altitude, then the flow is mostly driven by X-rays that penetrate past the ionization front and deliver heating at the base of the wind [180]. If the front sits below the sonic point, then the wind is EUV-driven [177, 180].

Detailed numerical calculations including EUV and X-ray heating have been carried out for hot Jupiters and have demonstrated that mass loss is negligible over their lifetimes [177, 180, 181]. One may visualize an efficiency of conversion of the incident photons into mechanical work. If all of the photons are used for radial flow, then we term it as being *energy-limited escape*. Hot Jupiters do not reside in this regime, because a significant fraction of the incident energy goes into a balance between ionization and recombination and also cooling by spectral lines [177, 180]. By contrast, the efficiency is close to being unity on Neptune- and Earth-like exoplanets, because of the lower escape speeds and shorter flow time scales needed to escape their gravitational potential wells—material effectively has less time to radiatively cool. Atmospheric escape is energy-limited to a good approximation, and it is believed to be responsible for sculpting the observed[2] mass-radius relations of these smaller exoplanets [181]. The energy-limited mass loss rate is

$$\dot{M} \sim \frac{\pi F_{\mathrm{X/EUV}} R^2}{E_g}, \tag{13.30}$$

where $F_{\mathrm{X/EUV}}$ is the X-ray or EUV flux from the star, R is the radius of the exoplanet and E_g is its specific potential energy. If atmospheric escape is not energy-limited, then the preceding expression needs to be multiplied by a dilution factor. The preceding expression arises from dimensional analysis.

13.5.2 Charge exchange

Stars produce winds containing charged particles that impinge upon the atmosphere of an exoplanet, a phenomenon that has been extensively studied for the solar wind. Specifically, energetic protons may swap electrons with atmospheric hydrogen atoms to produce a secondary population of hot hydrogen atoms known as *energetic neutral atoms* (ENAs) [104],

$$\mathrm{p}' + \mathrm{H} \rightarrow \mathrm{H}' + \mathrm{p}, \tag{13.31}$$

where primed quantities refer to the energetic populations. The preceding process is known as *charge exchange*. It has been observed to be at work in the interstellar medium shocked by supernova remnants and pulsar wind nebulae,

---

[2]Rather than the initial mass-radius relations, which are the outcomes of planet formation.

although the ENAs are known instead by the term *broad neutrals* as the thermal widths associated with their hydrogen lines are $\sim$ 100–1000 km s^{-1} in these objects [32, 42, 86].

In atmospheres, charge exchange creates ENAs with thermal speeds that easily exceed the escape speeds. It is a non-thermal process, because it occurs in the collisionless exosphere. Fundamentally, these ENAs originate from the stellar wind and are mainly a probe of the wind conditions[3] rather than those in the exosphere of the exoplanet [104], but they do demonstrate that the atmospheric escape of atomic hydrogen is occurring [56]. Charge exchange is believed to be at work in the exospheres of Mars, Earth and Venus [215].

Other non-thermal processes include *sputtering*, where fast ions impact the atoms or molecules and knock them out of the atmosphere.

13.6 PROBLEM SETS

13.6.1 Jeans escape

(a) Using equation (13.6), show that

$$\int f_{\mathrm{M}} \, d^3 v = n, \tag{13.32}$$

where f_{M} is the Maxwell-Boltzmann distribution.
(b) By calculating the first moment of the Maxwell-Boltzmann distribution,

$$F_{\mathrm{J}} = \int_0^{2\pi} \int_0^1 \int_0^\infty v^3 f_{\mathrm{M}} \, dv \, d\mu \, d\phi, \tag{13.33}$$

where $\mu \equiv \cos\theta$, derive the Jeans escape flux as stated in equation (13.7). Why is the integral over the polar angle only from $\mu = 0$ to $\mu = 1$ and not from $\mu = -1$ to $\mu = 1$?
(c) Show that the location of the exosphere may be approximated by

$$r_{\mathrm{exo}} \sim \frac{k_{\mathrm{B}} T}{mg} \ln\left(\frac{P_0 \sigma_{\mathrm{coll}}}{mg}\right), \tag{13.34}$$

where k_{B} is the Boltzmann constant, T is the temperature, m is the mean molecular mass, g is the surface gravity of the exoplanet, P_0 is its surface pressure and $\sigma_{\mathrm{coll}} \sim 10^{-15}$ cm^2 is the cross section for particle collisions. (It is an approximation because we have assumed hydrostatic equilibrium, both to relate the radial coordinate to pressure and for the expression of the pressure scale height.) Estimate the values of λ_{esc} for Mars, Earth and Venus.

---

[3]ENAs with thermal widths of 30–100 km s^{-1} are consistent with stellar wind temperatures $\sim 10^5$–10^6 K.

13.6.2   Non-isothermal, asymptotic breeze and accretion solutions

Derive the asymptotic breeze and accretion solutions associated with the non-isothermal Parker wind solution.

13.6.3   Numerical solution for non-isothermal Parker winds

Equation (13.27) is a transcendental equation for v/v_0, which is generally difficult to solve. However, in situations where the dependent variable appears as itself and a slowly-varying function, one may employ a trick[4] to obtain a numerical solution via iteration. Let $y \equiv (v/v_0)^2$. Equation (13.27) may be rewritten as

$$
\begin{aligned}
y = 1 &+ \frac{2}{\gamma_{\mathrm{ad}} + 1} \left[\frac{\gamma_{\mathrm{ad}} - 1}{2} \left(1 + \frac{1}{\lambda_{\mathrm{esc}}} \right) \ln (y) \right. \\
&+ \left. \frac{\lambda_{\mathrm{J}}}{\lambda_{\mathrm{esc}}} \left(\frac{1}{x} - 1 \right) + \frac{2 (\gamma_{\mathrm{ad}} - 1)}{\alpha \lambda_{\mathrm{esc}}} (x^\alpha - 1) \right].
\end{aligned}
\tag{13.35}
$$

Operationally, one computes an initial value for y by first ignoring the contribution of the logarithmic term. Subsequently, one then updates the value of y by inserting its previous value into the logarithmic term. With each iteration, one computes the fractional difference between the old and new values of y. Once this fractional difference is smaller than a numerical threshold (e.g., 10^{-6}), then a converged solution is obtained, which should occur within a few iterations. Write a `Python` program to execute this procedure. Reproduce the graphical solutions in Figure 13.2.

13.6.4   Jeans parameter for hot Jupiters

Estimate the value of the Jeans parameter for hot Jupiters. Are hot Jupiters better described by Jeans or hydrodynamic escape?

---

[4]This was first taught to me by Nick Gnedin. I have used this trick to perform similar calculations with a pocket calculator.

Chapter Fourteen

Outstanding Problems of Exoplanetary Atmospheres

In this chapter, I briefly describe and summarize my views on the present and future direction of the study of exoplanetary atmospheres. It is probably the only chapter of the textbook with a limited shelf life, as I expect most, if not all, of these puzzles to be resolved (or adequately addressed) in the coming decade. A pedagogical reason for listing these puzzles is to teach the student that, in professional research, asking the right question is as important as obtaining the right answer, and that not everything about exoplanetary atmospheres is completely understood.

At the time of writing of the textbook (2015–2016), astronomers have had nearly a decade of experience of performing photometry of exoplanetary atmospheres [239], both from the ground and from space (e.g., using the Spitzer Space Telescope) [31]. Early issues with data calibration and reduction had largely been resolved [4, 48, 69, 83, 123, 235] and the field was focusing on *spectrophotometry* [16, 17, 47, 132, 194, 220, 221, 222, 223, 224] and anticipating full-fledged spectroscopy with the James Webb Space Telescope [76] (then scheduled for launch in 2018) and 30- to 40-meter-class ground-based telescopes such as the European Extremely Large Telescope (E-ELT). Multi-wavelength phase curves had started to be collected [236], enabling exoplanetary atmospheres to be probed across longitude and depth.

This was also a period of transition for theory and simulation. In the early years, models typically possessed a level of sophistication far beyond what could be tested by the observations [27]. The situation had since evolved and models of atmospheric dynamics, chemistry and radiation were being confronted by a combination of transmission and emission spectra, optical and infrared phase curves, and geometric albedos [30, 207].

The advances in observational techniques led theorists to realize that a list of outstanding issues had to be resolved in order for the interpretation of data and our understanding of exoplanetary atmospheres to continue to mature [31]. Below, I describe my list of outstanding issues. They are not part of the main textbook, because they are not sufficiently understood to be included as standard fare.

1. Tidally-locked exoplanets with permanent daysides and nightsides are, by definition, poorly described by one-dimensional models, especially if one wishes to predict phase curves. Transmission spectra probe the entire *terminator* region, where the dayside transitions into the nightside and vice

versa. A transmission spectrum thus encodes information on the total flux emanating from fairly complicated parts of the atmosphere, where chemical and radiative equilibria may not hold. Yet, practitioners have mostly resorted to interpreting these transmission spectra using one-dimensional models. Is such an approach valid?

2. *General circulation models* (GCMs), which solve the equations of fluid dynamics on a sphere and were traditionally developed for the study of Earth, have been adapted to study exoplanets across a diverse set of conditions [96]. They may be used to create three-dimensional virtual laboratories that provide a coherent and holistic interpretation of different types of data of the same exoplanet [115, 217]. They may also be used to test if one-dimensional models are adequate for interpreting transmission spectra. However, to realize their full potential requires that a set of technical challenges are overcome, including the ambiguities associated with numerical dissipation[1] [87], sources of friction [44, 73, 198], shocks [90] and numerical conservation of quantities [193]. Furthermore, GCMs traditionally solve the primitive equations, which may not be general enough for all exoplanetary atmospheres.

3. Measurements of the radii of hot Jupiters reveal that more intensely irradiated objects tend to be larger [49]. This has come to be known as the *inflated hot Jupiter problem*. What is the mechanism behind this phenomenon [11, 12, 25, 26, 78, 80]? Do magnetic fields play a key role [14, 106, 188, 189, 256]? Curiously, the measured geometric albedos of hot Jupiters show no correlation with incident stellar flux, surface gravity or stellar metallicity [6, 91]. Are these phenomena related? A compelling theory may need to reconcile these observed phenomena.

4. As described in Chapter 5, spectral lines are poorly described by the Voigt profile in the far line wings. In the absence of reliable theory, practitioners have resorted to truncating the Voigt profile, for each line, at a fixed wavenumber interval (e.g., 100 cm^{-1}) [77]. Intuitively, we expect that such an ad hoc approach must be flawed, because a fixed wavenumber interval corresponds to a varying number of Doppler or Lorentz widths, depending on the temperature and the Einstein A-coefficient of the line. Elucidating a generalized line profile that includes the correct line-wing shape, due to pressure broadening, will have far-reaching implications for models of exoplanetary atmospheres.

5. Indispensable ingredients for models of exoplanetary atmospheres include reliable experimental data for spectroscopic line lists, Gibbs free energies of atoms and molecules, and rate coefficients for chemical kinetics—across a

---

[1]This is dissipation associated with the model grid and numerical scheme, rather than originating from any physical mechanism.

broad range of temperatures and pressures. Furthermore, pressure broadening of molecules due to a buffer gas (air broadening) is usually quantified for nitrogen and oxygen, because of the relevance to the atmosphere of Earth. This needs to be done for other buffer gases such as molecular hydrogen and helium. Making these measurements in the laboratory is a slow and thankless endeavor that often goes unappreciated. There needs to be a place for these researchers on the main stage of exoplanet research, instead of relegating them to the sidelines.

6. Atmospheric retrieval has been heralded as a modern way of interpreting the spectra of exoplanetary atmospheres, as it claims to back out the chemical abundances and thermal structure with minimal reliance on model assumptions. The weakness of such an approach is that the solutions obtained may be chemically or physically nonsensical [100]. Furthermore, traditional approaches for the Solar System that use layer-by-layer retrieval techniques [109] may not be suitable for interpreting the relatively sparse data of exoplanetary atmospheres, as there are formally at least as many parameters as there are layers in the model atmosphere. Using an insufficient number of layers may lead to numerical non-convergence. Attempts to reduce the number of parameters require specifying an ad hoc relationship between the parameters of neighboring layers, which are based on numerical, rather than physical, choices [138, 143]. For point-source exoplanetary atmospheres, where we have no realistic chance of making in-situ measurements, the retrievals need to be anchored, to some extent, by the laws of physics and chemistry. Should radiative and/or chemical equilibrium be enforced? Are the retrieved abundances chemically sensible? A related issue is to decode the identities of atoms and molecules in a formal way. In other words, the list of atoms and molecules to include in a retrieval calculation should be based on the information encoded in a spectrum, rather than being the *choice* of the modeler [249].

7. In deciphering the spectra of exoplanetary atmospheres, the carbon-to-oxygen ratio (C/O) is a quantity of interest because it potentially opens a window into the formation history of an exoplanet [133, 178]. Determining whether atmospheres are carbon-poor or carbon-rich has been a source of controversy [100], because one needs to infer C/O from molecular abundances, which are themselves the outcome of a retrieval calculation that may be afflicted by degeneracies. How does one robustly infer C/O from measured spectra? What are the key molecules involved in this inference? Is the formation history of an exoplanet uniquely pinned down by measuring the C/O of its atmosphere?

8. As described in Chapter 2, the presence of aerosols, condensates, clouds or hazes leads to degeneracies in our interpretation of chemical abundances and the temperature-pressure profile of an atmosphere. In the worst case scenario, aerosols/condensates/clouds/hazes produce flat transmis-

sion spectra [125, 131], which allow little to be inferred about the exoplanetary atmosphere. Furthermore, if the radius and mass of the exoplanet are unknown, then the presence of clouds implies that a family of mass-radius relationships will be consistent with the data [29, 139]. These degeneracies may be alleviated by using simplistic cloud/haze models that prescribe a number of key parameters, but ultimately a first-principles theory of formation is desired. Such a theory would form the seed particles from the atmospheric gas and self-consistently describe the chemistry of the solid and gaseous components.

9. Traditional models of the *habitable zone* aim to describe the range of distances of an exoplanet from its host star, such that liquid water may exist on its surface. The inner and outer boundaries of the habitable zone occur when the condensible and incondensible greenhouse gases become incondensible and condensible, respectively [191]. A key ingredient of such models is the assumption that the *carbonate-silicate cycle* functions to stabilize the long-term presence of carbon dioxide, a greenhouse gas, in the atmosphere [250]. Silicate rocks interact with water to transform carbon dioxide into calcium carbonate—a process known as *weathering*—which is then subducted into the mantle of the Earth. Outgassing and volcanism return the carbon dioxide into the atmosphere. Weathering acts more strongly when the temperature is high and water is present, which provides a self-limiting mechanism for the amount of carbon dioxide present.[2] It is natural to ask if there are generalizations of the carbonate-silicate cycle that function with different types of rocks, greenhouse gases and solvents, beyond what is found on Earth? Do these generalized cycles operate in the absence of plate tectonics? Answering these questions will allow us to construct generalized habitable zones for a wide variety of atmospheres [192].

10. The habitable zones of *red dwarfs* (also known as *M stars*) coincide with the range of distances where we expect the exoplanet to be tidally locked. If the distribution of heat from the permanent dayside to the nightside is insufficient, then runaway condensation of greenhouse gases may occur on the nightside, leading to *atmospheric collapse* [89, 114, 255]. What is the minimum mass of atmosphere needed to prevent this fate? How does it depend on different mixtures of greenhouse gases? Is tidal locking an inevitable outcome [137]? Given the abundance of red dwarfs in our cosmic neighborhood and their promise for characterizing exoplanetary atmospheres that approach being Earth-like, there is a need to address these puzzles.

---

[2]If Earth falls into a snowball state, it is believed that weathering shuts off, but outgassing continues to operate. Eventually, enough carbon dioxide is outgassed such that greenhouse warming provides a release from the snowball state.

11. Chapter 13 describes energy-limited atmospheric escape, where all of the available stellar flux is channeled into unbinding the atmosphere from the gravitational well of the exoplanet. In reality, only a fraction of the stellar flux goes into atmospheric escape. Is it possible to calculate this dilution or efficiency factor, from first principles, for *all* types of exoplanets [181]? A related goal is to thoroughly understand the relationship between the lower atmosphere (where the infrared radiation emanates from), the exosphere and everything in between.

12. The ultimate theoretical goal is to understand exoplanetary atmospheres so thoroughly that we may infer the presence of life from recording spectra and phase curves across galactic distances. The grand challenge is that the gases produced by life—even as we known it—are often also produced by geology. Disentangling the telltale signs of *biosignature gases* from the geological false positives becomes a difficult but necessary task [209, 210, 211, 212]. Central to this task is understanding how these gases interact with the existing atmosphere, which requires a complete theory of exoplanetary atmospheres.

Appendix A: Summary of Standard Notation

We provide a list of the notation used that is standard across all of the chapters. If a symbol is not listed here, it means that it may be used in different ways between the chapters—the breadth of the subjects covered do not allow us to totally avoid degenerate notation. I may also sometimes use standard notation to represent something else, but usually only in a fleeting fashion within a section of a chapter.

The challenge is that some of the symbols are entrenched as standard notation in specific subdisciplines. For example, g is universally used as the surface gravity by astronomers, while some atmospheric scientists use it to denote the cumulative sum of intervals in the k-distribution method. In this case, I made the judgment that the astronomers are in the majority and used y instead. As another example, geophysicists favor θ for the potential temperature, but this is standard notation, found in mathematical handbooks, for the polar angle or co-latitude in spherical geometry. Fluid dynamicists favor using ν for the kinematic viscosity, but this is treasured astronomical notation for the frequency. Since I rarely use frequency in this textbook—choosing instead to use wavenumber—it is only an issue in one of the problem sets. The use of G as Newton's gravitational constant is entrenched in the larger physics community, but chemists also use G to denote the Gibbs free energy. I have chosen to keep both conventions, such that the notation is degenerate, but this is not a problem as G is never used to represent both quantities within the same chapter. Another example of degenerate notation is ω, which I use to represent both the vorticity and wave frequency. Generally, x, y and z denote distances, but not always. In particular, x is my favorite "workhorse" symbol and I use it for other things, e.g., normalized wavenumber. When a chapter and its text are read in context, there will be no confusion over what x or y means.

If a specific value is not given, it means that it depends on input parameters. For physical units, cgs (centimeters, grams and seconds) units are usually used; "—" means that it is dimensionless, whereas "various" means that the quantity comes in various flavors and/or physical units. A non-standard unit of pressure that is used throughout the book is "bar." We note that

$$1 \text{ bar} = 10^6 \text{ dyn cm}^{-2} = 10^5 \text{ Pa}. \tag{1}$$

Do not confuse 1 bar with 1 atm: 1 bar = 0.98692 atm. Some references like to use the term "ergs" to denote the unit of energy (erg) in the plural. I find this way of writing it to be superfluous and unnecessarily confusing.

When we refer to a quantity per unit mass, the adjective "specific" is added. If the quantity is cast in per unit volume, then the qualifier "density" is used.

The physical constants are culled from the National Institute of Standards and Technology (NIST) Reference on Constants, Units and Uncertainties, while the astronomical constants are taken from the International Astronomical Union (IAU).

Table 1: Physical Constants or Constants of Nature

Symbol	Definition	Value	Units
k_{B}	Boltzmann constant	$1.38064852 \times 10^{-16}$	erg K^{-1}
σ_{SB}	Stefan-Boltzmann constant	5.670367×10^{-5}	cgs†
c	speed of light	$2.99792458 \times 10^{10}$	cm s^{-1}
c_s	speed of sound	various	cm s^{-1}
h	Planck constant	$6.626070040 \times 10^{-27}$	erg s
e	elementary charge	$4.80320425 \times 10^{-10}$	esu
m_p	mass of proton	$1.672621898 \times 10^{-24}$	g
m_{u}	atomic mass unit	$1.660539040 \times 10^{-24}$	g
m_e	mass of electron	$9.10938356 \times 10^{-28}$	g
N_{A}	Avogadro's constant	$6.022140857 \times 10^{23}$	—
G	gravitational constant	6.67408×10^{-8}	cm^3 s^{-2} g^{-1}
$\mathcal{R}_{\mathrm{univ}}$	universal gas constant	8.3144598×10^{7}	erg g^{-1} K^{-1}
—	Wien's law	2.8977729×10^{3}	μm K
—	eV to erg conversion	$1.6021766208 \times 10^{-12}$	erg eV^{-1}
$R_{\oplus}$	equatorial radius of Earth	6.3781×10^{8}	cm
$R_{\oplus}$	polar radius of Earth	6.3568×10^{8}	cm
R_{J}	equatorial radius of Jupiter	7.1492×10^{9}	cm
R_{J}	polar radius of Jupiter	6.6854×10^{9}	cm
$R_{\odot}$	solar radius	6.9566×10^{10}	cm
$M_{\oplus}$	mass of Earth	5.972365×10^{27}	g
M_{J}	mass of Jupiter	1.898187×10^{30}	g
$M_{\odot}$	solar mass	1.988547×10^{33}	g
$\mathcal{L}_{\odot}$	solar luminosity	3.828×10^{33}	erg s^{-1}
AU	astronomical unit	$1.49597870700 \times 10^{13}$	cm

Note: For shallow water models, "h" is used as the shallow water height.
Otherwise, it usually denotes the Planck constant.
$\dagger$: One way of expressing the physical units for the Stefan-Boltzmann
constant is erg cm^{-2} s^{-1} K^{-4}.

Table 2: Quantities Related to Thermodynamics and Chemistry

Symbol	Definition	Units
n_{dof}	number of degrees of freedom	—
κ_{ad}	adiabatic coefficient	—
γ_{ad}	adiabatic gas index	—
c_P	specific heat capacity (constant pressure)	erg g^{-1} K^{-1}
c_V	specific heat capacity (constant volume)	erg g^{-1} K^{-1}
$\mathcal{R} \equiv \mathcal{R}_{\mathrm{univ}}/\mu_{\mathrm{m}}$	specific gas constant	erg g^{-1} K^{-1}
G	specific Gibbs free energy	erg g^{-1}
ΔG	change in specific Gibbs free energy	erg g^{-1}
$\Delta \tilde{G}$	change in molar Gibbs free energy	kJ mol^{-1}
S	specific entropy	erg g^{-1}
V	specific volume	cm^3 g^{-1}
E_{int}	internal energy	various‡
E_{g}	potential energy	various‡
E_{k}	kinetic energy	various‡
E	total energy	various‡
K_{eq}	dimensionless equilibrium constant	—
K'_{eq}	dimensional equilibrium constant	various
K'	normalized equilibrium constant	—
k_{f}	forward rate coefficient	various
k_{r}	reverse rate coefficient	various
E_{f}	forward activation energy	erg g^{-1}
E_{r}	reverse activation energy	erg g^{-1}
C/O	carbon-to-oxygen ratio (by number)	—
$\tilde{n}_{\mathrm{X}}$	mixing ratio for molecular species X	—
$\tilde{n}_{\mathrm{Z}}$	elemental abundance for atomic species Z	—
K_{zz}	diffusion coefficient	cm^2 s^{-1}

Note: Typically, H is used for the specific enthalpy, but I avoid this notation as I use it for the pressure scale height. Instead, I explicitly list the functional form of the specific enthalpy; in one instance, I use Q to denote it.

$\ddagger$: The physical units are either erg cm^{-3} or erg g^{-1}, depending on whether it is a energy density or specific energy, respectively.

Table 3: Basic Quantities

Symbol	Definition	Units
t	time	s
z	vertical coordinate	cm
r	radial coordinate	cm
θ	polar angle / co-latitude	—
ϕ	azimuthal angle / longitude	—
$\mu \equiv \cos\theta$	cosine of co-latitude	—
μ_{m}	mean molecular weight	—
m	mean molecular mass	g
$\vec{v}$	velocity	cm s^{-1}
ρ	mass density	g cm^{-3}
n	number density	cm^{-3}
P	pressure	bar
T	temperature	K
$\vec{B}$	magnetic field strength	G
$\vec{J}$	current density	G s^{-1}
Q	heating	various
Θ	potential temperature	K
g	surface gravity of exoplanet	cm s^{-2}
Ω	angular rotational frequency of exoplanet	s^{-1}
R	radius of exoplanet	cm
M	mass of exoplanet	g
T_{eq}	equilibrium temperature of exoplanet	K
T_{irr}	irradiation temperature of exoplanet	K
T_{int}	interior/internal temperature of exoplanet	K
$g_\star$	stellar surface gravity	cm s^{-2}
$R_\star$	stellar radius	cm
$M_\star$	stellar mass	g
$T_\star$	stellar effective temperature	K
$\mathcal{L}_\star$	stellar luminosity	erg s^{-1}
l	characteristic length scale	cm
U	characteristic speed	cm s^{-1}
l_{mfp}	mean free path	cm
l_{D}	Debye length	cm

Table 4: Quantities Related to Fluid Dynamics

Symbol	Definition	Units
ν	kinematic viscosity	$\mathrm{cm^2\ s^{-1}}$
μ_ν	dynamic viscosity	$\mathrm{g\ cm^{-1}\ s^{-1}}$
c_0	gravity wave speed	$\mathrm{cm\ s^{-1}}$
ω	vorticity	$\mathrm{s^{-1}}$
Φ	potential vorticity	various
Ψ	streamfunction	$\mathrm{cm^2\ s^{-1}}$
$\mathcal{M}$	Mach number	—
$\mathcal{R}_0$	Rossby number	—
$\mathcal{R}_e$	Reynolds number	—
$\mathcal{R}_\mathrm{B}$	magnetic Reynolds number	—

Note: For shallow water models, "ω" is used as the frequency of the normal modes. Otherwise, we use it to denote the vorticity, $\vec{\omega} \equiv \nabla \times \vec{v}$.

Table 5: Quantities Related to Atmospheric Radiation

Symbol	Definition	Units
τ	optical depth[†]	—
κ	opacity[†]	$\mathrm{cm^2\ g^{-1}}$
$\tilde{m}$	column mass	$\mathrm{g\ cm^{-2}}$
ω_0	single scattering albedo[†]	—
g_0	scattering asymmetry factor[†]	—
A_B	Bond albedo	—
A_g	geometric albedo[†]	—
A_s	spherical albedo[†]	—
λ	wavelength	cm
$\tilde{\nu}$	wavenumber	$\mathrm{cm^{-1}}$
H	pressure scale height	cm
I	monochromatic[†] intensity	various units
J	monochromatic[†] total intensity	various units
B	Planck function[†]	various units
$F_\uparrow$	monochromatic[†] outgoing flux	various units
$F_\downarrow$	monochromatic[†] incoming flux	various units
F_-	monochromatic[†] net flux	various units
F_+	monochromatic[†] total flux	various units
$\mathcal{F}_\uparrow$	outgoing flux	various units
$\mathcal{F}_\downarrow$	incoming flux	various units
$\mathcal{F}_-$	net flux	various units
$\mathcal{F}_+$	total flux	various units
$\mathcal{T}$	transmission function[†]	—

†: Depends on wavelength, frequency or wavenumber.

Appendix B: Essential Formulae of Vector Calculus

The most central operator used in vector calculus is the *del* or *nabla operator*, represented by the symbol ∇. When operated on a scalar F, it produces its gradient in all three spatial dimensions, ∇F. For a vector $\vec{C}$, $\nabla.\vec{C}$ is its *divergence*. Its *curl* is $\nabla \times \vec{C}$, which measures the amount of "rotation" of the vector about a specific point.

In Cartesian coordinates, we have $\vec{C} = (C_x, C_y, C_z)$ and

$$
\begin{aligned}
\nabla F &= \frac{\partial F}{\partial x}\hat{x} + \frac{\partial F}{\partial y}\hat{y} + \frac{\partial F}{\partial z}\hat{z}, \\
\nabla.\vec{C} &= \frac{\partial C_x}{\partial x} + \frac{\partial C_y}{\partial y} + \frac{\partial C_z}{\partial z}, \\
\nabla \times \vec{C} &= \left(\frac{\partial C_z}{\partial y} - \frac{\partial C_y}{\partial z}\right)\hat{x} + \left(\frac{\partial C_x}{\partial z} - \frac{\partial C_z}{\partial x}\right)\hat{y} + \left(\frac{\partial C_y}{\partial x} - \frac{\partial C_x}{\partial y}\right)\hat{z},
\end{aligned}
\tag{2}
$$

with $\hat{x}$, $\hat{y}$ and $\hat{z}$ being the unit vectors in the x-, y- and z-directions, respectively.

In spherical coordinates (as is relevant for exoplanets), the issue is more subtle as the gradients of the unit vectors are sometimes non-zero. This implies that the gradient or Laplacian of a vector is *not* the sum of the gradient or Laplacian of its components. Stated without proof, the gradients of the unit vectors $\hat{r}$, $\hat{\theta}$ and $\hat{\phi}$ are

$$
\begin{aligned}
\frac{\partial \hat{r}}{\partial r} &= 0, & \frac{\partial \hat{r}}{\partial \theta} &= \hat{\theta}, & \frac{\partial \hat{r}}{\partial \phi} &= \sin\theta\,\hat{\phi}, \\
\frac{\partial \hat{\theta}}{\partial r} &= 0, & \frac{\partial \hat{\theta}}{\partial \theta} &= -\hat{r}, & \frac{\partial \hat{\theta}}{\partial \phi} &= \cos\theta\,\hat{\phi}, \\
\frac{\partial \hat{\phi}}{\partial r} &= 0, & \frac{\partial \hat{\phi}}{\partial \theta} &= 0, & \frac{\partial \hat{\phi}}{\partial \phi} &= -\sin\theta\,\hat{r} - \cos\theta\,\hat{\theta}.
\end{aligned}
\tag{3}
$$

These identities imply that we have [7]

$$
\begin{aligned}
\nabla.\vec{C} &= \frac{1}{r^2}\frac{\partial}{\partial r}\left(r^2 C_r\right) + \frac{1}{r\sin\theta}\left[\frac{\partial}{\partial \theta}\left(C_\theta \sin\theta\right) + \frac{\partial C_\phi}{\partial \phi}\right], \\
\nabla \times \vec{C} &= \frac{\hat{r}}{r\sin\theta}\left[\frac{\partial}{\partial \theta}\left(C_\phi \sin\theta\right) - \frac{\partial C_\theta}{\partial \phi}\right] \\
&\quad + \frac{\hat{\theta}}{r\sin\theta}\left[\frac{\partial C_r}{\partial \phi} - \sin\theta\frac{\partial}{\partial r}\left(rC_\phi\right)\right] + \frac{\hat{\phi}}{r}\left[\frac{\partial}{\partial r}\left(rC_\theta\right) - \frac{\partial C_r}{\partial \theta}\right],
\end{aligned}
\tag{4}
$$

and

$$\nabla^2\vec{C}\Big|_r = \nabla^2 C_r - \frac{2C_r}{r^2} - \frac{2}{r^2}\frac{\partial C_\theta}{\partial\theta} - \frac{2\cos\theta C_\theta}{r^2\sin\theta} - \frac{2}{r^2\sin\theta}\frac{\partial C_\phi}{\partial\phi},$$

$$\nabla^2\vec{C}\Big|_\theta = \nabla^2 C_\theta - \frac{C_\theta}{r^2\sin^2\theta} + \frac{2}{r^2}\frac{\partial C_r}{\partial\theta} - \frac{2\cos\theta}{r^2\sin^2\theta}\frac{\partial C_\phi}{\partial\phi}, \tag{5}$$

$$\nabla^2\vec{C}\Big|_\phi = \nabla^2 C_\phi - \frac{C_\phi}{r^2\sin^2\theta} + \frac{2}{r^2\sin\theta}\frac{\partial C_r}{\partial\phi} + \frac{2\cos\theta}{r^2\sin^2\theta}\frac{\partial C_\theta}{\partial\phi}.$$

Also, we have [7]

$$\nabla F = \hat{r}\frac{\partial F}{\partial r} + \frac{\hat{\theta}}{r}\frac{\partial F}{\partial\theta} + \frac{\hat{\phi}}{r\sin\theta}\frac{\partial F}{\partial\phi},$$

$$\nabla^2 F = \frac{1}{r^2}\frac{\partial}{\partial r}\left(r^2\frac{\partial F}{\partial r}\right) + \frac{1}{r^2\sin\theta}\frac{\partial}{\partial\theta}\left(\sin\theta\frac{\partial F}{\partial\theta}\right) + \frac{1}{r^2\sin^2\theta}\frac{\partial^2 F}{\partial\phi^2}. \tag{6}$$

The unit vectors in cartesian and spherical coordinates may be related using [7]

$$\hat{x} = \sin\theta\cos\phi\,\hat{r} + \cos\theta\cos\phi\,\hat{\theta} - \sin\phi\,\hat{\phi},$$

$$\hat{y} = \sin\theta\sin\phi\,\hat{r} + \cos\theta\sin\phi\,\hat{\theta} + \cos\phi\,\hat{\phi}, \tag{7}$$

$$\hat{z} = \cos\theta\,\hat{r} - \sin\theta\,\hat{\phi},$$

and

$$\hat{r} = \sin\theta\cos\phi\,\hat{x} + \sin\theta\sin\phi\,\hat{y} + \cos\theta\,\hat{z},$$

$$\hat{\theta} = \cos\theta\cos\phi\,\hat{x} + \cos\theta\sin\phi\,\hat{y} - \sin\theta\,\hat{z}, \tag{8}$$

$$\hat{\phi} = -\sin\phi\,\hat{x} + \cos\phi\,\hat{y}.$$

When the divergence of a vector $\vec{C}$ is integrated over a volume V, the *divergence theorem* (sometimes known as *Gauss's theorem*) states that [7]

$$\int \nabla.\vec{C}\ dV = \oint \vec{C}.d\vec{A}, \tag{9}$$

where $\vec{A}$ is the bounding surface of this volume. When the curl of a vector is integrated over a surface, *Stokes's theorem* states that one may equivalently integrate over the curve $\vec{l}$ bounding this surface [7],

$$\int \nabla\times\vec{C}.d\vec{A} = \oint \vec{C}.d\vec{l}. \tag{10}$$

Appendix C: Essential Formulae of Thermodynamics

It is not always appreciated that the Boltzmann constant (k_B), specific gas constant ($\mathcal{R}$), universal gas constant ($\mathcal{R}_{univ}$), specific heat capacity at constant pressure (c_P) and volume (c_V), adiabatic coefficient (κ_{ad}), adiabatic gas index (γ_{ad}) and number of degrees of freedom in a gas (n_{dof}) are inter-dependent quantities. First, the mean molecular weight (μ_m) and mean molecular mass (m) are related by

$$m = \mu_m m_u, \tag{11}$$

where m_u is the atomic mass unit and is approximately the mass of the hydrogen atom (m_H).

Tremendous confusion surrounds the definition of the universal gas constant. Strictly speaking, $\mathcal{R}_{univ} = 8.3144598 \times 10^7$ erg K^{-1} mol^{-1}, where "mol" is a mole. One mole of substance has exactly N_A atoms or molecules in it, where $N_A = 6.022140857 \times 10^{23}$ mol^{-1} is Avogadro's constant. By definition, we have

$$m_u N_A \equiv 1 \text{ g mol}^{-1}. \tag{12}$$

In other words, N_A particles, each with a mass of m_u, make up exactly 1 g of matter. The universal gas constant is defined as

$$\mathcal{R}_{univ} \equiv k_B N_A. \tag{13}$$

Let the pressure be P, the number density be n and the mass density be $\rho = nm$. The ideal gas law is

$$P = n k_B T = \frac{\rho \mathcal{R}_{univ} T}{\mu m_u N_A} = \frac{\rho \mathcal{R}'_{univ} T}{\mu}. \tag{14}$$

In cgs units, one may absorb the $m_u N_A$ factor into $\mathcal{R}_{univ}$, thereby expressing it as

$$\mathcal{R}'_{univ} \equiv \frac{\mathcal{R}_{univ}}{m_u N_A} = 8.3144598 \times 10^7 \text{ erg K}^{-1} \text{ g}^{-1}. \tag{15}$$

It cannot be over-emphasized that this only works in cgs units! In mks units, we have $m_u N_A = 10^{-3}$ kg mol^{-1} and therefore

$$\begin{aligned} \mathcal{R}'_{univ} &= 8.3144598 \times 10^3 \text{ J K}^{-1} \text{ kg}^{-1}, \\ \mathcal{R}_{univ} &= 8.3144598 \text{ J K}^{-1} \text{ mol}^{-1}. \end{aligned} \tag{16}$$

Only in cgs units may we liberally switch between the g^{-1} and mol^{-1} definitions of the universal gas constant. For example, Pierrehumbert [191] favors the use of $\mathcal{R}'_{univ}$ in cgs units.

If we now enjoy this liberty, then the specific and universal gas constants, as well as the Boltzmann constant, are related by

$$\mathcal{R} = \frac{\mathcal{R}_{\text{univ}}}{\mu_{\text{m}}} = \frac{k_{\text{B}}}{m}. \tag{17}$$

The adiabatic coefficient and adiabatic gas index may both be related to the number of degrees of freedom,

$$\kappa_{\text{ad}} = \frac{2}{2 + n_{\text{dof}}}, \quad \gamma_{\text{ad}} = 1 + \frac{2}{n_{\text{dof}}} = \frac{c_P}{c_V}. \tag{18}$$

If we consider only translation and rotation (and ignore vibration), then we have $n_{\text{dof}} = 3$ and 5 for monoatomic and diatomic gases, respectively. The adiabatic coefficient is also the dimensionless, dry adiabatic lapse rate, often termed the *adiabatic gradient*. Finally, the specific heat capacity at constant pressure may be expressed as

$$c_P = c_V + \mathcal{R} = \frac{\mathcal{R}}{\kappa_{\text{ad}}} = \frac{k_{\text{B}} \left(2 + n_{\text{dof}}\right)}{2m}. \tag{19}$$

The preceding expressions show that the specific heat capacities are functions of the composition and chemistry of the atmospheric gas.

Appendix D: Gibbs Free Energies of Various Molecules and Reactions

We collect the molar Gibbs free energies of formation, associated with various common molecules, from the JANAF tables (http://kinetics.nist.gov/janaf/).

Table 6: Molar Gibbs free energies of various chemical species ($P_0 = 1$ bar)

T	H_2O	CH_4	CO	CO_2	C_2H_2	C_2H_4
Name	water	methane	carbon monoxide	carbon dioxide	acetylene	ethylene
(K)	(kJ/mol)	(kJ/mol)	(kJ/mol)	(kJ/mol)	(kJ/mol)	(kJ/mol)
500	-219.051	-32.741	-155.414	-394.939	197.452	80.933
600	-214.007	-22.887	-164.486	-395.182	191.735	88.017
700	-208.812	-12.643	-173.518	-395.398	186.097	95.467
800	-203.496	-2.115	-182.497	-395.586	180.534	103.180
900	-198.083	8.616	-191.416	-395.748	175.041	111.082
1000	-192.590	19.492	-200.275	-395.886	169.607	119.122
1100	-187.033	30.472	-209.075	-396.001	164.226	127.259
1200	-181.425	41.524	-217.819	-396.098	158.888	135.467
1300	-175.774	52.626	-226.509	-396.177	153.588	143.724
1400	-170.089	63.761	-235.149	-396.240	148.319	152.016
1500	-164.376	74.918	-243.740	-396.288	143.078	160.331
1600	-158.639	86.088	-252.284	-396.323	137.861	168.663
1700	-152.883	97.265	-260.784	-396.344	132.665	177.007
1800	-147.111	108.445	-269.242	-396.353	127.487	185.357
1900	-141.325	119.624	-277.658	-396.349	122.327	193.712
2000	-135.528	130.802	-286.034	-396.333	117.182	202.070
2100	-129.721	141.975	-294.372	-396.304	112.052	210.429
2200	-123.905	153.144	-302.672	-396.262	106.935	218.790
2300	-118.082	164.308	-310.936	-396.209	101.830	227.152
2400	-112.252	175.467	-319.165	-396.142	96.738	235.515
2500	-106.416	186.622	-327.358	-396.062	91.658	243.880
2600	-100.575	197.771	-335.517	-395.969	86.589	252.246
2700	-94.729	208.916	-343.643	-395.862	81.530	260.615
2800	-88.878	220.058	-351.736	-395.742	76.483	268.987
2900	-83.023	231.196	-359.797	-395.609	71.447	277.363
3000	-77.163	242.332	-367.826	-395.461	66.421	285.743

Note: The molar Gibbs free energy associated with H_2 is 0 J mol^{-1} by definition.

We also list the molar Gibbs free energy associated with the hydrogen atom in units of kJ/mol, from 0 to 6000 K (in increments of 100 K) and at $P_0 = 1$ bar: 216.035, 212.450, 208.004, 203.186, 198.150, 192.957, 187.640, 182.220, 176.713, 171.132, 165.485, 159.782, 154.028, 148.230, 142.394, 136.522, 130.620, 124.689, 118.734, 112.757, 106.760, 100.744, 94.712, 88.664, 82.603, 76.530, 70.444, 64.349, 58.243, 52.129, 46.007, 39.877, 33.741, 27.598, 21.449, 15.295, 9.136, 2.973, -3.195, -9.366, -15.541, -21.718, -27.899, -34.082, -40.267, -46.454, -52.643, -58.834, -65.025, -71.218, -77.412, -83.606, -89.801, -95.997, -102.192, -108.389, -114.584, -120.780, -126.976, -133.172, -139.368. If we denote each of these numbers by $\tilde{G}_H$, then we have $\Delta\tilde{G}_0 = -2\tilde{G}_H$ for the net reaction in equation (7.1).

Table 7: Molar Gibbs free energies of net reactions ($P_0 = 1$ bar)

T	$\Delta\tilde{G}_{0,1}$	$\Delta\tilde{G}_{0,2}$	$\Delta\tilde{G}_{0,3}$	$\Delta\tilde{G}_{0,4}$
(K)	(kJ/mol)	(kJ/mol)	(kJ/mol)	(kJ/mol)
500	96.378	262.934	20.474	116.519
600	72.408	237.509	16.689	103.718
700	47.937	211.383	13.068	90.630
800	23.114	184.764	9.593	77.354
900	-1.949	157.809	6.249	63.959
1000	-27.177	130.623	3.021	50.485
1100	-52.514	103.282	-0.107	36.967
1200	-77.918	75.840	-3.146	23.421
1300	-103.361	48.336	-6.106	9.864
1400	-128.821	20.797	-8.998	-3.697
1500	-154.282	-6.758	-11.828	-17.253
1600	-179.733	-34.315	-14.600	-30.802
1700	-205.166	-61.865	-17.323	-44.342
1800	-230.576	-89.403	-20.000	-57.870
1900	-255.957	-116.921	-22.634	-71.385
2000	-281.308	-144.422	-25.229	-84.888
2100	-306.626	-171.898	-27.789	-98.377
2200	-331.911	-199.353	-30.315	-111.855
2300	-357.162	-226.786	-32.809	-125.322
2400	-382.38	-254.196	-35.275	-138.777
2500	-407.564	-281.586	-37.712	-152.222
2600	-432.713	-308.953	-40.123	-165.657
2700	-457.830	-336.302	-42.509	-179.085
2800	-482.916	-363.633	-44.872	-192.504
2900	-507.97	-390.945	-47.211	-205.916
3000	-532.995	-418.243	-49.528	-219.322

The first, second, third and fourth net reactions are listed in the following order:

$$CH_4 + H_2O \leftrightharpoons CO + 3H_2,$$
$$2CH_4 \leftrightharpoons C_2H_2 + 3H_2,$$
$$CO_2 + H_2 \leftrightharpoons CO + H_2O,$$
$$C_2H_4 \leftrightharpoons C_2H_2 + H_2.$$

$$(20)$$

The change in *molar* Gibbs free energy, at the reference pressure (P_0), is denoted by $\Delta \tilde{G}_0$ with self-explanatory subscripts for the four net reactions.

Appendix E: Python Scripts for Generating Figures

In this appendix, I have collected Python scripts used to make various figures in the textbook, so as to provide you with examples to learn how to code in Python. As a student, I found it helpful to learn from ready-made examples. I have selected examples that I feel are useful and also highlight important points.

We begin with something relatively simple: the Python program used to generate Figure 3.1. From the numpy library, it calls a pair of basic functions: exp (exponential function) and log (natural logarithm). It also calls arange, which allows one to define an array of double-precision numbers. The zeros command creates an empty array of zeros of length n. From the scipy library, the special (special functions) package is called, since we need to evaluate the exponential integral. The script itself has a simple organization: it consists of a single function (diffusivity), a main part to define the arrays and compute the diffusivity factor and a secondary part to plot the figure, which is saved in the encapsulated postscript (EPS) format.

```python
from numpy import exp,log,arange,zeros
from scipy import special
from matplotlib import pyplot as plt

# function to compute diffusivity factor
def diffusivity(x):
    eintegral = special.expn(1,x)
    term1 = ( 1.0 - x )/exp(x)
    term2 = x*x*eintegral
    term3 = log(term1+term2)
    result = -1.0*term3/x
    return result

dindex = arange(-2.0, 2.01, 0.01)
dtau = 10.0**dindex # array of optical depths
n = len(dtau)
dd = zeros(n)
for i in range(0,n):
    dd[i] = diffusivity(dtau[i])

# plot
plt.plot(dtau, dd, linewidth=5, color='k', linestyle='-')
plt.xscale('log')
plt.xlabel(r'$\Delta\tau$', fontsize=22)
```

```
plt.ylabel(r'${\cal D}$', fontsize=22)
plt.savefig('diffusivity.eps', format='eps')
```

The next Python program we will examine is somewhat more complex. It was used to compute the temperature-pressure profiles in Figures 4.3, 4.4 and 4.5. It contains several functions and performs three experiments associated with temperature-pressure profiles. As you will notice, I have hard-coded some of the inputs when calling up the tpprofile function, which is generally not recommended, but I wanted to avoid declaring too many inputs.

```
from numpy import arange,zeros
from scipy import special
from matplotlib import pyplot as plt

# function to compute T-P profile
def tpprofile(m,m0,tint,tirr,kappa_S,kappa0,kappa_cia,beta_S0,
              beta_L0,el1,el3):
   albedo = (1.0-beta_S0)/(1.0+beta_S0)
   kappa_L = kappa0 + kappa_cia*m/m0
   beta_S = kappa_S*m/beta_S0
   coeff1 = 0.25*(tint**4)
   coeff2 = 0.125*(tirr**4)*(1.0-albedo)
   term1 = 1.0/el1 + m*( kappa0 + 0.5*kappa_cia*m/m0 )
           /el3/(beta_L0**2)
   term2 = 0.5/el1 + special.expn(2,beta_S)*( kappa_S/kappa_L
           /beta_S0 - kappa_cia*m*beta_S0/el3/kappa_S/m0
           /(beta_L0**2) )
   term3 = kappa0*beta_S0*( 1./3.  - special.expn(4,beta_S) )
           /el3/kappa_S/(beta_L0**2)
   term4 = kappa_cia*(beta_S0**2)*( 0.5 - special.expn(3,beta_S) )
           /el3/m0/(kappa_S**2)/(beta_L0**2)
   result = ( coeff1*term1 + coeff2*(term2 + term3
            + term4) )**0.25
   return result

# input parameters (default values)
g = 1e3 # surface gravity
tint = 150.0 # internal temperature (K)
tirr = 1200.0 # irradiation temperature (K)
kappa_S0 = 0.01 # shortwave opacity
kappa0 = 0.02 # infrared opacity (constant component)
kappa_cia = 0.0 # CIA opacity normalization
```

```
beta_S0 = 1.0 # shortwave scattering parameter
beta_L0 = 1.0 # longwave scattering parameter
el1 = 3.0/8.0 # first longwave Eddington coefficient
el3 = 1.0/3.0 # second longwave Eddington coefficient

# define pressure and column mass arrays
logp = arange(-3,2.01,0.01)
pressure = 10.0**logp # pressure in bars
bar2cgs = 1e6 # convert bar to cgs units
p0 = max(pressure) # BOA pressure
m = pressure*bar2cgs/g # column mass
m0 = p0*bar2cgs/g # BOA column mass

# Experiment 1:  greenhouse effect and CIA
np = len(m)
tp0 = zeros(np)
tp1 = zeros(np)
tp2 = zeros(np)
tp3 = zeros(np)

# (set all Tint=0 except for fiducial)
for i in range(0,np):
   tp0[i] = tpprofile(m[i],m0,0.0,tirr,kappa_S0,
           kappa0,kappa_cia,beta_S0,beta_L0,el1,el3)
   tp1[i] = tpprofile(m[i],m0,0.0,tirr,kappa_S0,
           0.03,kappa_cia,beta_S0,beta_L0,el1,el3)
   tp2[i] = tpprofile(m[i],m0,0.0,tirr,kappa_S0,
           kappa0,1.0,beta_S0,beta_L0,el1,el3)
   tp3[i] = tpprofile(m[i],m0,tint,tirr,kappa_S0,
           kappa0,kappa_cia,beta_S0,beta_L0,el1,el3)

line1, =plt.plot(tp0, pressure, linewidth=4, color='k',
      linestyle='-')
line2, =plt.plot(tp1, pressure, linewidth=4, color='k',
      linestyle='--')
line3, =plt.plot(tp2, pressure, linewidth=4, color='k',
      linestyle=':')
plt.plot(tp3, pressure, linewidth=2, color='k', linestyle='-')
plt.yscale('log')
plt.xlim([800,1100])
plt.ylim([1e2,1e-3])
plt.xlabel('$T$ (K)', fontsize=18)
plt.ylabel('$P$ (bar)', fontsize=18)
plt.legend([line1,line2,line3],['fiducial',r'$\kappa_0=0.03$
          cm$^2$ g$^{-1}$',r'$\kappa_{\rm CIA}=1$ cm$^2$
```

```
          g$^{-1}$'],frameon=False,prop={'size':18})
plt.savefig('tp1.eps', format='eps')
plt.close()

# Experiment #2:  anti-greenhouse effect
tp4 = zeros(np)
tp5 = zeros(np)

for i in range(0,np):
   tp4[i] = tpprofile(m[i],m0,tint,tirr,0.02,kappa0,kappa_cia,
         beta_S0,beta_L0,el1,el3)
   tp5[i] = tpprofile(m[i],m0,tint,tirr,0.04,kappa0,kappa_cia,
         beta_S0,beta_L0,el1,el3)

line1, =plt.plot(tp3, pressure, linewidth=4, color='k',
      linestyle='-')
line2, =plt.plot(tp4, pressure, linewidth=4, color='k',
      linestyle='--')
line3, =plt.plot(tp5, pressure, linewidth=4, color='k',
      linestyle=':')
plt.yscale('log')
plt.xlim([800,1100])
plt.ylim([1e2,1e-3])
plt.xlabel('$T$ (K)', fontsize=18)
plt.ylabel('$P$ (bar)', fontsize=18)
plt.legend([line1,line2,line3],['fiducial',r'$\kappa_{\rm S_0}
         =0.02$ cm$^2$ g$^{-1}$',r'$\kappa_{\rm S_0}=0.04$
         cm$^2$ g$^{-1}$'],frameon=False,prop={'size':18})
plt.savefig('tp2.eps', format='eps')
plt.close()

# Experiment #3:  scattering greenhouse and anti-greenhouse
tp6 = zeros(np)
tp7 = zeros(np)
tp8 = zeros(np)
tp9 = zeros(np)

for i in range(0,np):
   tp6[i] = tpprofile(m[i],m0,tint,tirr,kappa_S0,kappa0,
         kappa_cia,0.75,beta_L0,el1,el3)
   tp7[i] = tpprofile(m[i],m0,tint,tirr,kappa_S0,kappa0,
         kappa_cia,beta_S0,0.75,el1,el3)
   tp8[i] = tpprofile(m[i],m0,tint,tirr,kappa_S0,kappa0,
         kappa_cia,0.5,beta_L0,el1,el3)
   tp9[i] = tpprofile(m[i],m0,tint,tirr,kappa_S0,kappa0,
```

244

```
            kappa_cia,beta_S0,0.5,el1,el3)

line1, =plt.plot(tp3, pressure, linewidth=4, color='k',
        linestyle='-')
line2, =plt.plot(tp7, pressure, linewidth=4, color='r',
        linestyle=':')
line3, =plt.plot(tp9, pressure, linewidth=4, color='r',
        linestyle='--')
line4, =plt.plot(tp6, pressure, linewidth=4, color='b',
        linestyle=':')
line5, =plt.plot(tp8, pressure, linewidth=4, color='b',
        linestyle='--')
plt.yscale('log')
plt.xlim([700,1450])
plt.ylim([1e2,1e-3])
plt.xlabel('$T$ (K)', fontsize=18)
plt.ylabel('$P$ (bar)', fontsize=18)
plt.legend([line1,line2,line3,line4,line5],['fiducial',
           r'$\beta_{\rm L_0}=0.75$',r'$\beta_{\rm L_0}=0.5$',
           r'$\beta_{\rm S_0}=0.75$',r'$\beta_{\rm S_0}=0.5$'],
           frameon=False,prop={'size':18})
plt.savefig('tp3.eps', format='eps')
plt.close()
```

Next, we have the Python program used to generate the line profiles in Figure 5.2. It involves evaluating an integral numerically using the integrate package in scipy.

```
# uses equation (4) from Zaghloul (2007, MNRAS, 375, 1043)
from numpy import exp,sin,cos,pi,sqrt,arange,zeros
from scipy import special,integrate
from matplotlib import pyplot as plt

# first part of Voigt H-function
def voigt1(a0,u0):
    term1 = exp(a0*a0)*special.erfc(a0)
    term2 = exp(-u0*u0)*cos(2.0*a0*u0)
    return term1*term2

# integrand for second part of Voigt H-function
def vintegrand(x,a0,u0):
    coeff = 2.0/sqrt(pi)
```

```python
   term1 = exp(x*x - u0*u0)
   term2 = sin(2.0*a0*(u0-x))
   return coeff*term1*term2

# full Voigt H-function
def voigt(a0,u0):
   term1 = voigt1(a0,u0)
   term2 = integrate.quad(vintegrand, 0, u0, args=(a0,u0) )
   return term1+term2[0]

# Doppler profile
def doppler(u0):
   result = exp(-u0*u0)
   return result

# Lorentz profile
def lorentz(a0,u0):
   result = a0/sqrt(pi)/( a0*a0 + u0*u0 )
   return result

x = arange(-6.0,6.01,0.01)
n = len(x)

y1 = zeros(n)
y2 = zeros(n)
y3 = zeros(n)
y4 = zeros(n)
y5 = zeros(n)
y6 = zeros(n)

a1 = 1e-2
a2 = 1.0

for i in range(0,n):
   y1[i] = voigt(a1,x[i])
   y2[i] = doppler(x[i])
   y3[i] = lorentz(a1,x[i])
   y4[i] = voigt(a2,x[i])
   y5[i] = doppler(x[i])
   y6[i] = lorentz(a2,x[i])

line1, =plt.plot(x, y1, linewidth=4, color='k', linestyle='-')
line2, =plt.plot(x, y2, linewidth=4, color='k', linestyle='--')
line3, =plt.plot(x, y3, linewidth=4, color='k', linestyle=':')
line4, =plt.plot(x, y4, linewidth=2, color='r', linestyle='-')
```

```
line5, =plt.plot(x, y5, linewidth=2, color='r', linestyle='--')
line6, =plt.plot(x, y6, linewidth=2, color='r', linestyle=':')
plt.yscale('log')
plt.xlim([-6,6])
plt.ylim([1e-4,1.0])
plt.xlabel(r'$x$', fontsize=25)
plt.ylabel('normalized line profile', fontsize=18)
plt.legend([line1,line2,line3,line4,line5,line6],['Voigt
          ($a_0=10^{-2}$)','Doppler','Lorentz','Voigt ($a_0=1$)',
          'Doppler','Lorentz'],frameon=False,prop={'size':10})
plt.savefig('line_profiles.eps', format='eps')
```

Finally, we have the Python program used to perform the atmospheric chemistry calculations in Figure 7.2. It involves using the polyroots function to solve the quintic equation for the mixing ratio of methane. By trial and error, I found that the root with the index 4 is the correct one, as it is both real (non-complex) and positive. This program may be straightforwardly modified to produce Figure 7.3.

```
from numpy import mean,arange,zeros,polynomial,array,interp,exp
from matplotlib import pyplot as plt

# function to compute first equilibrium constant
def kprime(my_temperature,pbar):
    runiv = 8.3144621 # J/K/mol
    temperatures = arange(500.0, 3100.0, 100.0)
    dg = [96378.0, 72408.0, 47937.0, 23114.0, -1949.0, -27177.0,
          -52514.0, -77918.0, -103361.0, -128821.0, -154282.0,
          -179733.0, -205166.0, -230576.0, -255957.0, -281308.0,
          -306626.0, -331911.0, -357162.0, -382380.0, -407564.0,
          -432713.0, -457830.0, -482916.0, -507970.0, -532995.0]
    my_dg = interp(my_temperature,temperatures,dg)
    result = exp(-my_dg/runiv/my_temperature)/pbar/pbar
    return result

# function to compute second equilibrium constant
def kprime2(my_temperature,pbar):
    runiv = 8.3144621 # J/K/mol
    temperatures = arange(500.0, 3100.0, 100.0)
    dg2 = [262934.0, 237509.0, 211383.0, 184764.0, 157809.0,
           130623.0, 103282.0, 75840.0, 48336.0, 20797.0, -6758.0,
           -34315.0, -61865.0, -89403.0, -116921.0, -144422.0,
```

```
               -171898.0, -199353.0, -226786.0, -254196.0, -281586.0,
               -308953.0, -336302.0, -363633.0, -390945.0, -418243.0]
     my_dg = interp(my_temperature,temperatures,dg2)
     result = exp(-my_dg/runiv/my_temperature)/pbar/pbar
     return result

# function to compute third equilibrium constant
def kprime3(my_temperature):
     runiv = 8.3144621 # J/K/mol
     temperatures = arange(500.0, 3100.0, 100.0)
     dg3 = [20474.0, 16689.0, 13068.0, 9593.0, 6249.0, 3021.0,
               -107.0, -3146.0, -6106.0, -8998.0, -11828.0, -14600.0,
               -17323.0, -20000.0, -22634.0, -25229.0, -27789.0,
               -30315.0, -32809.0, -35275.0, -37712.0,-40123.0,
               -42509.0, -44872.0, -47211.0, -49528.0]
     my_dg = interp(my_temperature,temperatures,dg3)
     result = exp(-my_dg/runiv/my_temperature)
     return result

# function to compute mixing ratio for methane
# (note:  n_o is oxygen abundance, n_c is carbon abundance)
def n_methane(n_o,n_c,temp,pbar):
     k1 = kprime(temp,pbar)
     k2 = kprime2(temp,pbar)
     k3 = kprime3(temp)
     a5 = 8.0*k1*k2*k2/k3
     a4 = 8.0*k1*k2/k3
     a3 = 2.0*k1/k3*( 1.0 + 8.0*k2*(n_o-n_c) ) + 2.0*k1*k2
     a2 = 8.0*k1/k3*(n_o-n_c) + 2.0*k2 + k1
     a1 = 8.0*k1/k3*(n_o-n_c)*(n_o-n_c) + 1.0 + 2.0*k1*(n_o-n_c)
     a0 = -2.0*n_c
     result = polynomial.polynomial.polyroots([a0,a1,a2,a3,a4,a5])
     return result[4] # picks out the correct root of cubic

# function to compute mixing ratio for methane
def n_water(n_o,n_c,temp,pbar):
     k2 = kprime2(temp,pbar)
     n_ch4 = n_methane(n_o,n_c,temp,pbar)
     result = 2.0*k2*n_ch4*n_ch4 + n_ch4 + 2.0*(n_o-n_c)
     return result

# function to compute mixing ratio for carbon monoxide
def n_cmono(n_o,n_c,temp,pbar):
     kk = kprime(temp,pbar)
     n_ch4 = n_methane(n_o,n_c,temp,pbar)
```

248

```python
    n_h2o = n_water(n_o,n_c,temp,pbar)
    result = kk*n_ch4*n_h2o
    return result

# function to compute mixing ratio for carbon dioxide
def n_cdio(n_o,n_c,temp,pbar):
    kk3 = kprime3(temp)
    n_h2o = n_water(n_o,n_c,temp,pbar)
    n_co = n_cmono(n_o,n_c,temp,pbar)
    result = n_co*n_h2o/kk3
    return result

# function to compute mixing ratio for acetylene
def n_acet(n_o,n_c,temp,pbar):
    kk2 = kprime2(temp,pbar)
    n_ch4 = n_methane(n_o,n_c,temp,pbar)
    result = kk2*n_ch4*n_ch4
    return result

# inputs
n_o = 5e-4 # elemental abundance of oxygen
n_c = 0.1*n_o # elemental abundance of carbon
main_label = 'C/O=0.1'
file_label = 'co0p1'

xx = arange(500.0, 3010.0, 10.0) # temperature array
n = len(xx)
n_ch4 = zeros(n)
n_h2o = zeros(n)
n_co = zeros(n)
n_co2 = zeros(n)
n_c2h2 = zeros(n)
n_ch4_2 = zeros(n)
n_h2o_2 = zeros(n)
n_co_2 = zeros(n)
n_co2_2 = zeros(n)
n_c2h2_2 = zeros(n)

pbar = 1e-2 # pressure in bars
pbar2 = 1e0

for i in range(0,n):
    n_h2o[i] = n_water(n_o,n_c,xx[i],pbar)
    n_ch4[i] = n_methane(n_o,n_c,xx[i],pbar)
    n_co[i] = n_cmono(n_o,n_c,xx[i],pbar)
```

```
    n_co2[i] = n_cdio(n_o,n_c,xx[i],pbar)
    n_c2h2[i] = n_acet(n_o,n_c,xx[i],pbar)
    n_h2o_2[i] = n_water(n_o,n_c,xx[i],pbar2)
    n_ch4_2[i] = n_methane(n_o,n_c,xx[i],pbar2)
    n_co_2[i] = n_cmono(n_o,n_c,xx[i],pbar2)
    n_co2_2[i] = n_cdio(n_o,n_c,xx[i],pbar2)
    n_c2h2_2[i] = n_acet(n_o,n_c,xx[i],pbar2)


# plot
line1, =plt.plot(xx, n_ch4, linewidth=1, color='k',
        linestyle='-')
line2, =plt.plot(xx, n_h2o, linewidth=1, color='m',
        linestyle=':')
line3, =plt.plot(xx, n_co, linewidth=1, color='c',
        linestyle='-.')
line4, =plt.plot(xx, n_co2, linewidth=1, color='g',
        linestyle='--')
line5, =plt.plot(xx, n_c2h2, linewidth=1, color='y',
        linestyle='-')
line6, =plt.plot(xx, n_ch4_2, linewidth=5, color='k',
        linestyle='-')
line7, =plt.plot(xx, n_h2o_2, linewidth=5, color='m',
        linestyle=':')
line8, =plt.plot(xx, n_co_2, linewidth=5, color='c',
        linestyle='-.')
line9, =plt.plot(xx, n_co2_2, linewidth=5, color='g',
        linestyle='--')
line10, =plt.plot(xx, n_c2h2_2, linewidth=5, color='y',
        linestyle='-')
plt.title(main_label, fontsize=15)
plt.yscale('log')
plt.xlim([500,3000])
plt.ylim([1e-22,1e0])
plt.xlabel(r'$T$ (K)', fontsize=18)
plt.ylabel(r'$\tilde{n}_{\rm X}$', fontsize=18)
plt.legend([line1,line2,line3,line4,line5,line6,line7,line8,
           line9,line10],[r'CH$_4$ ($P=10$ mbar)',r'H$_2$O'
             ($P=10$ mbar),r'CO ($P=10$ mbar)',r'CO$_2$
             ($P=10$ mbar)',r'C$_2$H$_2$ ($P=10$ mbar)',
           r'CH$_4$ ($P=1$ bar)',r'H$_2$O ($P=1$ bar)',
           r'CO ($P=1$ bar)',r'CO$_2$ ($P=1$ bar)',r'C$_2$H$_2$
             ($P=1$ bar)'],frameon=False,prop='size':10,loc=4)
plt.savefig('chemistry_'+file_label+'.eps', format='eps')
```

Bibliography

[1] Abel, M., & Frommhold, L. 2013, "Collision-induced spectra and current astronomical research," Canadian Journal of Physics, 91, 857–869

[2] Abramowitz, M., & Stegun, I.A. 1970, "Handbook of Mathematical Functions," ninth printing (New York: Dover Publications)

[3] Agol, E., Steffen, J., Sari, R., & Clarkson, W. 2005, "On detecting terrestrial planets with timing of giant planet transits," Monthly Notices of the Royal Astronomical Society, 359, 567–579

[4] Agol, E., Cowan, N.B., Knutson, H.A., Deming, D., Steffen, J.H., Henry, G.W., & Charbonneau, D. 2010, "The Climate of HD 189733b from Fourteen Transits and Eclipses Measured by Spitzer," Astrophysical Journal, 721, 1861–1877

[5] Andrews, S.M., & Williams, J.P. 2007, "High-Resolution Submillimeter Constraints on Circumstellar Disk Structure," Astrophysical Journal, 659, 705–728

[6] Angerhausen, D., DeLarme, E., & Morse, J.A. 2015, "A Comprehensive Study of Kepler Phase Curves and Secondary Eclipses: Temperatures and Albedos of Confirmed Kepler Giant Planets," Publications of the Astronomical Society of the Pacific, 127, 1113–1130

[7] Arfken, G.B., & Weber, H.J. 1995, "Mathematical Methods for Physicists," fourth edition (San Diego: Academic Press)

[8] Armstrong, B.H. 1969, "The Radiative Diffusivity Factor for the Random Malkmus Band," Journal of the Atmospheric Sciences, 26, 741–743

[9] Atkins, P.W., & de Paula, J. 2006, "Physical Chemistry," eighth edition (New York: Freeman)

[10] Balbus, S.A., & Terquem, C. 2001, "Linear Analysis of the Hall Effect in Protostellar Disks," Astrophysical Journal, 552, 235–247

[11] Baraffe, I., Chabrier, G., Barman, T.S., Allard, F., & Hauschildt, P.H. 2003, "Evolutionary models for cool brown dwarfs and extrasolar giant planets. The case of HD 209458," Astronomy & Astrophysics, 402, 701–712

[12] Baraffe, I., Chabrier, G., & Barman, T. 2010, "The physical properties of extra-solar planets," Reports on Progress in Physics, 73, 016901

[13] Barman, T.S., Macintosh, B., Konopacky, Q.M., & Marois, C. 2011, "Clouds and Chemistry in the Atmosphere of Extrasolar Planet HR 8799b," Astrophysical Journal, 733, 65

[14] Batygin, K., & Stevenson, D.J. 2010, "Inflating Hot Jupiters with Ohmic Dissipation," Astrophysical Journal Letters, 714, L238–L243

[15] Batygin, K., & Brown, M.E. 2016, "Evidence for a Distant Giant Planet in the Solar System," Astronomical Journal, 151, 22

[16] Bean, J.L., Miller-Ricci Kempton, E., & Homeier, D. 2010, "A ground-based transmission spectrum of the super-Earth exoplanet GJ 1214b," Nature, 468, 669–672

[17] Bean, J.L., et al. 2011, "The Optical and Near-infrared Transmission Spectrum of the Super-Earth GJ 1214b: Further Evidence for a Metal-rich Atmosphere," Astrophysical Journal, 743, 92

[18] Benneke, B., & Seager, S. 2012, "Atmospheric Retrieval for Super-Earths: Uniquely Constraining the Atmospheric Composition with Transmission Spectroscopy," Astrophysical Journal, 753, 100

[19] Betts, A.K. 1986, "A new convective adjustment scheme. Part I: Observational and theoretical basis," Quarterly Journal of the Royal Meteorological Society, 112, 677–691

[20] Betts, A.K., & Miller, M.J. 1986, "A new convective adjustment scheme. Part II: Single column tests using GATE wave, BOMEX, ATEX and arctic air-mass data sets," Quarterly Journal of the Royal Meteorological Society, 112, 693–709

[21] Bondi, H. 1952, "On Spherically Symmetrical Accretion," Monthly Notices of the Royal Astronomical Society, 112, 195–204

[22] Borucki, W.J., & Summers, A.L. 1984, "The photometric method of detecting other planetary systems," Icarus, 58, 121–134

[23] Brown, T.M. 2001, "Transmission Spectra as Diagnostics of Extrasolar Giant Planet Atmospheres," Astrophysical Journal, 553, 1006–1026

[24] Burrows, A., & Sharp, C.M. 1999, "Chemical Equilibrium Abundances in Brown Dwarf and Extrasolar Giant Planet Atmospheres," Astrophysical Journal, 512, 843–863

[25] Burrows, A. 2003, "A Theory for the Radius of the Transiting Giant Planet HD 209458b," Astrophysical Journal, 594, 545–551

[26] Burrows, A., Hubeny, I., Budaj, J., & Hubbard, W.B. 2007a, "Possible Solutions to the Radius Anomalies of Transiting Giant Planets," Astrophysical Journal, 661, 502–514

[27] Burrows, A., Hubeny, I., Budaj, J., Knutson, H.A., & Charbonneau, D. 2007b, "Theoretical Spectral Models of the Planet HD 209458b with a Thermal Inversion and Water Emission Bands," Astrophysical Journal Letters, 668, L171–L174

[28] Burrows, A., Ibgui, L., & Hubeny, I. 2008, "Optical Albedo Theory of Strongly Irradiated Giant Planets: The Case of HD 209458b," Astrophysical Journal, 682, 1277–1282

[29] Burrows, A., Heng, K., & Nampaisarn, T. 2011, "The Dependence of Brown Dwarf Radii on Atmospheric Metallicity and Clouds: Theory and Comparison with Observations," Astrophysical Journal, 736, 47

[30] Burrows, A. 2014a, "Spectra as windows into exoplanet atmospheres," Proceedings of the National Academy of Sciences, 111, 12601–12609

[31] Burrows, A. 2014b, "Highlights in the study of exoplanet atmospheres," Nature, 513, 345–352

[32] Bychkov, K.V., & Lebedev, V.S. 1979, "On the origin of High-velocity Gas Hα-emission from the Cygnus Loop and IC 443," Astronomy & Astrophysics, 80, 167–169

[33] Cahoy, K.L., Marley, M.S., & Fortney, J.J. 2010, "Exoplanet Albedo Spectra and Colors as a Function of Planet Phase, Separation, and Metallicity," Astrophysical Journal, 724, 189–214

[34] Champagne, F.H. 1978, "The fine-scale structure of the turbulent velocity field," Journal of Fluid Mechanics, 86, 67–108

[35] Chandrasekhar, S. 1960, "Radiative Transfer" (New York: Dover Publications)

[36] Chapman, S. 1946, "Some Thoughts on Nomenclature," Nature, 157, 405

[37] Charbonneau, D., Brown, T.M., Latham, D.W., & Mayor, M. 2000, "Detection of Planetary Transits Across a Sun-like Star," Astrophysical Journal Letters, 529, L45–L48

[38] Charbonneau, D., Brown, T.M., Noyes, R.W., & Gilliland, R.L. 2002, "Detection of an Extrasolar Planet Atmosphere," Astrophysical Journal, 568, 377–384

[39] Charbonneau, D., et al. 2005, "Detection of Thermal Emission from an Extrasolar Planet," Astrophysical Journal, 626, 523–529

[40] Charbonneau, D. 2009, "The Rise of the Vulcans," Proceedings of the International Astronomical Union, eds. F. Pont, D. Sasselov and M. Holman, 253, 1–8

[41] Charbonneau, D., et al. 2009, "A super-Earth transiting a nearby low-mass star," Nature, 462, 891–894

[42] Chevalier, R.A., Kirshner, R.P., & Raymond, J.C. 1980, "The Optical Emission from a Fast Shock Wave with Application to Supernova Remnants," Astrophysical Journal, 235, 186–195

[43] Cho, J.Y.-K., Menou, K., Hansen, B.M.S., & Seager, S. 2003, "The Changing Face of the Extrasolar Giant Planet HD 209458b," Astrophysical Journal Letters, 587, L117-L120

[44] Cho, J.Y.-K., Polichtchouk, I., & Thrastarson, H.Th. 2015, "Sensitivity and variability redux in hot-Jupiter flow simulations," Monthly Notices of the Royal Astronomical Society, 454, 3423–3431

[45] Cooper, C.S., & Showman, A.P. 2006, "Dynamics and Disequilibrium Carbon Chemistry in Hot Jupiter Atmospheres, with Application to HD 209458b," Astrophysical Journal, 649, 1048–1063

[46] Deming, D., Harrington, J., Seager, S., & Richardson, L.J. 2006, "Strong Infrared Emission from the Extrasolar Planet HD 189733b," Astrophysical Journal, 644, 560–564

[47] Deming, D., et al. 2013, "Infrared Transmission Spectroscopy of the Exoplanets HD 209458b and XO-1b Using the Wide Field Camera-3 on the Hubble Space Telescope," Astrophysical Journal, 774, 95

[48] Deming, D., et al. 2015, "Spitzer Secondary Eclipses of the Dense, Modestly-irradiated, Giant Exoplanet HAT-P-20b Using Pixel-level Decorrelation," Astrophysical Journal, 805, 132

[49] Demory, B.-O., & Seager, S. 2011, "Lack of Inflated Radii for Kepler Giant Planet Candidates Receiving Modest Stellar Irradiation," Astrophysical Journal Supplements, 197, 12

[50] de Wit, J., Gillon, M., Demory, B.-O., & Seager, S. 2012, "Towards consistent mapping of distant words: secondary-eclipse scanning of the exoplanet HD 189733b," Astronomy & Astrophysics, 548, A128

[51] Doyle, L.R., et al. 2011, "Kepler-16: A Transiting Circumbinary Planet," Science, 333, 1602–1606

[52] Draine, B.T., & Lee, H.M. 1984, "Optical Properties of Interstellar Graphite and Silicate Grains," Astrophysical Journal, 285, 89–108

[53] Draine, B.T., & Lee, H.M. 1987, "Optical Properties of Interstellar Graphite and Silicate Grains: Erratum," Astrophysical Journal, 318, 485

[54] Draine, B.T., & McKee, C.F. 1993, "Theory of Interstellar Shocks," Annual Review of Astronomy & Astrophysics, 31, 373–432

[55] Draine, B.T. 2011, "Physics of the Interstellar and Intergalactic Medium" (Princeton: Princeton University Press)

[56] Ehrenreich, D., et al. 2015, "A giant comet-like cloud of hydrogen escaping the warm Neptune-mass exoplanet GJ 436b," Nature, 522, 459–461

[57] Eisenberg, D., & Crothers, D. 1979, "Physical Chemistry with Applications to the Life Sciences" (California: Benjamin/Cummings)

[58] Ertel, H. 1942, "Ein neuer hydrodynamischer Erhaltungssatz," Die Naturwissenschaften, 30, 543–544

[59] Ertel, H., & Rossby, C.-G. 1949, "A new conservation theorem of hydrodynamics," Geofisica Pura e Applicata, 14, 189–193

[60] Fabrycky, D.C. 2010, "Non-Keplerian Dynamics of Exoplanets," in Exoplanets, ed. S. Seager, 217–238 (Tucson: University of Arizona Press)

[61] Forget, F., & Pierrehumbert, R.T. 1997, "Warming Early Mars with Carbon Dioxide Clouds That Scatter Infrared Radiation," Science, 278, 1273–1276

[62] Fortney, J.J. 2005, "The effect of condensates on the characterization of transiting planet atmospheres with transmission spectroscopy," Monthly Notices of the Royal Astronomical Society, 364, 649–653

[63] Fortney, J.J., Lodders, K., Marley, M.S., & Freedman, R.S. 2008, "A Unified Theory for the Atmospheres of the Hot and Very Hot Jupiters: Two Classes of Irradiated Atmospheres," Astrophysical Journal, 678, 1419–1435

[64] Fortney, J.J., Shabram, M., Showman, A.P., Lian, Y., Freedman, R.S., Marley, M.S., & Lewis, N.K. 2010, "Transmission Spectra of Three-Dimensional Hot Jupiter Model Atmospheres," Astrophysical Journal, 709, 1396–1406

[65] Fu, Q., & Liou, K.N. 1992, "On the Correlated k-Distribution Method for Radiative Transfer in Nonhomogeneous Atmospheres," Journal of the Atmospheric Sciences, 49, 2139–2156

[66] Gail, H.-P., & Sedlmayr, E. 2014, "Physics and Chemistry of Circumstellar Dust Shells" (New York: Cambridge University Press)

[67] Galperin, B., Young, R.M.B., Sukoriansky, S., Dikovskaya, N., Read, P.L., Lancaster, A.J., & Armstrong, D. 2014, Icarus, 229, 295–320

[68] García Muñoz, A. 2007, "Physical and chemical aeronomy of HD 209458b," Planetary and Space Science, 55, 1426–1455

[69] Gibson, N.P., Aigrain, S., Roberts, S., Evans, T.M., Osborne, M., & Pont, F. 2012, "A Gaussian process framework for modelling instrumental systematics: application to transmission spectroscopy," Monthly Notices of the Royal Astronomical Society, 419, 2683–2694

[70] Gill, A.E. 1980, "Some simple solutions for heat-induced tropical circulation," Quarterly Journal of the Royal Meteorological Society, 106, 447–462

[71] Gilman, P.A. 2000, "Magnetohydrodynamic 'Shallow Water' Equations for the Solar Tachocline," Astrophysical Journal Letters, 544, L79–L82

[72] Glassman, I., Yetter, R.A., & Glumac, N.G. 2015, "Combustion," fifth edition (Massachusetts: Elsevier)

[73] Goodman, J. 2009, "Thermodynamics of Atmospheric Circulation on Hot Jupiters," Astrophysical Journal, 693, 1645–1649

[74] Goody, R.M., & Yung, Y.L. 1989, "Atmospheric Radiation: Theoretical Basis," third edition (New York: Oxford University Press)

[75] Gould, A., & Loeb, A. 1992, "Discovering Planetary Systems Through Gravitational Microlenses," Astrophysical Journal, 396, 104–114

[76] Greene, T.P., Line, M.R., Montero, C., Fortney, J.J., Lustig-Yaeger, J., & Luther, K. 2016, "Characterizing Transiting Exoplanet Atmospheres with JWST," Astrophysical Journal, 817, 17

[77] Grimm, S.L. & Heng, K. 2015, "HELIOS-K: An Ultrafast, Open-source Opacity Calculator for Radiative Transfer," Astrophysical Journal, 808, 182

[78] Guillot, T., & Showman, A.P. 2002, "Evolution of '51 Pegasus b-like' planets," Astronomy & Astrophysics, 385, 156–165

[79] Guillot, T. 2010, "On the radiative equilibrium of irradiated planetary atmospheres," Astronomy & Astrophysics, 520, A27

[80] Hansen, B.M.S., & Barman, T. 2007, "Two Classes of Hot Jupiters," Astrophysical Journal, 671, 861–871

[81] Hansen, B.M.S. 2008, "On the absorption and redistribution of energy in irradiated planets," Astrophysical Journal Supplements, 179, 484–508

[82] Harrington, J., Hansen, B.M., Luszcz, S.H., Seager, S., Deming, D., Menou, K., Cho, J.Y.-K., & Richardson, L.J. 2006, "The Phase-Dependent Infrared Brightness of the Extrasolar Planet υ Andromedae b," Science, 314, 623–626

[83] Harrington, J., Luszcz, S., Seager, S., Deming, D., & Richardson, L.J. 2007, "The hottest planet," Nature, 447, 691–693

[84] Helling, Ch., & Woitke, P. 2006, "Dust in brown dwarfs. V. Growth and evaporation of dirty dust grains," Astronomy & Astrophysics, 455, 325–338

[85] Heng, K., & Spitkovsky, A. 2009, "Magnetohydrodynamic Shallow Water Waves: Linear Analysis," Astrophysical Journal, 703, 1819–1831

[86] Heng, K. 2010, "Balmer-Dominated Shocks: A Concise Review," Publications of the Astronomical Society of Australia, 27, 23–44

[87] Heng, K., Menou, K., & Phillipps, P.J. 2011, "Atmospheric circulation of tidally locked exoplanets: a suite of benchmark tests for dynamical solvers," Monthly Notices of the Royal Astronomical Society, 413, 2380–2402

[88] Heng, K., Hayek, W., Pont, F., & Sing, D.K. 2012, "On the effects of clouds and hazes in the atmospheres of hot Jupiters: semi-analytical temperature-pressure profiles," Monthly Notices of the Royal Astronomical Society, 420, 20–36

[89] Heng, K., Kopparla, P. 2012, "On the Stability of Super-Earth Atmospheres," Astrophysical Journal, 754, 60

[90] Heng, K. 2012, "On the Existence of Shocks in Irradiated Exoplanetary Atmospheres," Astrophysical Journal Letters, 761, L1

[91] Heng, K., & Demory, B.-O. 2013, "Understanding Trends Associated with Clouds in Irradiated Exoplanets," Astrophysical Journal, 777, 100

[92] Heng, K. 2013, "Why Does Nature Form Exoplanets Easily?" American Scientist, 101, 184–187

[93] Heng, K., & Workman, J. 2014, "Analytical Models of Exoplanetary Atmospheres. I. Atmospheric Dynamics via the Shallow Water System," Astrophysical Journal Supplements, 213, 27

[94] Heng, K., Mendonça, J.M., & Lee, J.-M. 2014, "Analytical Models of Exoplanetary Atmospheres. II. Radiative Transfer via the Two-stream Approximation," Astrophysical Journal Supplements, 215, 4

[95] Heng, K. 2014, "The Nature of Scientific Proof in the Age of Simulations," American Scientist, 102, 174–177

[96] Heng, K., & Showman, A.P. 2015, "Atmospheric Dynamics of Hot Exoplanets," Annual Review of Earth & Planetary Sciences, 43, 509–540

[97] Heng, K., Wyttenbach, A., Lavie, B., Sing, D.K., Ehrenreich, D., & Lovis, C. 2015, "A Non-isothermal Theory for Interpreting Sodium Lines in Transmission Spectra of Exoplanets," Astrophysical Journal Letters, 803, L9

[98] Heng, K., & Winn, J. 2015, "The Next Great Exoplanet Hunt," American Scientist, 103, 196–203

[99] Heng, K., Lyons, J.R., & Tsai, S.-M. 2016, "Atmospheric Chemistry for Astrophysicists: A Self-Consistent Formalism and Analytical Solutions for Arbitrary C/O," Astrophysical Journal, 816, 96

[100] Heng, K., & Lyons, J.R. 2016, "Carbon Dioxide in Exoplanetary Atmospheres: Rarely Dominant Compared to Carbon Monoxide and Water in Hot, Hydrogen-dominated Atmospheres," Astrophysical Journal, 817, 149

[101] Henyey, L.G., & Greenstein, J.L. 1941, "Diffuse Radiation in the Galaxy," Astrophysical Journal, 93, 70–83

[102] Herzberg, G. 1952, "Spectroscopic Evidence of Molecular Hydrogen in the Atmospheres of Uranus and Neptune," Astrophysical Journal, 115, 337–340

[103] Holman, M.J., & Murray, N.W. 2005, "The Use of Transit Timing to Detect Terrestrial-Mass Extrasolar Planets," Science, 307, 1288–1291

[104] Holmström, M., Ekenbäck, A., Selsis, F., Penz, T., Lammer, H., & Wurz, P. 2008, "Energetic neutral atoms as the explanation for the high-velocity hydrogen around HD 209458b," Nature, 451, 970–972

[105] Holton, J.R. 2004, "Introduction to Dynamic Meteorology," fourth edition (Massachusetts: Elsevier)

[106] Huang, X., & Cumming, A. 2012, "Ohmic Dissipation in the Interiors of Hot Jupiters," Astrophysical Journal, 757, 47

[107] Hubeny, I., Burrows, A., & Sudarsky, D. 2003, "A Possible Bifurcation in Atmospheres of Strongly Irradiated Stars and Planets," Astrophysical Journal, 594, 1011–1018

[108] Hubeny, I., & Mihalas, D. 2015, "Theory of Stellar Atmospheres" (New Jersey: Princeton University Press)

[109] Irwin, P.G.J., et al. 2008, "The NEMESIS planetary atmosphere radiative transfer and retrieval tool," Journal of Quantitative Spectroscopy & Radiative Transfer, 109, 1136–1150

[110] Jackson, J.D. 1999, "Classical Electrodynamics," third edition (New York: Wiley)

[111] Jacobson, M.Z. 2005, "Fundamentals of Atmospheric Modeling" (New York: Cambridge University Press)

[112] Jeans, J.H. 1916, "The Dynamical Theory of Gases," second edition (London: Cambridge University Press)

[113] Johnston, H.S. 1966, "Gas Phase Reaction Rate Theory" (New York: Ronald Press Company)

[114] Joshi, M.M., Haberle, R.M., & Reynolds, R.T. 1997, "Simulations of the Atmospheres of Synchronously Rotating Terrestrial Planets Orbiting M Dwarfs: Conditions for Atmospheric Collapse and the Implications for Habitability," Icarus, 129, 450–465

[115] Kataria, T., Showman, A.P., Fortney, J.J., Stevenson, K.B., Line, M.R., Kreidberg, L., Bean, J.L., & Désert, J.-M. 2015, "The Atmospheric Circulation of the Hot Jupiter WASP-43b: Comparing Three-dimensional Models to Spectrophotometric Data," Astrophysical Journal, 801, 86

[116] Kirkpatrick, J.D. 2005, "New Spectral Types L and T," Annual Review of Astronomy & Astrophysics, 43, 195–245

[117] Kitzmann, D. 2016, "Revisiting the Scattering Greenhouse Effect of CO_2 Ice Clouds," Astrophysical Journal Letters, 817, L18

[118] Klotz, I.M., & Rosenberg, R.M. 2008, "Chemical Thermodynamics: Basic Concepts and Methods," seventh edition (New Jersey: Wiley)

[119] Koskinen, T.T., Aylward, A.D., Smith, C.G.A., & Miller, S. 2007, "A Thermospheric Circulation Model for Extrasolar Giant Planets," Astrophysical Journal, 661, 515–526

[120] Koskinen, T.T., Aylward, A.D., & Miller, S. 2009, "The Upper Atmosphere of HD 17156b," Astrophysical Journal, 693, 868–885

[121] Knutson, H.A., et al. 2007, "A map of the day-night contrast of the extrasolar planet HD 189733b," Nature, 447, 183–186

[122] Knutson, H.A., et al. 2009a, "Multiwavelength Constraints on the Day-Night Circulation Patterns of HD 189733b," Astrophysical Journal, 690, 822–836

[123] Knutson, H.A., Charbonneau, D., Cowan N.B., Fortney, J.J., Showman, A.P., Agol, E., & Henry, G.W. 2009b, "The 8 μm Phase Variation of the Hot Saturn HD 149026b," Astrophysical Journal, 703, 769–784

[124] Knutson, H.A., Howard, A.W., & Isaacson, H. 2010, "A Correlation Between Stellar Activity and Hot Jupiter Emission Spectra," Astrophysical Journal, 720, 1569–1576

[125] Knutson, H.A., Benneke, B., Deming, D., & Homeier, D. 2014, "A featureless transmission spectrum for the Neptune-mass exoplanet GJ 436b," Nature, 505, 66–68

[126] Kolmogorov, A.N. 1991a, "The Local Structure of Turbulence in Incompressible Viscous Fluid for Very Large Reynolds Numbers," translated by V. Levin, Proceedings of the Royal Society of London A (Mathematical and Physical Sciences), 434, 9–13

[127] Kolmogorov, A.N. 1991b, "Dissipation of energy in the locally isotropic turbulence," translated by V. Levin, Proceedings of the Royal Society of London A (Mathematical and Physical Sciences), 434, 15–17

[128] Konopacky, Q.M., Barman, T.S., Macintosh, B.A., & Marois, C. 2013, "Detection of Carbon Monoxide and Water Absorption Lines in an Exoplanet Atmosphere," Science, 339, 1398–1401

[129] Kraichnan, R.H. 1967, "Inertia Ranges in Two-Dimensional Turbulence," Physics of Fluids, 10, 1417–1423

[130] Kraichnan, R.H. 1971, "Inertia-range transfer in two- and three-dimensional turbulence," Journal of Fluid Mechanics, 47, 525–535

[131] Kreidberg, L., et al. 2014a, "Clouds in the atmosphere of the super-Earth exoplanet GJ 1214b," Nature, 505, 69–72

[132] Kreidberg, L., et al. 2014b, "A Precise Water Abundance Measurement for the Hot Jupiter WASP-43b," Astrophysical Journal Letters, 793, L27

[133] Kuchner, M.J., & Seager, S. 2005, "Extrasolar Carbon Planets," arXiv:astro-ph/0504214

[134] Kundu, P.K., & Cohen, I.M. 2004, "Fluid Mechanics," third edition (San Diego: Elsevier)

[135] Lacis, A.A., & Oinas, V. 1991, "A Description of the Correlated k Distribution Method for Modeling Nongray Gaseous Absorption, Thermal Emission, and Multiple Scattering in Vertically Inhomogeneous Atmospheres," Journal of Geophysical Research, 96, 9027–9063

[136] Laor, A., & Draine, B.T. 1993, "Spectroscopic Constraints on the Properties of Dust in Active Galactic Nuclei," Astrophysical Journal, 402, 441–468

[137] Leconte, J., Wu, H., Menou, K., & Murray, N. 2015, "Asynchronous rotation of Earth-mass planets in the habitable zone of lower-mass stars," Science, 347, 632–635

[138] Lee, J.-M., Fletcher, L.N., & Irwin, P.G.J. 2012, "Optimal estimation retrievals of the atmospheric structure and composition of HD 189733b from secondary eclipse spectroscopy," Monthly Notices of the Royal Astronomical Society, 420, 170–182

[139] Lee, J.-M., Heng, K., & Irwin, P.G.J. 2013, "Atmospheric Retrieval Analysis of the Directly Imaged Exoplanet HR 8799b," Astrophysical Journal, 778, 97

[140] Lesur, G., & Papaloizou, J.C.B. 2010, "The subcritical baroclinic instability in local accretion disc models," Astronomy & Astrophysics, 513, A60

[141] Li, H., Finn, J.M., Lovelace, R.V.E., & Colgate, S.A. 2000, "Rossby Wave Instability of Thin Accretion Disks. II. Detailed Linear Theory," Astrophysical Journal, 533, 1023–1034

[142] Liepmann, H., & Roshko, A. 1957, "Elements of Gas Dynamics" (New York: Wiley)

[143] Line, M.R., et al. 2013, "A Systematic Retrieval Analysis of Secondary Eclipse Spectra. I. A Comparison of Atmospheric Retrieval Techniques," Astrophysical Journal, 775, 137

[144] Line, M.R., Knutson, H., Wolf, A.S., & Yung, Y.L. 2014, "A Systematic Retrieval Analysis of Secondary Eclipse Spectra. II. A Uniform Analysis of Nine Planets and their C to O Ratios," Astrophysical Journal, 783, 70

[145] Lissauer, J.J., et al. 2011, "A closely packed system of low-mass, low-density planets transiting Kepler-11," Nature, 470, 53–58

[146] Liu, B., & Showman, A.P. 2013, "Atmospheric Circulation of Hot Jupiters: Insensitivity to Initial Conditions," Astrophysical Journal, 770, 42

[147] Lodders, K., & Fegley, B. 2002, "Atmospheric Chemistry in Giant Planets, Brown Dwarfs, and Low-Mass Dwarf Stars. I. Carbon, Nitrogen, and Oxygen," Icarus, 155, 393–424

[148] Longuet-Higgins, M.S. 1967, "On the trapping of wave energy round islands," Journal of Fluid Mechanics, 29, 781–821

[149] Longuet-Higgins, M.S. 1968, "The Eigenfunctions of Laplace's Tidal Equations over a Sphere," Philosophical Transactions of the Royal Society of London (Series A), 262, 511–607

[150] Lovis, C., & Fischer, D.A. 2010, "Radial Velocity Techniques for Exoplanets," in Exoplanets, ed. S. Seager, 27–53 (Tucson: University of Arizona Press)

[151] Madhusudhan, N., & Seager, S. 2009, "A Temperature and Abundance Retrieval Method for Exoplanet Atmospheres," Astrophysical Journal, 707, 24–39

[152] Madhusudhan, N., & Seager, S. 2010, "On the Inference of Thermal Inversions in Hot Jupiter Atmospheres," Astrophysical Journal, 725, 261–274

[153] Madhusudhan, N., & Seager, S. 2011, "High Metallicity and Non-equilibrium Chemistry in the Dayside Atmosphere of hot-Neptune GJ 436b," Astrophysical Journal, 729, 41

[154] Madhusudhan, N., Burrows, A., & Currie, T. 2011, "Model Atmospheres for Massive Gas Giants with Thick Clouds: Application to the HR 8799 Planets and Predictions for Future Detections," Astrophysical Journal, 737, 34

[155] Madhusudhan, N. 2012, "C/O Ratio as a Dimension for Characterizing Exoplanetary Atmospheres," Astrophysical Journal, 758, 36

[156] Majeau, C., Agol, E., & Cowan, N.B. 2012, "A Two-dimensional Infrared Map of the Extrasolar Planet HD 189733b," Astrophysical Journal Letters, 747, L20

[157] Manabe, S., Smagorinsky, J., & Strickler, R.F. 1965, "Simulated Climatology of a General Circulation Model with a Hydrologic Cycle," Monthly Weather Review, 93, 769–798

[158] Mandel, K., & Agol, E. 2002, "Analytic Light Curves for Planetary Transit Searches," Astrophysical Journal Letters, 580, L171

[159] Mao, S., & Paczyński, B. 1991, "Gravitational Microlensing by Double Stars and Planetary Systems," Astrophysical Journal Letters, 374, L37–L40

[160] Marley, M.S., Gelino, C., Stephens, D., Lunine, J.I., & Freedman, R. 1999, "Reflected Spectra and Albedos of Extrasolar Giant Planets. I. Clear and Cloudy Atmospheres," Astrophysical Journal, 513, 879–893

[161] Marley, M.S., Saumon, D., Cushing, M., Ackerman, A.S., Fortney, J.J., & Freedman, R. 2012, "Masses, Radii, and Cloud Properties of the HR 8799 Planets," Astrophysical Journal, 754, 135

[162] Marley, M.S., Ackerman, A.S., Cuzzi, J.N., & Kitzmann, D. 2013, "Clouds and Hazes in Exoplanet Atmospheres," in Comparative Climatology of Terrestrial Planets, eds. S.J. Mackwell, A.A. Simon-Miller, J.W. Harder and M.A. Bullock, 367–391 (Tucson: University of Arizona Press)

[163] Marois, C., Macintosh, B., Barman, T., Zuckerman, B., Song, I., Patience, J., Lafrenière, D., Doyon, R. 2008, "Direct Imaging of Multiple Planets Orbiting the Star HR 8799," Science, 322, 1348–1352

[164] Marois, C., Zuckerman, B., Konopacky, Q.M., Macintosh, B., & Barman, T. 2010, "Images of a fourth planet orbiting HR 8799," Nature, 468, 1080–1083

[165] Matsuno, T. 1966, "Quasi-Geostrophic Motions in the Equatorial Area," Journal of the Meteorological Society of Japan, 44, 25–42

[166] Mayor, M., & Queloz, D. 1995, "A Jupiter-mass companion to a solar-type star," Nature, 378, 355–359

[167] McLaughlin, D.B. 1924, "Some Results of a Spectrographic Study of the Algol System," Astrophysical Journal, 60, 22–31

[168] Meador, W.E., & Weaver, W.R. 1980, "Two-Stream Approximations to Radiative Transfer in Planetary Atmospheres: A Unified Description of Existing Methods and a New Improvement," Journal of the Atmospheric Sciences, 37, 630–643

[169] Menou, K., Cho, J.Y.-K., Seager, S., & Hansen, B.M.S. 2003, "Weather Variability of Close-in Extrasolar Giant Planets," Astrophysical Journal Letters, 587, L113-L116

[170] Menou, K. 2012a, "Magnetic Scaling Laws for the Atmospheres of Hot Giant Exoplanets," Astrophysical Journal, 745, 138

[171] Menou, K. 2012b, "Thermo-resistive Instability of Hot Planetary Atmospheres," Astrophysical Journal Letters, 754, L9

[172] Mihalas, D. 1970, "Stellar Atmospheres" (San Francisco: Freeman and Company)

[173] Mihalas, D., & Weibel-Mihalas, B. 1999, "Foundations of Radiation Hydrodynamics," second edition (New York: Dover Publications)

[174] Moore, W.J. 1972, "Physical Chemistry," fourth edition (New Jersey: Prentice-Hall)

[175] Moses, J.I., et al. 2011, "Disequilibrium Carbon, Oxygen, and Nitrogen Chemistry in the Atmospheres of HD 189733b and HD 209458b," Astrophysical Journal, 737, 15

[176] Moses, J.I., et al. 2013, "Compositional Diversity in the Atmospheres of Hot Neptunes, with Application to GJ 436b," Astrophysical Journal, 777, 34

[177] Murray-Clay, R.A., Chiang, E.I., & Murray, N. 2009, "Atmospheric Escape from Hot Jupiters," Astrophysical Journal, 693, 23–42

[178] Öberg, K.I., Murray-Clay, R., & Bergin, E.A. 2011, "The Effects of Snow-lines on C/O in Planetary Atmospheres," Astrophysical Journal Letters, 743, L16

[179] Ou, S., Ji, J., Liu, L., & Peng, X. 2007, "Disk-planet Interaction Simulations. I. Baroclinic Generation of Vortensity and Nonaxisymmetric Rossby Wave Instability," Astrophysical Journal, 667, 1220–1228

[180] Owen, J.E., & Jackson, A.P. 2012, "Planetary evaporation by UV and X-ray radiation: basic hydrodynamics," Monthly Notices of the Royal Astronomical Society, 425, 2931–2947

[181] Owen, J.E., & Wu, Y. 2013, "Kepler Planets: A Tale of Evaporation," Astrophysical Journal, 775, 105

[182] Ozawa, H., Ohmura, A., Lorenz, R.D., & Pujol, T. 2003, "Reviews of Geophysics," 41, 4

[183] Paczyński, B. 1986, "Gravitational Microlensing by the Galactic Halo," Astrophysical Journal, 304, 1–5

[184] Parker, E.N. 1960, "The Hydrodynamic Theory of Solar Corpuscular Radiation and Stellar Winds," Astrophysical Journal, 132, 821–866

[185] Parker, E.N. 1965, "Dynamical Theory of the Solar Wind," Space Science Reviews, 4, 666–708

[186] Peixóto, J.P., & Oort, A.H. 1984, "Physics of climate," Reviews of Modern Physics, 56, 365–429

[187] Penner, S.S. 1952, "Emission and Absorption of Radiation by Spectral Lines with Doppler Contour," Journal of Chemical Physics, 20, 507–510

[188] Perna, R., Menou, K., & Rauscher, E. 2010a, "Magnetic Drag on Hot Jupiter Atmospheric Winds," Astrophysical Journal, 719, 1421–1426

[189] Perna, R., Menou, K., & Rauscher, E. 2010b, "Ohmic Dissipation in the Atmospheres of Hot Jupiters," Astrophysical Journal, 724, 313–317

[190] Pierrehumbert, R.T., & Swanson, K.L. 1995, "Baroclinic Instability," Annual Review of Fluid Mechanics, 27, 419–467

[191] Pierrehumbert, R.T. 2010, "Principles of Planetary Climate" (New York: Cambridge University Press)

[192] Pierrehumbert, R., & Gaidos, E. 2011, "Hydrogen Greenhouse Planets Beyond the Habitable Zone," Astrophysical Journal Letters, 734, L13

[193] Polichtchouk, I., Cho, J.Y.-K., Watkins, C., Thrastarson, H.Th., Umurhan, O.M., & de la Torre Juárez, M. 2014, "Intercomparison of general circulation models for hot extrasolar planets," 229, 355–377

[194] Pont, F., Sing, D.K., Gibson, N.P., Aigrain, S., Henry, G., & Husnoo, N. 2013, "The prevalence of dust on the exoplanet HD 189733b from Hubble and Spitzer observations," Monthly Notices of the Royal Astronomical Society, 432, 2917–2944

[195] Press, W.H., Teukolsky, S.A., Vetterling, W.T., & Flannery, B.P. 2007, "Numerical Recipes: the Art of Scientific Computing," third edition (New York: Cambridge University Press)

[196] Prinn, R.G., & Barshay, S.S. 1977, "Carbon Monoxide on Jupiter and Implications for Atmospheric Convection," Science, 198, 1031–1034

[197] Queloz, D., Eggenberger, A., Mayor, M., Perrier, C., Beuzit, J.L., Naef, D., Sivan, J.P., & Udry, S. 2000, "Detection of a spectroscopic transit by the planet orbiting the star HD 209458," Astronomy & Astrophysics, 359, L13–L17

[198] Rauscher, E., & Menou, K. 2012, "The Role of Drag in the Energetics of Strongly Forced Exoplanet Atmospheres," Astrophysical Journal, 745, 78

[199] Rodgers, C.D. 2000, "Inverse Methods for Atmospheric Sounding: Theory and Practice" (Singapore: World Scientific)

[200] Rossiter, R.A. 1924, "On the Detection of an Effect of Rotation During Eclipse in the Velocity of the Brighter Component of Beta Lyrae, and on the Constancy of Velocity of this System," Astrophysical Journal, 60, 15–21

[201] Roston, G.D., & Obaid, F.S. 2005, "Exact analytical formula for Voigt spectral line profile," Journal of Quantitative Spectroscopy & Radiative Transfer, 94, 255–263

[202] Rothman, L.S., et al. 1996, "The HITRAN Molecular Spectroscopic Database and HAWKS (HITRAN Atmospheric Workstation): 1996 Edition," Journal of Quantitative Spectroscopy & Radiative Transfer, 60, 665–710

[203] Rothman, L.S., et al. 2010, "HITEMP, the high-temperature molecular spectroscopic database," Journal of Quantitative Spectroscopy & Radiative Transfer, 111, 2139–2150

[204] Russell, H.N. 1916, "On the Albedo of the Planets and their Satellites," Astrophysical Journal, 43, 173–196

[205] Seager, S., & Sasselov, D.D. 2000, "Theoretical Transmission Spectra During Extrasolar Giant Planet Transits," Astrophysical Journal, 537, 916–921

[206] Seager, S., & Mallén-Ornelas, G. 2003, "A Unique Solution of Planet and Star Parameters from an Extrasolar Planet Transit Light Curve," Astrophysical Journal, 585, 1038–1055

[207] Seager, S., & Deming, D. 2010, "Exoplanet Atmospheres," Annual Review of Astronomy and Astrophysics, 48, 631–672

[208] Seager, S. 2010, "Exoplanet Atmospheres" (New Jersey: Princeton University Press)

[209] Seager, S., Schrenk, M., & Bains, W. 2012, "An Astrophysical View of Earth-Based Metabolic Biosignature Gases," Astrobiology, 12, 61–82

[210] Seager, S., Bains, W., & Hu, R. 2013a, "A Biomass-based Model to Estimate the Plausibility of Exoplanet Biosignature Gases," Astrophysical Journal, 775, 104

[211] Seager, S., Bains, W., & Hu, R. 2013b, "Biosignature Gases in H_2-dominated Atmospheres on Rocky Exoplanets," Astrophysical Journal, 777, 95

[212] Seager, S. 2014, "The future of spectroscopic life detection on exoplanets," Proceedings of the National Academy of Sciences, 111, 12634–12640

[213] Shakura, N.I., & Sunyaev, R.A. 1973, "Black holes in binary systems. Observational appearance," Astronomy & Astrophysics, 24, 337–355

[214] Sharp, C.M., & Burrows, A. 2007, "Atomic and Molecular Opacities for Brown Dwarf and Giant Planet Atmopsheres," Astrophysical Journal Supplements, 168, 140–166

[215] Shizgal, B.D., & Arkos, G.G. 1996, "Nonthermal Escape of the Atmospheres of Venus, Earth, and Mars," Review of Geophysics, 34, 483–505

[216] Showman, A.P., & Guillot, T. 2002, "Atmospheric circulation and tides of '51 Pegasus b-like' planets," Astronomy & Astrophysics, 385, 166–180

[217] Showman, A.P., Fortney, J.J., Lian, Y., Marley, M.S., Freedman, R.S., Knutson, H.A., & Charbonneau, D. 2009, "Atmospheric Circulation of Hot Jupiters: Coupled Radiative-Dynamical General Circulation Model Simulations of HD 189733b and HD 209458b," Astrophysical Journal, 699, 564–584

[218] Showman, A.P., & Polvani, L.M. 2010, "The Matsuno-Gill model and equatorial superrotation," Geophysical Research Letters, 37, L18811

[219] Showman, A.P., & Polvani, L.M. 2011, "Equatorial Superrotation on Tidally Locked Exoplanets," Astrophysical Journal, 738, 71

[220] Sing, D.K., et al. 2011, "Hubble Space Telescope transmission spectroscopy of the exoplanet HD 189733b: high-altitude atmospheric haze in the optical and near-ultraviolet with STIS," Monthly Notices of the Royal Astronomical Society, 416, 1443–1455

[221] Sing, D.K., et al. 2012, "GTC OSIRIS transiting exoplanet atmospheric survey: detection of sodium in XO-2b from differential long-slit spectroscopy," Monthly Notices of the Royal Astronomical Society, 426, 1663–1670

[222] Sing, D.K., et al. 2013, "HST hot-Jupiter transmission spectral survey: evidence for aerosols and lack of TiO in the atmosphere of WASP-12b," Monthly Notices of the Royal Astronomical Society, 436, 2956–2973

[223] Sing, D.K., et al. 2015, "HST hot-Jupiter transmission spectral survey: detection of potassium in WASP-31b along with a cloud deck and Rayleigh scattering," Monthly Notices of the Royal Astronomical Society, 446, 2428–2443

[224] Sing, D.K., et al. 2016, "A continuum from clear to cloudy hot-Jupiter exoplanets without primordial water depletion," Nature, 529, 59–62

[225] Slater, J.C. 1939, "Introduction to Chemical Physics" (New York: McGraw-Hill)

[226] Smith, W.R., & Missen, R.W. 1982, "Chemical Reaction Equilibrium Analysis: Theory and Algorithms" (New York: Wiley)

[227] Snellen, I.A.G., de Kok, R.J., de Mooij, E.J.W., & Albrecht, S. 2010, "The orbital motion, absolute mass and high-altitude winds of exoplanet HD 209458b," Nature, 465, 1049–1051

[228] Snellen, I.A.G., Brandl, B.R., de Kok, R.J., Brogi, M., Birkby, J., & Schwarz, H. 2014, "Fast spin of the young extrasolar planet β Pictoris b," Nature, 509, 63–65

[229] Spiegel, D.S., Silverio, K., & Burrows, A. 2009, "Can TiO Explain Thermal Inversions in the Upper Atmospheres of Irradiated Giant Planets?" Astrophysical Journal, 699, 1487–1500

[230] Spiegel, D.S., Burrows, A., & Milsom, J.A. 2011, "The Deuterium-burning Mass Limit for Brown Dwarfs and Giant Planets," Astrophysical Journal, 727, 57

[231] Spitkovsky, A., Levin, Y., & Ushomirsky, G. 2002, "Propagation of Thermonuclear Flames on Rapidly Rotating Neutron Stars: Extreme Weather During Type I X-Ray Bursts," Astrophysical Journal, 566, 1018–1038

[232] Spitzer, L. 1978, "Physical Processes in the Interstellar Medium" (New York: Wiley)

[233] Steinfeld, J.I., Francisco, J.S., & Hase, W.L. 1989, "Chemical Kinetics and Dynamics" (New Jersey: Prentice-Hall)

[234] Stevens, D.J., & Gaudi, B.S. 2013, "A Posteriori Transit Probabilities," Publications of the Astronomical Society of the Pacific, 125, 933–950

[235] Stevenson, K.B., et al. 2012, "Transit and Eclipse Analyses of the Exoplanet HD 149026b Using BLISS Mapping," Astrophysical Journal, 754, 136

[236] Stevenson, K.B., et al. 2014, "Thermal structure of an exoplanet atmosphere from phase-resolved emission spectroscopy," Science, 346, 838–841

[237] Sudarsky, D., Burrows, A., & Pinto, P. 2000, "Albedo and Reflection Spectra of Extrasolar Giant Planets," Astrophysical Journal, 538, 885–903

[238] Swendsen, R.H. 2012, "An Introduction to Statistical Mechanics and Thermodynamics" (New York: Oxford University Press)

[239] Tinetti, G., et al. 2007, "Water vapour in the atmosphere of a transiting extrasolar planet," Nature, 448, 169–171

[240] Toon, O.B., McKay, C.P., & Ackerman, T.P. 1989, "Rapid Calculation of Radiative Heating Rates and Photodissociation Rates in Inhomogeneous Multiple Scattering Atmospheres," Journal of Geophysical Research, 94, 16287–16301

[241] Triaud, A.H.M.J., et al. 2010, "Spin-orbit angle measurements for six southern transiting planets," Astronomy & Astrophysics, 524, A25

[242] Umurhan, O.M. 2008, "A shallow-water theory for annular sections of Keplerian disks," Astronomy & Astrophysics, 489, 953–962

[243] Upadhyay, S.K. 2006, "Chemical Kinetics and Reaction Dynamics" (Dordrecht: Springer)

[244] Vallis, G.K. 2006, "Atmospheric and Oceanic Fluid Dynamics" (New York: Cambridge University Press)

[245] van Zeggeren, F., & Storey, S.H. 1970, "The Computation of Chemical Equilibria" (New York: Cambridge University Press)

[246] Vidal-Madjar, A., Lecavelier des Etangs, A., Désert, J.-M., Ballester, G.E., Ferlet, R., Hébrard, G., & Mayor, M. 2003, "An extended upper atmosphere around the extrasolar planet HD 209458b," Nature, 422, 143–146

[247] Visscher, C., & Moses, J.I. 2011, "Quenching of Carbon Monoxide and Methane in the Atmospheres of Cool Brown Dwarfs and Hot Jupiters," Astrophysical Journal, 738, 72

[248] Volkov, A.N., Johnson, R.E., Tucker, O.J., & Erwin, J.T. 2011, "Thermally Driven Atmospheric Escape: Transition from Hydrodynamic to Jeans Escape," Astrophysical Journal Letters, 729, L24

[249] Waldmann, I.P., Tinetti, G., Rocchetto, M., Barton, E.J., Yurchenko, S.N., & Tennyson, J. 2015, "Tau-Rex I: A Next Generation Retrieval Code for Exoplanetary Atmospheres," Astrophysical Journal, 802, 107

[250] Walker, J.C.G., Hays, P.B., & Kasting, J.F. 1981, "A negative feedback mechanism for the long-term stabilization of Earth's surface temperature," Journal of Geophysical Research, 86, 9776-9782

[251] Washington, W.M., & Parkinson, C.L. 2005, "An Introduction to Three-Dimensional Climate Modeling," second edition (Sausalito: University Science Books)

[252] Winn, J.N. 2010, "Exoplanet Transits and Occultations," in Exoplanets, ed. S. Seager, 55–77 (Tucson: University of Arizona Press)

[253] Wolszczan, A., & Frail, D.A. 1992, "A planetary system around the millisecond pulsar PSR1257+12," Nature, 355, 145–147

[254] Wolszczan, A. 1994, "Confirmation of Earth-Mass Planets Orbiting the Millisecond Pulsar PSR B1257+12," Science, 264, 538–542

[255] Wordsworth, R. 2015, "Atmospheric Heat Redistribution and Collapse on Tidally Locked Rocky Planets," Astrophysical Journal, 806, 180

[256] Wu, Y., & Lithwick, Y. 2013, "Ohmic Heating Suspends, Not Reverses, the Cooling Contraction of Hot Jupiters," Astrophysical Journal, 763, 13

[257] Wyttenbach, A., Ehrenreich, D., Lovis, C., Udry, S., & Pepe, F. 2015, "Spectrally resolved detection of sodium in the atmosphere of HD 189733b with the HARPS spectrograph," Astronomy & Astrophysics, 577, A62

[258] Yelle, R.V. 2004, "Aeronomy of extra-solar giant planets at small orbital distances," Icarus, 170, 167–179

[259] Zaghloul, M.R. 2007, "On the calculation of the Voigt line profile: a single proper integral with a damped sine integrand," Monthly Notices of the Royal Astronomical Society, 375, 1043–1048

[260] Zaqarashvili, T.V., Oliver, R., Ballester, J.L., Carbonell, M., Khodachenko, M.L., Lammer, H., Leitzinger, M., & Odert, P. 2011, "Rossby waves and polar spots in rapidly rotating stars: implications for stellar wind evolution," Astronomy & Astrophysics, 532, A139

[261] Zel'dovich, Y.B., & Raizer, Y.P. 2002, "Physics of Shock Waves and High-Temperature Hydrodynamic Phenomena" (New York: Dover)

[262] Zellem, R.T., et al. 2014, "The 4.5 μm Full-orbit Phase Curve of the Hot
       Jupiter HD 209458b," Astrophysical Journal, 790, 53

Index